Lin Zongshou, Xing Weihong, Chen Wei
Cementitious Materials Science

Also of interest

Lin Zongshou, Xing Weihong, Chen Wei

Cementitious Materials Science

Theories and Applications

DE GRUYTER

WUTP 武汉理工大学出版社 Wuhan University of Technology Press

Authors
Lin Zongshou
School of Materials Science and Engineering
Wuhan University of Technology
122 Luoshi Road, 430070
Wuhan, China
Email: 13807182067@163.com
Tel: +86 138 0718 2067

Xing Weihong
School of Materials Science and Engineering
Wuhan University of Technology
122 Luoshi Road, 430070
Wuhan, China
Email: whxing@whut.edu.cn
Tel: +86 139 7129 1670

Chen Wei
State Key Laboratory of Silicate Materials for Architectures
Wuhan University of Technology
122 Luoshi Road, 430070
Wuhan, China
Email: chen.wei@whut.edu.cn
Tel: +86 138 7105 7299

ISBN 978-3-11-057209-4
e-ISBN (PDF) 978-3-11-057210-0
e-ISBN (EPUB) 978-3-11-057216-2

Library of Congress Control Number: 2018958584

Bibliographic information published by the Deutsche Nationalbibliothek
The Deutsche Nationalbibliothek lists this publication in the Deutsche Nationalbibliografie; detailed
bibliographic data are available on the Internet at http://dnb.dnb.de.

© 2019 Wuhan University of Technology Press, Wuhan and Walter de Gruyter GmbH, Berlin/Boston
Typesetting: Integra Software Services Pvt. Ltd.
Printing and binding: CPI books GmbH, Leck
Cover image: Peshkova / iStock / Getty Images Plus

www.degruyter.com

Preface

Cementitious Materials Science was first printed and published by the China Construction Industry Press in 1980, as a university teaching material for undergraduates with majors of construction engineering materials and products, cementitious materials and products and inorganic nonmetallic materials. This textbook was then published by the Wuhan Polytechnic University Press (now it is the Wuhan University of Technology Press) with a comprehensive revision made by Professor Yuan Runzhang in 1989. In 1996, the book was revised and modified again by Professor Yuan Runzhang, and it was published by the Wuhan University of Technology Press and reprinted many times later. In 2014, the textbook, *Cementitious Materials Science*, was updated by Professor Lin Zongshou; the contents of each chapter was updated according to the requirements of the teaching program for the major of cementitious materials, including the latest national standards and research results. The new edition of the book was published by the Wuhan University of Technology Press. In 2017, the copyright of this book was bought by German De Gruyter Press. The book is, therefore, re-edited by Professor Lin Zonghou.

The chapters of this edition on "Cementitious materials" have reserved the conventional classification of the original version. Starting from the common basic principles and rules of cementitious materials, and based on the characteristics and applications of various cementitious materials, the relationship between the composition, structure and properties of cementitious materials is systematically expatiated more deeply, with the emphasis of general-purpose Portland cement. In addition, it is also expatiated in the book about the basic laws of hydration and hardening process of cementitious materials; the relationship between the structural and engineering properties of hardened cementitious materials; the relationship between the forming process of hardened cementitious materials and process parameters; the relationships among the structure, properties and application conditions of hardened cementitious materials and so on. However, the production equipment and process parameters for each type of cementitious materials are not elaborated in this book. Hopefully, readers can master the basic principles and properties of various cementitious materials, and innovate and develop them in applications.

It must be pointed out that readers should analytically read some of the theories introduced in this book and test/develop them in practice because the cementitious material science itself is still developing with imperfect theories. It should also be mentioned that some of the data quoted in this book are from different authors obtained under different conditions, and the data quoted in the book are used to illustrate some principles and laws. It is expected that readers should not apply the data in the book mechanically in the actual work, but should apply the basic principles and laws in the book according to the actual situation and should solve practical problems through further experiments. Note that the principles and rules of

https://doi.org/10.1515/9783110572100-201

materials science are established through experiments. It is very important to understand that the application of these principles and laws to a specific production practice will also be carried out by experiments.

This book is a university teaching material for the major of cementitious materials and products; it can also be used as a reference for technicians in the fields of building material products factories and civil engineering, such as cement, concrete and other cementitious materials. Readers should have the basic knowledge of physical chemistry and phase analysis when reading this book.

The detailed work on writing and edition of this book is divided as follows: Professor Lin Zongshou is the chief editor of this book, and has compiled the *Preface, Introduction* and Chapters 1 and 2; Dr Xing Weihong works as a subeditor, and has compiled Chapters 3, 4, 5 and Professor Chen Wei is a subeditor, and has compiled Chapter 6.

The English translation of this book is implemented as follows: Dr Yu Rui has translated the *Preface, Introduction* and Sections 1.1–1.5; Dr Liu Zhichao has translated Sections 1.6.1–1.6.3, 1.7.6–1.7.12 and *Problems* of Chapter 1; Dr Rao Meijuan has translated Sections 1.6.5.7–1.6.5.8 and Chapter 5; Dr Hu Chuanlin has translated Sections 1.6.4–1.6.5.6 and 1.7.1–1.7.5; Dr Liu Yunpeng has translated Sections 2.1–2.3; Dr Yang Lu has translated Sections 2.4–2.10 and *Problems* of Chapter 2; Dr Jian Shouwei has translated Chapter 3; Dr Xing Weihong has translated Chapter 4 and Professor Chen Wei has translated Chapter 6 and *References*.

Proofreading of English version of this book is implemented as follows:

Professor Chen Wei has proofread *Introduction* and Sections 1.1–1.3; Dr Rao Meiju has proofread Sections 1.4–1.5;

Dr Yu Rui has proofread Sections 1.6.1–1.6.3, 1.6.5.7–1.6.5.8 and 1.7.6–1.7.12 and *Problems* of Chapter 1;

Dr Xing Weihong has proofread Sections 1.6.4–1.6.5.6 and 1.7.1–1.7.5 and Chapters 2 and 3;

Dr Liu Yunpeng has proofread Chapter 4; Dr Jian Shouwei has proofread Chapters 5 and 6 and *References*.

The book is refereed by Professor Shui Zhonghe. The editor expresses his sincere appreciation for the hard work done by English translators and the careful review done by chief referee. Thanks to De Gruyter Press in improving the quality of the textbook. In addition, the editor also expresses his appreciation to all readers for supporting this book.

Additionally, it will be appreciated if researchers, teachers, students or other readers can put forward valuable suggestions for further improvements of this book.

Contents

Brief introduction to the author

Lin Zongshou was born in Fuding, Fujian Province, in 1957. He graduated from Tongji University in December, 1981. In June 1985, he got his master's degree in Wuhan University of Technology. In 1990, he returned to China from Japan after studying in Tokyo Institute of Technology.

He was a representative in the Ninth and Tenth Chinese National People's Congress (NPC), a member of the Eleventh Chinese People's Political Consultative Conference (CPPCC) committee and a member of the CPPCC population resources and environment committee, and he was awarded the Chinese National "May 1" Labor Medal. He is an expert and has the special allowance of the Chinese State Council. He is currently a professor and doctoral supervisor at Wuhan University of Technology.

Professor Lin is mainly engaged in research of chemical and technological processes of cement. He has 33 patents and five computer software copyrights, and has published more than 120 papers. He has chiefly compiled "Inorganic Nonmetallic Materials Technology," "Cement Technology," "Cementitious Materials Science" and other teaching materials. He has written books of "Hundred Thousand Questions of Cement" with a set of 10 volumes, "Excess-Sulfate Phosphogypsum Slag Cement and Concrete," "Causes and Countermeasures of Sand-out in Cement" and "Slag-based Ecological Cement." He has established Wuhan Yisheng Science and Technology Co., Ltd., and presided the Chinese National "863" Program and several other projects. He is awarded for a new Chinese national key product, two first prizes in scientific and technological progress in Hubei Province and an outstanding achievement prize in comprehensive utilization of national natural resources.

https://doi.org/10.1515/9783110572100-202

1 Introduction

1.1 Definition and classification of cementitious materials

On the basis of the physical and chemical effects, in the process of the gradual transformation of plastic slurry into a solid stone, the substances that can bind other materials to form a whole structure with mechanical strengths are collectively referred as cementitious materials, also known as glued materials.

Cementitious materials can be generally divided into two categories: inorganic and organic types. All kinds of resin and asphalt are organic cementitious materials. The inorganic cementitious material can be divided into hydraulic and nonhydraulic types based on its hardening conditions. The hydraulic cementitious material, after mixing with water, can be hardened both in air and in water and generate strength, which is normally called cement, such as Portland cement, aluminate cement, sulphoaluminate cement and so on. Instead of hardening in water, the nonhydraulic cementitious materials could be hardened in air or other conditions. The one that can only be hardened in the air are called air-hardening cementitious materials, such as lime, gypsum and magnesium cementitious materials.

On the basis of the classification of hydration products, the inorganic cementitious materials can be divided into silicates that mostly produce calcium silicate hydrate, aluminate that mostly produce calcium aluminate hydrate, sulphoaluminate that mostly produce calcium sulphoaluminate hydrate and so on. In addition, there are also some other classification methods.

1.2 The role of cementitious materials in national economy

Cementitious materials can not only be widely used in industrial constructions, civil engineering, transportation, water conservancy, agriculture and forestry, national defense, seaport, urban and rural constructions, aerospace industry, nuclear industry and other new industrial constructions, but also replace steel and timber to produce sleeper, electrical pole, pressure pipe, cement carrier and various structures used for marine development and so on. Besides, it is also an indispensable material for a series of large-scale modern technical facilities and national defense engineering. Therefore, as one of the most important raw materials, cementitious materials have attracted a significant attention.

In the foreseeable decades, cement, concrete and other cementitious materials are still the main building materials. With the development of science and technology, the ability of mankind to transform nature has been continuously improved, and the scale has gradually expanded, which simultaneously has put forward a series of new requirements for the cementing materials. Therefore, the cementitious material

https://doi.org/10.1515/9783110572100-001

has a broad development prospects. The reasons for the continuous development of cementitious materials should be attributed to the following characteristics:
(1) abundant raw materials, raw materials locally available, low production costs;
(2) good durability, high adaptability, can be used in water, sea, hot or cold environments;
(3) advanced fire-resistance capacity;
(4) less maintenance work, low depreciation costs;
(5) as a substrate, it is capable to combine or composite with other materials, such as fiber-reinforced cementitious materials, polymer-reinforced cementitious materials, fiber–polymer–cementitious material multiple composites and so on. Based on the aforementioned points, a large category of new composite materials could be developed;
(6) beneficial for reusing industrial wastes.

The applications of cementitious materials have played an important role in all aspects of industry and civil engineering, and the cementitious material industry is an industry that cannot be ignored in the national economy. With the development of modern science and technology, new technologies in other fields will also inevitably penetrate into the cementitious material industry. The traditional cementitious material industry is bound to spring up the new technology revolution and varietal development with the rapid growth of science and technology. Simultaneously, its application areas will also be broadened, which could further strengthen its key role in the national economy.

1.3 The brief history of cementitious materials

The development of cementitious materials can be traced back to the prehistoric period of mankind. It has gone through several stages such as natural clay, gypsum-lime, lime-volcanic ash and hydraulic cementitious materials made from artificial ingredients. The ancient Egyptians used the Nile's slurry to make bricks without calcination. To increase strength and reduce shrinkage, sand and grass are also mixed into the mud. About 3,000–2,000 BC, the ancient Egyptians began to use calcined gypsum as a building cementitious material, and calcined gypsum has already been used in the construction of the ancient pyramids in Egypt.

Unlike the ancient Egyptians, the Greeks preferred to use the lime generated from calcined limestone in the construction of buildings. In 146 BC, the ancient Roman Empire conquered the ancient Greece, inheriting the tradition of producing and using lime in ancient Greece. The ancient Romans used lime by first dissolving it with water, mixing it with sand and then building the structures with the mortar. Some of the ancient Roman architecture, which was built with lime mortar, was very strong and remains even until today.

The ancient Romans improved the utilization technology of lime. Not only the sand was mixed in the lime, but also the ground volcanic ash. In the areas without volcanic ash, it was replaced by the ground brick, which had the same effect as that of volcanic ash. The mortar was much better in strength and water resistance than the lime–sand two-component mortar, and it was more durable in its regular buildings and underwater construction. Some people called the "lime–ash–sand" three-component mortar "Roman mortar."

The development of Chinese construction cementitious materials is unique and historical. As early as 5,000–3,000 BC, the period of Yangshao culture in the Neolithic, Chinese have already used "white ash" to daub the cave, the ground and the four walls of the excavation, which can result in a smooth and hard structure. The "white ash" was named because of its white powder appearance, which was made from grinding natural ginger stone. The ginger stone is a kind of limestone with high silica content. It is often mixed in the loess, and it is the calcareous concretion in loess. The "white ash" is the earliest building cementitious material that has ever found in ancient China.

In the sixteenth century BC, Shang Dynasty, the cave buildings were rapidly replaced by wooden structure. At this time, in addition to using "white ash" to daub the ground, the yellow mud was used to build walls. From 403 BC to 221 BC, the Warring Stage period, the grass was mixed with yellow mud to build walls and glue wall tiles. In the history of Chinese architecture, the "white ash" was eliminated long time ago. However, the yellow mud and yellow mud mixed with grasses as cementitious materials were utilized until modern society.

In the seventh century BC, lime appeared in the Zhou Dynasty, which was mainly made from the calcined shell of the large clam. The main component of clam shell is calcium carbonate. When CO_2 is exhaustively eliminated, the remaining substance is lime. It has been found in the Zhou Dynasty that the produced lime has good moisture absorption and resistance properties, which is the reason why it has been widely used in Chinese history for a long period.

In the fifth century, South and North Dynasties, a kind of building material named "three-mixture-soil" appeared, which was composed of lime, clay and fine sand. In the Ming Dynasty, the "three-mixture-soil" was composed of lime, pottery powders and gravels. In the Qing Dynasty, except for the lime, clay and fine sand-based "three-mixture-soil," the lime, slag and sand were utilized to produce a new "three-mixture-soil." Based on the modern point of view, "three-mixture-soil" can be treated as a kind of concrete, in which the lime, loess or other volcanic ash materials were utilized as cementitious materials and fine sand, gravel or slag played the role of fillers. There are many similarities between the "three-mixture-soil" and the three-component mortar (Roman mortar). After tamping, the "three-mixture-soil" has relatively high strength and good waterproofness capacity. It was used to build dams in the Qing Dynasty.

One of a striking feature for the development of cementitious materials in Chinese ancient architecture is the application of cementitious materials composed of organic materials and lime, such as "lime-glutinous rice," "lime-tung oil," "lime-blood," "lime-bletilla," "lime-glutinous rice-alum" and so on. In addition, in the application of "three-mixture-soil," the glutinous rice and blood were also added.

The development process of the cementitious materials in ancient Chinese architecture originated from "white ash" and yellow mud, which further transferred to the lime and "three-mixture-soil," and finally developed to the cementitious materials of lime doped with organic materials. The cementitious material in ancient Chinese architecture had its glorious histories. Compared to the development of that in ancient western architecture, due to the widespread adoption of the cementitious material of lime and organic materials, the Chinese ancient cementitious materials were even better.

In the second half of the eighteenth century, hydraulic lime and Roman cement were developed, which were made from calcined limestone containing clay. Based on this foundation, a natural cement was developed from calcined and finely ground natural cement rock (a limestone with clay content at 20–25%). Subsequently, it gradually developed to grind and mix the limestone with a certain amount of clay, and produced the hydraulic lime based on the calcination of artificial ingredients. This is actually the prototype for Portland cement production.

In the early nineteenth century (1810–1835), according to artificial ingredients, high-temperature calcination and grinding, the hydraulic cementitious materials can be produced. The calcination temperature has reached the melting points of several raw materials, which can also be treated as sintering. In 1824, the British Joseph Aspdin first obtained the patent for the product. Since the produced cementitious material has similar appearance and color in the hardened state as that limestone produced on Portland Island, it was originally called Portland cement. In our country, it was called silicate cement. Due to the fact that it has relatively high silicate content, the Portland cement can harden in water and show relatively high strength. An example of the first large-scale application of Portland cement was the Thames Tunnel Construction, which was built in 1825–1843.

The appearance of Portland cement has played an important role in engineering construction. With the demands of modern science and industrial development, in the early twentieth century, various cements for different applications were gradually produced. In the recent half of the century, sulphoaluminate cement, fluoroaluminate cement, aluminoferrite cement and other types of cement were successively developed, which promote further development of Portland cement to more categories. At the same time, new cognition has been acquired on ancient cementitious materials such as lime and gypsum, which enlarge their application fields and development speed. Today, cementitious materials have entered a stage of vigorous development.

1.4 The development of cementitious materials science

In the broad field of productive and scientific practice for cementitious materials, considerable knowledge has been accumulated, especially with the formation and development of materials science, new and profound changes are taking place in the understanding of cementitious materials. The characteristics and trends of this change can be summarized as follows:

(1) The understanding of cementitious materials is gradually deepened, from macro-scopic to microscopic, which gradually reveals the relationship between its performance and the internal structure. Thus, this provides a theoretical basis for developing new varieties of cementitious materials and expanding their areas of applications.

(2) The understandings for cementitious materials productive process and their hydration and hardening process have gradually improved from experience and phenomenon to theory and essence, which provides a theoretical basis for effectively controlling the productive process of cementitious materials products and adopting new technologies and new methods. There is no doubt that cementitious materials are being formed progressively as an important part of material science.

(3) The traditional cementitious materials were produced by natural raw materials. In future, no natural resources would be used for the production of cementitious materials, while various industrial solid wastes could be the main raw materials to produce near-zero emission environment-friendly cementitious materials with simple manufacturing process and little energy consumption.

The main research contents of cementitious materials can be summarized as follows:
(1) the relationship between composition, structure and cementitious properties of cementitious materials;
(2) the law of hydration and hardening process and structure formation of cementi-tious materials;
(3) the relationship between composition and structure of harden cementitious materials and their engineering properties;
(4) preparation of cementitious materials with specified properties and structure and technological approaches for the production of environment-friendly cementitious materials.

We are convinced that with the development of cementitious materials science, the cementitious materials and products industry will have a new neap in the future.

2 Common Portland cement

Generally, the cement can be defined as a hydraulic powder material. When cement is mixed with proper amount of water, a plastic paste can be formed, which can harden both in the air and in water; and the sand, stone and other materials can be firmly combined together.

There are many types of cement, based on their application purpose and property. Three general types are common Portland cement, special cement and characteristic cement. Common Portland cement is widely used in civil engineering, including Portland cement, ordinary Portland cement, Portland blast furnace slag cement, Portland pozzolana cement, Portland fly-ash cement and composite Portland cement. Special cement refers to cement serving special purposes such as oil well cement and masonry cement. Characteristic cement exhibits unique performances such as high early strength Portland cement, sulfate-resistant Portland cement, moderate heat Portland cement, sulfoaluminate expansive cement and self-stressing aluminate cement.

Based on the main hydraulic minerals, cement can be classified as silicate cement, aluminate cement, sulfoaluminate cement, fluoroaluminate cement and the cement produced from industrial solid waste and local materials. At present, there are more than 100 different types of cements.t

2.1 The invention and production of cement

2.1.1 The invention of cement

In the mid-eighteenth century, the lighthouses in England were mainly made of wood and "Roman mortar." Due to the fact that the used materials cannot withstand the corrosion and erosion from the sea, the lighthouse was often damaged. Hence, J. Smeaton, who was honored as the Father of British civil engineering, undertook the task of building new lighthouses. In 1756, in the process of lighthouse construction, it was observed that after adding some water to the calcinated and ground limestone containing clay, the formed mortar can harden slowly and has much higher strength than "Roman mortar" in the sea, which could simultaneously resist the erosions from seawater. The lime that is made of clay and limestone was called hydraulic lime. The discovery of Smeaton was a great leap in the knowledge accumulation and played an important role in the invention of Portland cement. In 1796, the British researcher J. Parker milled clayey limestone (called Sepa Tria) into pellets, and calcined them at a higher temperature than that of lime, and finally ground them again to produce cement. Parker called this cement "Roman Cement" and patented it. "Roman Cement" set quickly and can be used for engineering in water, which was widely used by British until the invention of "Portland Cement."

https://doi.org/10.1515/9783110572100-002

Almost at the same time when the "Roman cement" was produced, the French produced cement in the area of Boulogne by using the marl that has similar chemical compositions as that of modern cement. The natural marl, whose chemical composition is close to the modern cement, is called cement-used limestone, and the cement made from the limestone is called natural cement. Natural cement was also produced in America by using the cementitious rock in the areas of Rosendale and Louisville. For a long time in the 1880s, and later, natural cement was widely used in the United States and had played an important role in the construction industry.

The British researcher J. Foster mixed chalk and clay with a mass ratio of 2:1, and grounded the mixture with water to form slurry. Then, the slurry was subjected into the hopper to precipitate, and the precipitated fraction is dried in the atmosphere. After that, the dried substance was calcined in the limekiln at a certain temperature for complete volatilization of CO_2 in the material. Finally, the calcined product is yellowish and was ground to cement after cooling. The obtained cement was named "British cement" by Forster, which was also the British No. 4679 patent in October 22, 1882. Due to the lower calcination temperature, the quality of "British cement" was much lower than the "Roman cement." Although it has not been widely used, its manufacturing method was the prototype of the modern cement, which was another major leap in cement knowledge accumulation.

In October 21, 1824, J. Aspdin, a plasterer in Leeds, obtained the 5022nd "Portland cement" (silicate cement) patent, and became an inventor of cement. The "Portland cement" method that described in his patent certificate is: "The limestone was crushed into fine powder, blended with a certain amount of clay, mixed with waste to form slurry by artificial or mechanical stirring. Then the mud was placed on the plate, heated and dried. The dried material was hit into pieces, and fed into the limekiln to calcine until all the CO_2 is escaped. After that, the calcined product was cooled, crushed and ground to generate the cement. When mixing the cement with a small amount of water, the produced mortar with proper consistency could be applied to various work situations." The color of the cement after hydration and hardening was similar to the building stone in Portland, England, hence it was named "Portland cement."

The "Portland cement" manufacturing method described in the patent certificate of Aspdin is similar to Foster's "British cement," since the calcination temperatures were both up to the complete volatilization of CO_2 within the material. Based on the general knowledge of cement production, the quality of the produced "Portland cement" under this temperature should be lower than "British cement." However, "Portland cement" was more competitive than "British cement" in the market. In 1838, when the Thames tunnel construction was rebuilt, the "Portland cement" was finally chosen even its price is much higher than that of "British cement." It is a personal assumption that Aspdin did not show the "Portland cement" production technology completely on the patent certificate for the sake of confidentiality. He actually had more knowledge of cement production than that described in the patent.

Aspdin must have used higher calcination temperature in the manufacturing; otherwise, the cement after hardening would not have the same color as the stone in Portland, and it would not have won the competition in the market.

However, according to the contents described in the patent certificate and the relevant information, Aspdin failed to grasp the exact calcination temperature of "Portland cement" and the correct ratio of the raw materials, which cause that the quality of the product was very unstable, and even some buildings were collapsed due to the low quality of the produced cement.

Another cement research genius at the time of Aspdin in Britain was I. C. Johnson. He was the manager of the British Swan Valley White Company, specialized in the manufacture of "Roman cement" and "British cement." In 1845, in an experiment, Johnson accidentally found that the calcined cement bulk with a certain amount of glass results in good hydraulicity after grinding. In addition, it was also noticed that the cement would crack if the calcined bulk contained lime. According to these unexpected findings, Johnson identified two basic conditions for cement production: First, the temperature of the kiln must be high enough to ensure that the calcined bulk contains a certain amount of dark green glass. Second, the ratio of raw materials must be correct and fixed, the calcined bulk cannot contain excessive lime and the cement should not crack after hardening. These conditions ensured the quality of "Portland cement" and solved the problem of quality instability that Aspdin faced. Since then, the basic parameters of modern cement production were determined.

The earliest Chinese cement factory was the Green Island Cement Factory, a foreign-funded enterprise in Macao, which was founded in 1886. Tangshan cement factory was the earliest Chinese national cement enterprises, established by Chinese in 1889, which was three years later than Macao Green Island Cement Factory. Later, cement factories were built in Dalian, Shanghai and Guangzhou successively. According to the pronunciation of English word "cement," the productions of these factories were named "Ximiantu," "Shimintu," "Shuimenting" and "foreign ash". The name "Cement" was first used by "Hubei cement factory" located in Huangshi harbor, Daye county, Hubei province, which was the predecessor of Huaxin cement co., Ltd. It was built in May 2, 1909, and was capable to produce cement with 180–200 t/day.

2.1.2 The process of cement

According to the preparation methods of raw materials, the process of cement production can be divided into two categories: the dry method and the wet method. In the dry method, the raw materials are dried and crushed into powder, and then added into the kiln clinker. When an appropriate amount of water is added to the raw materials to generate pellets, then add into the kiln or Lepol kiln clinker to calcine. This is also called semidry method. In the wet method, raw materials with water are ground into slurry, and then added into the rotary kiln.

The suspension preheater, which was invented in the 1950s, experienced great development in the 1960s and greatly reduced the heat consumption of the clinker. In the 1970s, the kiln outside decomposition technology appeared, which significantly improved the output of clinker, and the heat consumption had decreased obviously. At the same time, the development of homogenization and pre-homogenization of the raw materials, and the continuous advancement of drying and grinding equipment improved the clinker quality. Hot air of the cooler was used for kiln outside the decomposition furnace, exhaust gas of kiln was used for drying raw materials and coal powder, the waste gas of kiln tail and the waste heat of the cooler were successfully used to regenerate power, so that the waste heat was fully utilized. About nearly a decade, with further optimization of new dry process production technology, further reduction of environmental load, various alternative materials and fuels, and the degradation and utilization of wastes, cement industry is treating the new dry production technology as the foundation and transforms to the ecological environment materials industry. The process of cement kiln outside the decomposition technology in dry method is shown in Figure 2.1.

In Figure 2.1, the limestone was broken into gravels, and then poured into the raw material warehouse after the homogenization executing in limestone pre-homogenization yard. The sandstone is transported into the factory, broken by the crusher, and then transported into the raw material warehouse. Iron powder is transported into the factory by car and directly stored in the raw material warehouse. According to the specified proportion calculation, the limestone, sandstone and iron powder are added into the raw mill for grinding to produce the raw material powders. Then the raw material powders were fed into the raw material warehouse. The homogenized raw materials are added into the preheater, decomposition furnace and rotary kiln to produce clinkers. The calcined clinkers are transported into the cooler for cooling, and the cooled clinkers are stored in the clinker warehouse. Gypsum is transported into the factory by car and stored into a gypsum warehouse after breaking. The fly ash is directly transported into the fly ash warehouse after being transported into the factory. After pressed by the roller press, the clinker is mixed with gypsum and fly ash at a certain proportion, and the mixture is added into the cement mill for grinding to produce the final cement. The produced cement is stored in the cement warehouse, and then packed up, or shipped in bulk by car, or transported by bulk cement vessels. The coal is transported into coal yard to be homogenized, and then the coal powder is prepared from coal mill, which can be used for calcination of rotary kiln and decomposing furnace. The hot gas required for the drying of coal mill is originated from the cooler. The high-temperature exhaust gas from the kiln tail preheater is cooled by the humidification tower, some of which is used for raw mill to dry raw materials and then discharged after dust collection, and the remaining part is discharged directly through dust collection. The hot gas generated after cooling the clinker can be partly used as secondary wind directly entering into the kiln to help coal powder combustion, while the other part is transported by tertiary air duct to the kiln tail decomposition furnace

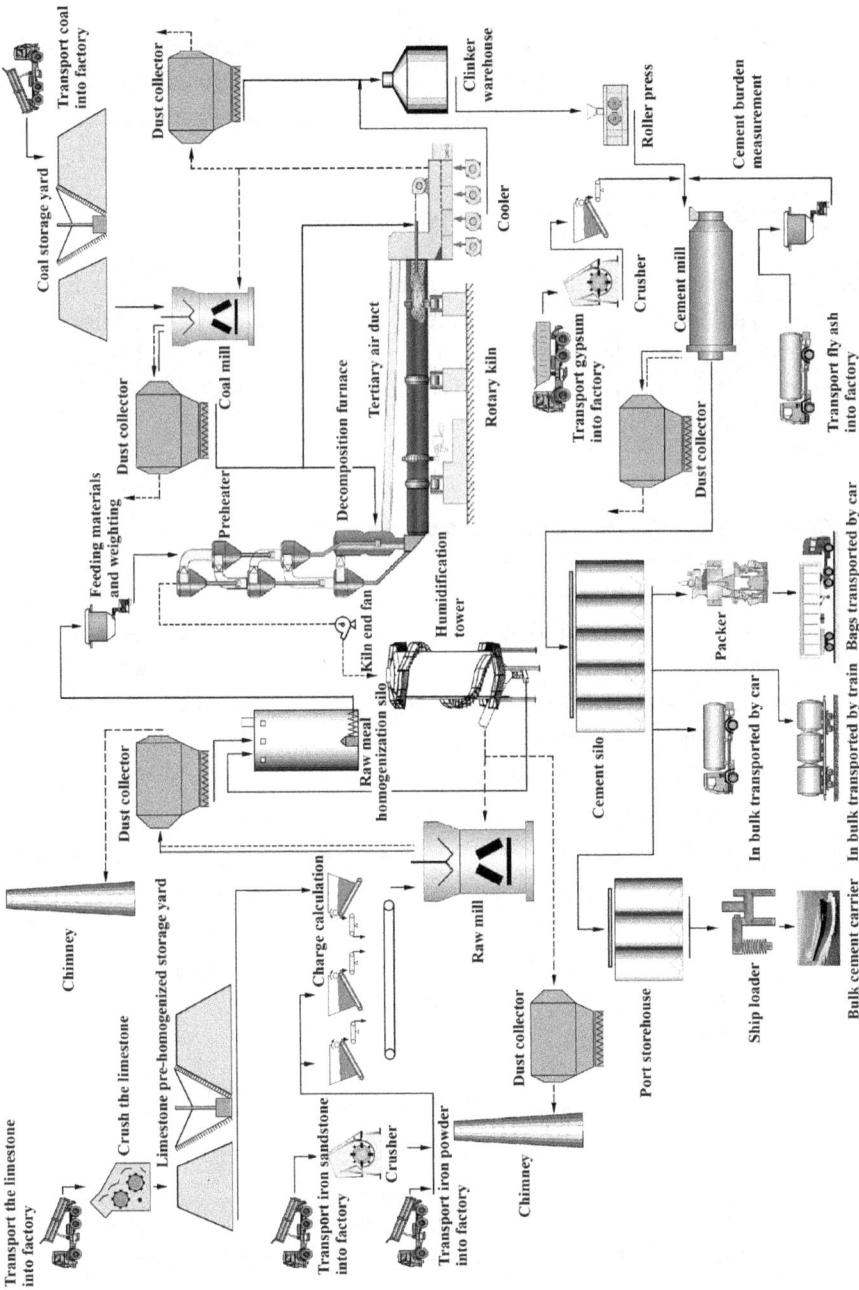

Figure 2.1: The process of cement kiln outside decomposition technology in dry method.

to help coal powder combustion. The excessive gas is used for drying the coal mill or been discharged by the dust system of the kiln. In the cogeneration pre-decomposition kiln system, some of the heat gas generated from the cooler and the exhaust gas of the kiln tail preheater are discharged after providing steam for boiler generator.

2.2 Composition of Portland cement clinker

When the raw materials containing CaO (abbreviation: C), SiO_2 (abbreviation: S), Al_2O_3 (abbreviation: A), Fe_2O_3 (abbreviation: F) are ground into powders with a certain proportion, and then calcined to partially melt, the obtained hydraulic cementitious material composed of calcium silicate can be called as Portland cement clinker, abbreviated as clinker.

2.2.1 Chemical and mineral compositions of clinker

2.2.1.1 Chemical composition

Portland cement clinker is mainly composed of four oxides: CaO, SiO_2, Al_2O_3 and Fe_2O_3 (normally more than 95 wt%). When the content of the main oxides [$w(CaO)+w(SiO_2)+w(Al_2O_3)$] is converted to 100% ($Fe_2O_3$ content is calculated in Al_2O_3), the composition of Portland cement clinker can be described by the graphic area of $C_3S–C_2S–C_3A$ triangle shown in Figure 2.2.

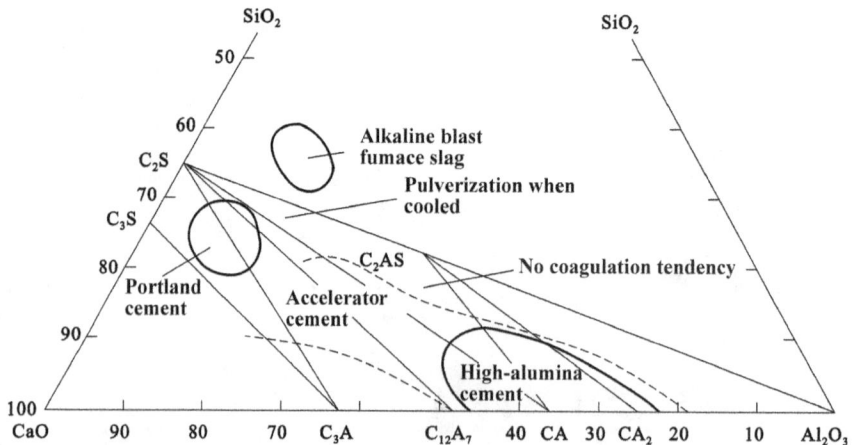

Figure 2.2: Cement areas of $CaO-SiO_2-Al_2O_3$ system.

In the modern Portland cement clinker, the fluctuation ranges of the main oxides are as follows: $w(CaO)$ is 62–67%; $w(SiO_2)$ is 20–24%; $w(Al_2O_3)$ is 4–7%; $w(Fe_2O_3)$ is 2.5–6.0%.

In some cases, due to the differences in cement variety, the composition of raw materials and the manufacturing process, the oxide contents may not be within the above-mentioned range. For instance, in white Portland cement, the content of Fe_2O_3 must be less than 0.5%, while the content of SiO_2 can be higher than 24%, or even up to 27%. Except for the four main oxides, there are also MgO (abbreviation: M), SO_3 (abbreviation: $\bar{S}$), K_2O (abbreviation: K), Na_2O (abbreviation: M), TiO_2 (abbreviation: T) and P_2O_5 (abbreviation: P).

2.2.1.2 Mineral composition
In Portland cement clinker, CaO, SiO_2, Al_2O_3 and Fe_2O_3 do not exist as separate oxides, but as multimineral aggregates produced by two or more oxides after high-temperature chemical reaction. Its crystal is small and the size of the clinker is usually 30–60 μm. There are four main minerals:
(1) Tricalcium silicate: $3CaO \cdot SiO_2$, represented as C_3S
(2) Dicalcium silicate: $2CaO \cdot SiO_2$, represented as C_2S
(3) Tricalcium aluminate: $3CaO \cdot Al_2O_3$, represented as C_3A
(4) Ferrite solid solution: $4CaO \cdot Al_2O_3 \cdot Fe_2O_3$ are represented typically as C_4AF.

In addition, there are also small amount of free calcium oxide (f-CaO), periclase (crystalline magnesium oxide), alkali-containing minerals and glass within the clinkers. Figure 2.3 presents a petrographic photograph of Portland cement clinker under a reflecting microscope. The black polygonal particles are C_3S; the round particles with black and white twin striation are C_2S, and between these two kinds of crystals are white mesophase C_4AF (light color) with strong reflection and black mesophase C_3A (dark) with weak reflection.

40 μm

Figure 2.3: Petrographic photograph of clinker under reflecting microscope.

Generally, the content of C_3S and C_2S in clinker is about 75 wt%, which is called the silicate mineral. The theoretical content of C_3A and C_4AF accounts for about 22 wt%. During the calcination of cement clinker, C_3A and C_4AF, MgO and alkali will be gradually melted at 1,250–1,280 °C to form liquid phases, promoting the formation of C_3S. Hence, it is called the fluxing mineral.

(1) Tricalcium silicate

C_3S is the main mineral of Portland cement clinker. The content is usually about 50%, and sometimes even up to above 60%. The pure C_3S is stable only under the temperature of 1,250–2,065 °C; when the temperature is above 2,065 °C, it incongruently melted to CaO and liquid phase; when the temperature is below 1,250 °C, it decomposes into C_2S and CaO, but the reaction is slow, so pure C_3S can be a metastable state at room temperature. C_3S has three types of crystals and seven variants:

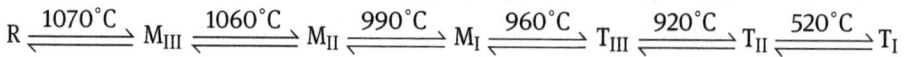

$$R \xrightleftharpoons{1070°C} M_{III} \xrightleftharpoons{1060°C} M_{II} \xrightleftharpoons{990°C} M_{I} \xrightleftharpoons{960°C} T_{III} \xrightleftharpoons{920°C} T_{II} \xrightleftharpoons{520°C} T_{I}$$

The R type belongs to the trigonal system, M type belongs to the monoclinic system and T type belongs to the triclinic system. The crystal structures of these variants are similar. In the Portland cement clinker, it usually does not exist as pure C_3S, while it always contains a small amount of MgO, Al_2O_3, Fe_2O_3 and so on. The mentioned composites can form solid solution, which is known as Alite or A mineral. Alite usually belongs to M type or R type.

Pure C_3S is white, with a density of $3.14\,g/cm^3$, and its crystal cross section is hexagonal or prismatic. The Alite single crystal of monoclinic system is hexagonal schistose or tabular. It often exists as an inclusion of C_2S and CaO in Alite.

C_3S presents the normal setting time, relatively fast hydration rate and high heat release amount. It has relatively high early strength, while its strength improvement at later age is also significant. Its strength at 28 days can reach 70–80% of that obtained after hydrating for one year, which is the highest among the four minerals. However, the hydration heat release for C_3S is high, and its water resistance is relatively poor.

(2) Dicalcium silicate

The content of C_2S in clinker is generally around 20%, which is one of the main minerals of Portland cement clinker. Moreover, it usually does not exist as pure C_2S, while it always contains a small amount of MgO, Al_2O_3, Fe_2O_3 and so on. The mentioned composites can form solid solution, which is known as Blite or B mineral. Below 1,450 °C, the pure C_2S has the following polycrystalline transformation:

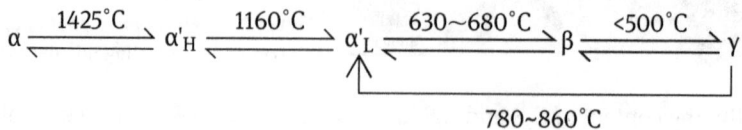

$$\alpha \xrightleftharpoons{1425°C} \alpha'_H \xrightleftharpoons{1160°C} \alpha'_L \xrightleftharpoons{630\sim680°C} \beta \xrightleftharpoons{<500°C} \gamma$$
$$\alpha'_L \xrightarrow{780\sim860°C} \gamma$$

(H: high temperature; L: low temperature)

At room temperature, α, α'_H, α'_L and β are not stable, which have a tendency to transform into γ. In the clinker, there are less α and α'. When the calcination temperature is relatively high and the cooling rate is relatively fast, β can exist in the clinker, since the solid solution contains a small amount of MgO, Al_2O_3, Fe_2O_3 and so on. The normally named C_2S or B mineral refers to β-C_2S.

The strength of α-C_2S and α'-C_2S is excellent, while γ-C_2S shows little hydraulicity. In the process of clinker calcination, when there is poor ventilation, strong reducing atmosphere, low sintering temperature, insufficient liquid phase and slow cooling rate, under $500\,°C$, β-C_2S with a density of $3.28\,g/cm^3$ can be easily transformed into γ-C_2S with a density of $2.97\,g/cm^3$, while the volume expansion is 10% resulting in clinker pulverization. If the liquid phase is too much, the fluxing mineral can be formed into glass, the β-C_2S can be included by glass, it can jump the transition temperature that β-C_2S transfer to γ-C_2S by rapid cooling.

Pure C_2S is white in color. C_2S is brown when it contains Fe_2O_3. The hydration speed of Blite is slow, and only about 20% can be hydrated after 28 days. Its setting and hardening are also slow, early strength is low, but the later strength growth rate is high, while the later strength can reach the level similar as Alite after a year. The hydration release heat of C_2S is small and its water resistance is advanced.

(1) Mesophase
The material filled between Alite and Bailey is collectively referred to as the mesophase. In the process of calcination, the mesophase is melted into liquid phase. During cooling, some liquid phases crystallize and the remaining solidify into glass.

(i) Tricalcium aluminate
The crystal structure of C_3A can be cube, octahedral or dodecahedron. Its shape varies with the cooling rate in the cement clinker. Clinker with high content of Al_2O_3 and slow cooling rate may crystallize into the complete large crystals. Generally, it could come into the glass phase or in irregular microcrystalline precipitation. The potential content of C_3A in clinker is 7–15%. Pure C_3A is a colorless crystal with a density of $3.04\,g/cm^3$ and a melting temperature of $1,533\,°C$. Under the reflecting microscope, it is drip-like with fast cooling rate, and rectangular or cylindrical with slowing cooling rate. Due to its poor reflective ability and dark gray color, C_3A is also called black mesophase.

The hydration rate of C_3A is relatively rapid. It can release relatively large amount of heat and coagulate quickly. If the gypsum and other retarders are not added, then the cement would condense and harden fast, and the strength within 3 days can be put out, but the absolute value is not high, and it hardly increase later, or even shrink. The cement with high content of C_3A has large shrinkage deformation and poor sulfate resistance.

(ii) Ferrite solid solution
The potential content of Ferrite solid solution in clinker is 10–18%. Its component is complex in clinker. It was considered as a component of C_2F–C_3A_3F continuous solid

solution; and also considered as a part of C_6A_2F–C_6AF_2 continuous solid solution. In the general Portland cement clinker, its composition is close to C_4AF, so the C_4AF is usually used to represent Ferrite phase in clinker. If the $w(Al_2O_3)/w(Fe_2O_3)$ is lower than 0.64 in clinker, the C_2F can be produced with certain hydraulic.

The early hydration rate of C_4AF is between C_3A and C_3S, but the subsequent development is not as good as C_3S. The early strength is similar to C_3A, and the later strength can increase continuously, which is similar to C_2S. Its impact resistance and sulfate resistance are great, and the heat of hydration is lower than that of C_3A, but the clinker with high content of C_4AF is difficult to be grinded. In road cement and sulfate-resistant cement, the high content of C_4AF is referable.

(iii) Glass
In the actual production, due to the faster cooling rate, some liquid phases are too late to crystallize and become a supercooled liquid, namely the glass. In glass, the particles are arranged disorderly, and the composition is also uncertain. Its main components are Al_2O_3, Fe_2O_3, CaO and a small amount of MgO and alkali.

In the calcination process, C_3A and C_4AF are melted into liquid phase, which can promote the formation of C_3S, it is their important function. If there are few fulxing minerals in the material, it is easy to be burned, the CaO cannot be easily absorbed and the f-CaO in the clinker increases, which affect the quality of the clinker, reducing the production of the kiln and increasing the consumption of the fuel; if the melting mineral is too much, the material lump is easily formed in the kiln, the ring is formed inside the rotary kiln, the furnace accretion is formed in the vertical kiln and so on, which seriously affecting the normal production of rotary kiln and vertical kiln.

(2) f-CaO and periclase
f-CaO refers to calcium oxide, which has not been combined with high-temperature calcination, also known as free lime. The high-temperature calcined f-CaO structure is relatively dense, and the hydration is very slow. It usually takes 3 days to react obviously. During hydration, f-CaO creates calcium hydroxide and the volume is increased by 97.9%, which causes local swelling stress in the hardened cement paste. With the increase in f-CaO content, the flexural strength decreases first, then the strength is reduced after 3 days, it could result in poor soundness. Therefore, in the clinker calcination, the f-CaO content should be strictly controlled. It is generally controlled below 1.5% in rotary kiln, while below 3.0% in vertical kiln in China. Because part of the free oxide of clinker in vertical kiln is the raw material, which has not been calcined by high temperature. The hydration of the f-CaO is fast and is less destructive to hardened cement paste.

Magnesite refers to the free state of MgO crystals. MgO has a low chemical affinity with SiO_2 and Fe_2O_3, so it generally does not participate in the chemical reactions during calcination. It is found in the following three forms in the clinker:

(i)   Dissolved in C_3A and C_3S to form solid solution
(ii) Dissolved in glass
(iii) In free form of magnesite

It is believed that MgO content of the first two forms is about 2% in clinker, which is not destructive to the hardened cement paste. In the form of magnesite, due to its low hydration rate, it starts hydration obviously after half a year to a year. The resulting hydrate is $Mg(OH)_2$ and the volume was expanded to 148%. Therefore, it also leads to poor stability. The extent of magnesia expansion is related to its crystal size and content. The bigger the size, the higher the content and the greater the expansion. In the production, fast cooling measurement should be carried out to reduce the size of the magnesite crystal.

2.2.2 Modulus value of clinker

Portland cement clinker is composed of two or more oxides, so it can better reflect the effect of clinker mineral compositions and properties to control the ratio (i.e., modulus value) between the oxides in cement production than the content of the oxides separately. As a result, the modulus value indicating the relative content between oxides is often used as an index in production control.

2.2.2.1 Limestone saturation coefficient

Among the four main oxides in clinker, CaO is the alkaline oxide and the other three are acidic oxides. They are combined with each other to form four main clinker minerals, such as C_3S, C_2S, C_3A and C_4AF. It is not difficult to understand that once the content of CaO exceeds the demand of all acidic oxides, it would inevitably exist in the form of f-CaO, which will lead to poor cement stability and damage if with high content. So theoretically, there is a limit lime content. A. Guttmann and Gille believed that the highest alkaline minerals formed by acidic oxide are C_3S, C_3A and C_4AF, and the theoretical limit content of lime can be calculated. For the convenience of calculation, C_4AF is rewritten as "C_3A" and "CF," the resulting "C_3A" is added to C_3A and the demanded lime content of each 1% acidic oxide, respectively:

the demanded lime content for 1% Al_2O_3 to form C_3A: $\frac{3M_r\,(CaO)}{M_r\,(Al_2O_3)} = \frac{3 \times 56.08}{101.96} = 1.65$;

the demanded lime content of 1% Fe_2O_3 to form CF: $\frac{M_r\,(CaO)}{M_r\,(Fe_2O_3)} = \frac{56.08}{159.70} = 0.35$;

the demanded lime content of 1% SiO_2 to form C_3S: $\frac{3M_r\,(CaO)}{M_r\,(SiO_2)} = \frac{3 \times 56.08}{60.09} = 2.8$.

The theoretical limit content of lime can be obtained by multiplying the demanded lime content of per 1% acidic oxides by the corresponding content of acidic oxide:

$$w(CaO) = 2.8w(SiO_2) + 1.65w(Al_2O_3) + 0.35w(Fe_2O_3) \tag{2.1}$$

Kindle and Junker announced that in the actual production, Al_2O_3 and Fe_2O_3 can always react with CaO (saturated with CaO), but only SiO_2 may not be completely saturated to generate C_3S, and there is a portion of C_2S. Otherwise, f-CaO will appear in the clinker. Hence, a coefficient less than 1 should be multiplied before $w(SiO_2)$, that is, the lime saturation coefficient (KH). Therefore, the free lime content is calculated as

$$w(CaO) = KH \times 2.8w(SiO_2) + 1.65w(Al_2O_3) + 0.35w(Fe_2O_3) \tag{2.2}$$

The upper formula is rewritten as follows:

$$KH = \frac{w(CaO) - 1.65w(Al_2O_3) - 0.35w(Fe_2O_3)}{2.8w(SiO_2)} \tag{2.3}$$

In formula (2.3), the numerator shows the content of CaO to form calcium silicate (C_3S+C_2S), and denominator shows the content of CaO demanded by all SiO_2 to form C_3S in theory. Therefore, the KH in clinker is the ratio of the content of CaO to form calcium silicate [$w(C_3S)+w(C_2S)$] to the content of CaO that demanded by all of the SiO_2 to form C_3S in theory. It means the degree of formation of C_3S, when SiO_2 is saturated with CaO in the clinker. Theoretically, when KH = 1, the remaining CaO in the clinker reacts with all C_2S to form C_3S, and there is no C_2S and only C_3S, C_3A and C_4AF are in the clinker. When [$w(CaO)-1.65w(Al_2O_3)-0.35w(Fe_2O_3)$] (the content of CaO combined with SiO_2) is just enough for SiO_2 to entirely form C_2S without C_3S, [$w(CaO)-1.65w(Al_2O_3)-0.35w(Fe_2O_3)$] should be

$$\frac{2M_r(CaO) \times w(SiO_2)}{M_r(SiO_2)} = \frac{2 \times 56.08 \times w(SiO_2)}{60.09} = 1.8665w(SiO_2)$$

Then,

$$KH = \frac{1.8665w(SiO_2)}{2.8w(SiO_2)} = 0.6666$$

That is to say, when KH = 0.6666 in the clinker, there are only C_2S, C_3A, C_4AF without C_3S. Therefore, KH is between 0.6666 and 1.0 in the actual clinker.

KH actually represents the ratio of C_3S to C_2S in percentage (mass fraction, wt%) in the clinker. The higher the value of KH, the higher the proportion of C_3S in the silicate minerals, and the better the strength of clinker. Therefore, increasing the KH is beneficial to improve the quality of cement clinker. But if KH is too high, the calcination of clinker is difficult, so it is necessary to increase the calcination temperature and prolong the calcination time. Otherwise, the f-CaO will appear; at the same time, it will lead to low output of the kiln and high heat consumption, and the working conditions of kiln liner will be deteriorated.

Formula (2.3) is applicable to the clinker when $w(Al_2O_3)/w(Fe_2O_3) \geq 0.64$; if $w(Al_2O_3)/w(Fe_2O_3) < 0.64$, the clinker minerals are composed of C_3S, C_2S, C_4AF and C_2F. Similarly, C_4AF is rewritten as "C_2A" and "C_2F," and the rewritten "C_2F" is added to C_2F, thus

$$KH = \frac{w(CaO) - 1.1w(Al_2O_3) - 0.7w(Fe_2O_3)}{2.8w(SiO_2)} \quad (2.4)$$

Considering that there are f-CaO, free silica and gypsum in the clinker, the formula can be rewritten as follows:

When $w(Al_2O_3)/w(Fe_2O_3) \geq 0.64$,

$$KH = \frac{w(CaO) - w(f - CaO) - 1.65w(Al_2O_3) - 0.35w(Fe_2O_3) - 0.7w(SO_3)}{2.8[w(SiO_2) - w(f - SiO_2)]} \quad (2.5)$$

When $w(Al_2O_3)/w(Fe_2O_3) < 0.64$,

$$KH = \frac{w(CaO) - w(f - CaO) - 1.1w(Al_2O_3) - 0.7w(Fe_2O_3) - 0.7w(SO_3)}{2.8[w(SiO_2) - w(f - SiO_2)]} \quad (2.6)$$

It is noteworthy that KH is normally adopted to control the lime content in clinker in Chinese cement factory (except some imported from abroad). But the modulus value formula used for controlling the lime content in clinker is not the same around the world. Some commonly used parameters are as follows:

(1) Hydraulic modulus
In 1868, the German W. Michaëlis proposed the hydraulic modulus as a coefficient to control optimum lime content in the clinker. It is the ratio of alkaline oxide to acidic oxide mass fraction (%) in clinker, expressed as HM (or m). The formula is as follows:

$$HM = \frac{w(CaO)}{w(SiO_2) + w(Al_2O_3) + w(Fe_2O_3)} \quad (2.7)$$

where $w(CaO)$, $w(SiO_2)$, $w(Al_2O_3)$, $w(Fe_2O_3)$ refer to the content (mass fraction, %) of CaO, SiO_2, Al_2O_3, Fe_2O_3 in the clinker.

At present, the hydraulic modulus is used in Japan, and the HM *value* is usually between 1.8 and 2.4.

(2) Lea and Parker lime saturation factor
According to the study of the quaternary phase diagram of CaO-Al_2O_3-SiO_2-Fe_2O_3, F.M. Lea and T.W. Parker suggested that although C_3S, C_3A and C_4AF can be formed in Portland cement clinker, the maximum allowable lime content should not be determined directly according to these mineral compositions. Because the clinker cannot achieve equilibrium cooling during the actual cooling process, it is possible to precipitate f-CaO in the uncombined lime in clinker. So, it is necessary to limit the lime content to a lower value. Accordingly, they proposed the lime saturation factor (LSF).

In order to facilitate the derivation of the formula, first, the influence of the cooling process on the clinker composition of the $CaO-C_2S-C_{12}A_7$ system is discussed. The phase transition of clinker in the $CaO-C_3S-C_3A$ ternary system during cooling process is shown in Figure 2.4.

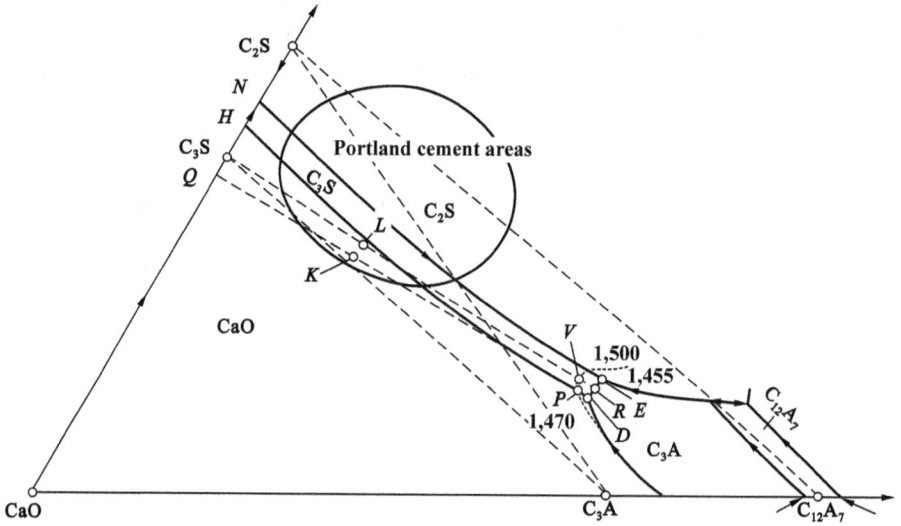

Figure 2.4: The Portland cement area of $CaO-C_2S-C_{12}A_7$ system.

Point D is the reaction point of the $CaO-C_3S-C_3A$ ternary system. When the liquid phase is cooled to this point, the CaO in the solid phase is sunk back to precipitate C_3S and C_3A. Original components K and L are on each side of the C_3S-D line. They are heated to firing temperature 1,500 °C to be balanced and then cooled.

When the composition K is heated to 1,500 °C, the liquid phase composition is the crossover point P of 1,500 °C isotherm in C_3S initial phase area and $CaO-C_3S$ boundary HD. The solid phase composition is the crossover Q of the PK extension line and the $CaO-C_3S$ line. Solid phase Q consists of C_3S and CaO. When cooled, the liquid phase P was moved to D. At point D, CaO was sunk back to precipitate C_3S and C_3A. As the reaction proceeds, the product layer that the CaO pass through becomes thicker continuously, and the resistance becomes stronger. Only when the cooling process is very slow, it is then possible to sink back the CaO entirely. Otherwise, CaO residue must be there.

When the composition L is heated to 1,500 °C, the solid phase composition is C_3S, liquid phase composition is crossover point V of C_3S-L extension line and 1,500 °C isotherm. At this point, no matter how the cooling process is, there will be no CaO residue.

Therefore, the maximum lime content cannot exceed the C_3S-D line to ensure that there is no f-CaO in clinker in $CaO-C_2S-C_{12}A_7$ system.

The clinker phases variation in the CaO-C_2S-$C_{12}A_7$-C_4AF quaternary system during the cooling process is shown in Figure 2.5.

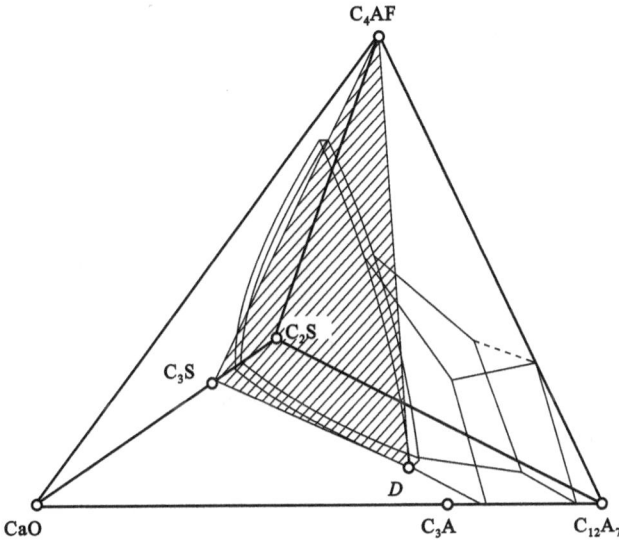

Figure 2.5: The maximum lime content in clinker of CaO-C_2S-$C_{12}A_7$-C_4AF system.

Based on the same mechanism, in order to ensure there is no f-CaO, the maximum lime content should not exceed the C_3S-D-C_4AF plane. The equation of the plane can be written by the composition of C_3S, D and C_4AF. The components of C_3S, D and C_4AF are given in Table 2.1.

Table 2.1: The components of points C_3S, D and C_4AF.

Composition point	Components (mass fraction, %)			
	$w(CaO)$	$w(Al_2O_3)$	$w(SiO_2)$	$w(Fe_2O_3)$
C_3S	73.6	0	26.4	0
D	59.7	32.8	7.5	0
C_4AF	46.1	21.0	0	32.9

The determinant of the plane equation is written from the contents of Al_2O_3, SiO_2 and Fe_2O_3 (mass fraction, %) from the three points:

$$\begin{vmatrix} w(Al_2O_3) & w(SiO_2) - 26.4 & w(Fe_2O_3) \\ 32.8 & 7.5 - 26.4 & 0 \\ 21.0 & 0 - 26.4 & 32.9 \end{vmatrix} = 0 \qquad (2.8)$$

Formula (2.8) can be rewritten as

$$3.78w\,(SiO_2) + 2.18w\,(Al_2O_3) + 1.65w\,(Fe_2O_3) = 100 \qquad (2.9)$$

According to the quaternary system, this can be further written as

$$w\,(CaO) + w\,(SiO_2) + w\,(Al_2O_3) + w\,(Fe_2O_3) = 100 \qquad (2.10)$$

From formulas (2.9) and (2.10), the maximum lime content in CaO-C_2S-$C_{12}A_7$-C_4AF system yields

$$w\,(CaO) = 2.78w\,(SiO_2) + 1.18w\,(Al_2O_3) + 0.65w\,(Fe_2O_3) \qquad (2.11)$$

The ratio of actual lime content to the maximum lime content in clinker is the Lea and Parker LSF:

$$LSF = \frac{100w(CaO)}{2.78w(SiO_2) + 1.18w(Al_2O_3) + 0.65w(Fe_2O_3)} \qquad (2.12)$$

The LSF of Portland cement clinker usually fluctuates between 85 and 95, which is mostly used by the European and American countries at present.

2.2.2.2 Silica modulus

The silica modulus is the abbreviation of the silica/oxygen ratio (also called silica modulus), which indicates the ratio of the percentage of SiO_2 to the percentage of Al_2O_3 and Fe_2O_3 in clinker, abbreviated as SM (in Russian: n):

$$SM = \frac{w(SiO_2)}{w(Al_2O_3) + w(Fe_2O_3)} \qquad (2.13)$$

The SM in Portland cement clinker is usually between 1.7 and 2.7. But the SM of white Portland cement can be up to 4.0 or even higher. In addition to reflecting the ratio of SiO_2 content to that of Al_2O_3 and Fe_2O_3, SM also indicates the ratio of silicate minerals to fluxing minerals in the clinker, which reflects the quality and burnability of the clinker. SM varies with the ratio of silicate minerals to fluxing minerals. If SM in the clinker is too high, the calcination of clinker is difficult and it would be difficult to form C_3S because at high temperatures, liquid phase decreases significantly. If the content of CaO is low, the content of C_2S would be too much, and the clinker would be easily pulverized. If the SM is too low, strength of clinker would decrease because of less silicate minerals, and due to the excessive liquid phase, the material lump, ring and furnace accretion would be easy to appear which will affect the operation of the kiln.

2.2.2.3 Alumina modulus

Alumina modulus is short for alumina ratio (also called iron modulus, IM), which represents the mass ratio of Al_2O_3 and Fe_2O_3 in the clinker. It can be abbreviated as IM (in Russian: p). The calculation formula is

$$IM = \frac{w(Al_2O_3)}{w(Fe_2O_3)} \qquad (2.14)$$

The IM of Portland cement clinker is usually between 0.9 and 1.7. The IM of sulfate-resistant Portland cement or low heat cement can be as low as 0.7. IM indicates the ratio of the content of Al_2O_3 to Fe_2O_3 in clinker, which also suggests the proportional relationship between C_3S and C_4AF. IM is also related to the condensation speed of clinker and liquid phase viscosity, thus affecting the degree of difficulty in clinker calcination.

When the IM in clinker is high, the content of C_3A and the liquid phase viscosity are high. The material is difficult to calcine, so the setting time of cement is short. But when the IM is low, the liquid phase viscosity is low and the particles in the liquid phase are easy to diffuse, which is beneficial to form C_3S. When the sintering range is narrow, the material lump is easy to be formed in kiln, which affects the operation of kiln.

At present, KH, SM and aluminum modulus (IM) are used in China. In order to calcine smoothly, ensure the quality of clinker and stability of mineral composition. The three modulus values should be chosen by the raw materials, fuel and equipment and other details in the factory. The three values should be coordinated with each other properly. A certain modulus value cannot be individually emphasized. In general, three rate values cannot be both high and low at the same time.

2.2.3 Calculation and conversion of mineral composition of clinker

The mineral composition of clinker can be determined by petrographic analysis and X-ray quantitative analysis. It can also be calculated according to the chemical composition. However, the calculated values based on chemical composition are theoretically possible minerals, which are called "potential minerals." In the case of stable production conditions, the true mineral composition of clinker has a certain correlation with the calculated mineral composition, which is able to indicate the effect of mineral composition on clinker and cement performance. Hence, it is still widely used in China.

The petrographic analysis is based on the percentage of each mineral in the unit area measured under a microscope, and then their percentage multiplies by the density of the corresponding minerals to obtain each mineral content. The calculation of mineral density is shown in Table 2.2.

Table 2.2: The density of mineral used for calculation.

Mineral	C_3S	C_2S	C_3A	C_4AF	Glass	MgO
Density (g/cm³)	3.13	3.28	3.00	3.77	3.00	3.58

The results of this mineral determination method are more consistent with the actual situation, but when the mineral crystals are smaller, the errors may result from overlapping.

X-ray analysis is based on the ratio of the characteristic peak intensity of each mineral in the clinker and the characteristic peak intensity of the single mineral to obtain its content. This method has less error, but when the content is too low, the measurement is not accurate. The error of infrared spectral analysis is relatively small. Quantitative analysis of clinker minerals can also be performed by electron probe analyses, X-ray spectrum analyses and so on.

2.2.3.1 The method of KH

To calculate conveniently, first, the ratio of relative molecular mass is listed:

$$C_3S: \frac{M_r(C_3S)}{M_r(CaO)} = 4.07; \quad C_2S = \frac{M_r(CaO)}{M_r(SiO_2)} = 1.87; \quad C_4AF: \frac{M_r(C_4AF)}{M_r(Fe_2O_3)} = 3.04;$$

$$C_3S: \frac{M_r(C_3A)}{M_r(Al_2O_3)} = 2.65; \quad CaSO_4: \frac{M_r(CaSO_4)}{M_r(SO_3)} = 1.7; \quad \frac{M_r(Al_2O_3)}{M_r(Fe_2O_3)} = 0.64$$

The content of CaO for the reaction with SiO_2 is set as C_s, and the content of SiO_2 for the reaction with CaO is set as S_c, then:

$$C_s = w \text{ (CaO)} - [1.65w \text{ (Al}_2O_3) + 0.35w \text{ (Fe}_2O_3) + 0.75w \text{ (SO}_3)]$$
$$= 2.8KH \bullet S_c \tag{2.15}$$

$$S_c = w(SiO_2) \tag{2.16}$$

In the case of general calcination, first, CaO and SiO_2 react to form C_2S, and the remaining CaO reacts with the partial C_2S to produce C_3S. The content of C_3S can be calculated from the remaining CaO content ($C_s - 1.87S_c$):

$$w(C_3S) = 4.07 \ (C_s - 1.87S_c)$$
$$= 4.07C_s - 7.06S_c$$
$$= 4.07 \ (2.8KH \bullet S_c - 7.60S_c) \tag{2.17}$$
$$= 3.8 \ (3KH - 2) \ w(SiO_2)$$

Because

$$C_s + S_c = w \ (C_3S) + w(C_2S) \tag{2.18}$$

Therefore,

$$w(C_2S) = C_s + S_c - w \ (C_3S) = C_s + S_c - \ (4.07C_s - 7.60S_c)$$
$$= 8.60S_c - 3.07C_s = 8.60S_c - 3.07 \ (2.8KH - S_c) \tag{2.19}$$
$$= 8.60 \ (1 - H) \ w \ (SiO_2)$$

When the C_3A content is calculated, the content of Al_2O_3 [$0.64w(Fe_2O_3)$] consumed by the formation of C_4AF should be deducted from the total content of Al_2O_3, and C_3A content can be calculated from the remaining Al_2O_3 [$w(Al_2O_3) - 0.64w(Fe_2O_3)$]:

$$w(C_3A) = 2.65\,[w\,(Al_2O_3) - 0.64w\,(Fe_2O_3)] \tag{2.20}$$

where $0.64w(Fe_2O_3)$ refers to the content of Al_2O_3 consumed by the formation of C_4AF:

0.64 – the ratio $\frac{M_r(Al_2O_3)}{M_r(Fe_2O_3)} = 0.64$.

According to $\frac{M_r(C_4AF)}{M_r(Fe_2O_3)} = 3.04$, the content of C_4AF can be calculated:

$$w(C_4AF) = 3.04M_r(Fe_2O_3) \tag{2.21}$$

The content of $CaSO_4$ can be calculated from the SO_3 content:

$$w(CaSO_4) = 1.71\,w\,(SO_3) \tag{2.22}$$

Similarly, the clinker mineral composition can be calculated when IM < 0.64:

$$w(C_3S) = 3.8\,(3KH - 2)\,w(SiO_2) \tag{2.23}$$

$$w(C_2S) = 8.60\,(1 - KH)\,w(SiO_2) \tag{2.24}$$

$$w(C_4AF) = 4.766\,w(Al_2O_3) \tag{2.25}$$

$$w(C_2F) = 1.70w(Fe_2O_3) - 2.666\,w(Al_2O_3) \tag{2.26}$$

$$w(CaSO_4) = 1.70\,w(SO_3) \tag{2.27}$$

2.2.3.2 The Bogue method

The R.H. Bogue method is also called the algebraic method. It is based on the material balance. First, the formula between clinker chemical composition, mineral composition and clinker modulus value is listed from the simultaneous equations, and then solve the equations. The calculation formula of clinker mineral composition can be done. In fact, as long as a set of data among the clinker chemical composition, modulus values and mineral composition are known, the other two sets of the data can be calculated. That is to say, the three can be converted into each other.

(1) When the clinker mineral composition is known, calculate the chemical component of clinker

The content of each mineral and oxide in the clinker is represented as $w(C_3S)$, $w(C_2S)$, $w(C_3A)$, $w(C_4AF)$, $w(C_2F)$, $w(CaSO_4)$ and $w(CaO)$, $w(SiO_2)$, $w(Al_2O_3)$, $w(Fe_2O_3)$, $w(SO_3)$, respectively. The chemical composition (mass fraction, %) of the five minerals and $CaSO_4$ is shown in Table 2.3.

Table 2.3: The chemical composition of the five minerals and $CaSO_4$ (mass fraction, %).

Oxides	C_3S	C_2S	C_3A	C_4AF	C_2F	$CaSO_4$
CaO	73.69	65.12	62.27	46.16	41.26	41.19
SiO_2	26.31	34.88				
Al_2O_3			37.73	20.98		
Fe_2O_3				32.86	58.74	
SO_3						58.81

According to the data shown in Table 2.3, the following formula of each oxide in clinker can be calculated:

When IM ≥ 0.64:

$$w(\text{CaO}) = 0.7369w(\text{C}_3\text{S}) + 0.6512w(\text{C}_2\text{S}) + 0.6227w(\text{C}_3\text{A})$$
$$+ 0.4616w(\text{C}_4\text{AF}) + 0.4119w(\text{CaSO}_4) \tag{2.28}$$

$$w(\text{SiO}_2) = 0.2361w(\text{C}_3\text{S}) + 0.3488w(\text{C}_2\text{S}) \tag{2.29}$$

$$w(\text{Al}_2\text{O}_3) = 0.3773w(\text{C}_3\text{A}) + 0.2098w(\text{C}_4\text{AF}) \tag{2.30}$$

$$w(\text{Fe}_2\text{O}_3) = 0.3286w(\text{C}_4\text{AF}) \tag{2.31}$$

$$w(\text{SO}_3) = 0.5881w(\text{CaSO}_4) \tag{2.32}$$

When IM <0.64:

$$w(\text{CaO}) = 0.7369w(\text{C}_3\text{S}) + 0.6512w(\text{C}_2\text{S}) + 0.4616w(\text{C}_4\text{AF})$$
$$+ 0.4126\,w(\text{C}_2\text{F}) + 0.4119w(\text{CaSO}_4) \tag{2.33}$$

$$w(\text{SiO}_2) = 0.2631w(\text{C}_3\text{S}) + 0.3488w(\text{C}_2\text{S}) \tag{2.34}$$

$$w(\text{Al}_2\text{O}_3) = 0.2098w(\text{C}_4\text{AF}) \tag{2.35}$$

$$w(\text{Fe}_2\text{O}_3) = 0.3286w(\text{C}_4\text{AF}) + 0.5874w(\text{C}_2\text{F}) \tag{2.36}$$

$$w(\text{SO}_3) = 0.5881w(\text{CaSO}_4) \tag{2.37}$$

(2) When the chemical component is known, calculate the mineral composition of clinker

Solve the equation set of formulas (2.28)–(2.32) and equation set of formulas (2.33)–(2.37), the percentage of each mineral is calculated as follows:

When IM ≥ 0.64:

$$w(C_3S) = 4.07w(CaO) - 7.5986\,w(SiO_2) - 6.7171w(Al_2O_3)$$
$$- 1.4286\,w(Fe_2O_3) - 2.8506w(SO_3)$$
$$(2.38)$$

$$w(C_2S) = 8.5986\,w(SiO_2) + 5.0667w(Al_2O_3) + 1.0776w(Fe_2O_3)$$
$$- 3.07w(CaO) + 2.1502w(SO_3)$$
$$(2.39)$$

$$w(C_3A) = 2.6504w(Al_2O_3) - 1.6922w(Fe_2O_3) \tag{2.40}$$

$$w(C_4AF) = 3.0432w(Fe_2O_3) \tag{2.41}$$

$$w(CaSO_4) = 1.7004w(SO_3) \tag{2.42}$$

When IM < 0.64:

$$w(C_3S) = 4.07w(CaO) - 7.5986\,w(SiO_2) - 4.4769\,w(Al_2O_3)$$
$$- 2.8588\,w(Fe_2O_3) - 2.8506\,w(SO_3)$$
$$(2.43)$$

$$w(C_2S) = 8.5986\,w(SiO_2) + 3.3769\,w(Al_2O_3) + 2.1564\,w(Fe_2O_3)$$
$$- 3.07\,w(CaO) + 2.1502\,w(SO_3)$$
$$(2.44)$$

$$w(C_4AF) = 4.7662\,w(Al_2O_3) \tag{2.45}$$

$$w(C_2F) = 1.7024w(Fe_2O_3) - 2.6663\,w(Al_2O_3) \tag{2.46}$$

$$w(CaSO_4) = 1.7004w(SO_3) \tag{2.47}$$

(3) When the clinker mineral composition is known, calculate the modulus values of clinker

(i) When the clinker minerals consist of C₃S, C₂S, C₃A, C₄AF and CaSO₄:

$$KH = \frac{w(C_3S) + 0.8837w(C_2S)}{w(C_3S) + 1.3256w(C_2S)} \tag{2.48}$$

$$KH^- = \frac{w(C_3S) + 0.8837w(C_2S) - 1.3572w(f - CaO) - 0.9507w(SO_3)}{w(C_3S) + 1.3256w(C_2S) - 3.8w(f - SiO_2)} \tag{2.49}$$

$$SM = \frac{w(C_3S) + 1.325w(C_2S)}{1.434w(C_3A) + 2.046w(C_4AF)} \tag{2.50}$$

$$IM = \frac{1.15w(C_3A)}{w(C_4AF)} + 0.64 \tag{2.51}$$

$$LST = \frac{73.69w(C_3S) + 65.12w(C_2S) + 62.27w(C_3A) + 46.16w(C_4AF) + 41.19w(CaSO_4)}{0.7367w(C_3S) + 0.9766w(C_2S) + 0.4452w(C_3A) + 0.4612w(C_4AF)}$$
$$(2.52)$$

$$\mathrm{HM} = \frac{0.7369w(\mathrm{C_3S}) + 0.6512w(\mathrm{C_2S}) + 0.6227w(\mathrm{C_3A}) + 0.4616w(\mathrm{C_4AF}) + 0.4119w(\mathrm{CaSO_4})}{0.2631w(\mathrm{C_3S}) + 0.3488w(\mathrm{C_2S}) + 0.3773w(\mathrm{C_3A}) + 0.5384w(\mathrm{C_4AF})}$$

(2.53)

(ii) When the clinker minerals consist of C_3S, C_2S, C_2F, C_4AF and $CaSO_4$:

$$\mathrm{KH} = \frac{w(\mathrm{C_3S}) + 0.8837w(\mathrm{C_2S})}{w(\mathrm{C_3S}) + 1.3256w(\mathrm{C_2S})}$$

(2.54)

$$\mathrm{KH^-} = \frac{w(\mathrm{C_3S}) + 0.8837w(\mathrm{C_2S}) - 1.3572w(\mathrm{f-CaO}) - 0.9507w(\mathrm{SO_3})}{w(\mathrm{C_3S}) + 1.3256w(\mathrm{C_2S}) - 3.8w(\mathrm{f-SiO_2})}$$

(2.55)

$$\mathrm{SM} = \frac{0.2631w(\mathrm{C_3S}) + 0.3488w(\mathrm{C_2S})}{0.5384w(\mathrm{C_4AF}) + 0.5874w(\mathrm{C_2F})}$$

(2.56)

$$\mathrm{IM} = \frac{0.2098w(\mathrm{C_4AF})}{0.3286w(\mathrm{C_4AF}) + 0.5874w(\mathrm{C_2F})}$$

(2.57)

$$\mathrm{LST} = \frac{73.69w(\mathrm{C_3S}) + 65.12w(\mathrm{C_2S}) + 46.12w(\mathrm{C_4AF}) + 41.26w(\mathrm{C_4F}) + 41.19w(\mathrm{CaSO_4})}{0.7367w(\mathrm{C_3S}) + 0.9766w(\mathrm{C_2S}) + 0.4612w(\mathrm{C_4AF}) + 0.3818w(\mathrm{C_4F})}$$

(2.58)

$$\mathrm{HM} = \frac{0.7369w(\mathrm{C_3S}) + 0.6512w(\mathrm{C_2S}) + 0.4616w(\mathrm{C_4AF}) + 0.4126w(\mathrm{C_2F}) + 0.4119w(\mathrm{CaSO_4})}{0.2631w(\mathrm{C_3S}) + 0.3488w(\mathrm{C_2S}) + 0.5834w(\mathrm{C_4AF}) + 0.5874w(\mathrm{C_2F})}$$

(2.59)

(4) When the modulus values of clinker are known, calculate the chemical component of clinker

(i) When KH, SM, IM of clinker are known, calculate the mineral composition and chemical component:

$$w(\mathrm{Fe_2O_3}) = \frac{\Sigma}{(2.8\mathrm{KH} + 1)\ (\mathrm{IM} + 1)\ \mathrm{SM} + 2.65\mathrm{IM} + 1.35}$$

(2.60)

$$w(\mathrm{Al_2O_3}) = \mathrm{IM} \cdot w(\mathrm{Fe_2O_3})$$

(2.61)

$$w(\mathrm{SiO_2}) = \mathrm{SM} \cdot [w(\mathrm{Al_2O_3}) + w(\mathrm{Fe_2O_3})]$$

(2.62)

$$w(\mathrm{CaO}) = \Sigma - [w(\mathrm{SiO_2}) + w(\mathrm{Al_2O_3}) + w(\mathrm{Fe_2O_3})]$$

(2.63)

In the formulas, Σ represents the estimated value of total content of the four main oxides of SiO_2, Al_2O_3, Fe_2O_3 and CaO (97–99%).

(ii) When LSF, SM and IM of clinker are known, calculate the mineral composition and chemical component:

$$w(\mathrm{Fe_2O_3}) = \frac{100\Sigma}{2.8\mathrm{LST} \cdot \mathrm{SM}(\mathrm{IM} + 1) + 1.18\mathrm{LST} \cdot \mathrm{IM} + 0.65\mathrm{LST} + 100\ (\mathrm{SM} + 1)(\mathrm{IM} + 1)}$$

(2.64)

$$w(Al_2O_3) = IM \cdot w(Fe_2O_3) \tag{2.65}$$

$$w(SiO_2) = SM(IM + 1) \cdot w(Fe_2O_3) \tag{2.66}$$

$$w(CaO) = \sum - [w(SiO_2) + w(Al_2O_3) + w(Fe_2O_3)] \tag{2.67}$$

In the formula, Σ refers to the estimated value of total content of the four main oxides of SiO_2, Al_2O_3, Fe_2O_3 and CaO (97–99%).

(iii) **When HM, SM and IM of clinker are known, calculate the mineral composition and chemical component:**

$$w(Fe_2O_3) = \frac{\Sigma}{(HM+1)\,(SM \cdot IM + SM + IM + 1)} \tag{2.68}$$

$$w(Al_2O_3) = IM \cdot w(Fe_2O_3) \tag{2.69}$$

$$w(SiO_2) = SM[w(Al_2O_3) + w(Fe_2O_3)] \tag{2.70}$$

$$w(CaO) = \sum - [w(SiO_2) + w(Al_2O_3) + w(Fe_2O_3)] \tag{2.71}$$

In the formula, Σ concerns the estimated value of total content of the four main oxides of SiO_2, Al_2O_3, Fe_2O_3 and CaO (97–99%).

2.2.3.3 The difference between actual mineral composition and calculated mineral composition of clinker

The calculation of mineral composition of Portland cement clinker is based on the assumption that the clinker is cooled in balance and the mineral is pure C_3S, C_2S, C_3A and C_4AF. Its calculation results are not completely consistent with actual mineral composition, sometimes there are large differences.

(1) The influence of solid solution

The calculated minerals are pure C_3S, C_2S, C_3A and C_4AF, but the actual minerals are solid solutions and a few solid solutions with other oxides, such as Alite, Blite and ferrite solid solution. For example, if the composition of Alite is considered as $54CaO \cdot 15SiO_2 \cdot MgO \cdot Al_2O_3$ (can be abbreviated as $C_{54}S_{15}MA$), in the formula of calculating $w(C_3S)$, the coefficient in front of $w(SiO_2)$ is not 3.80 but 4.30, so the actual content will be increased by 11%. But the content of C_3S is decreased due to partial Al_2O_3 is dissolved into Alite.

(2) The influence of cooling condition

If the Portland cement clinker cooling rate is slow and the process is balanced crystallization, the liquid phase is almost crystallized to C_3A, C_4AF and other minerals. But in the industrial production condition, the cooling rate is faster, so the liquid phase can partially or even mostly become glass. At this point, the content of C_3A and C_4AF is lower than the calculated value, while the content of C_3S may increase and the content of C_2S may decrease.

(3) The influence of alkalis and other trace components

$K_2O\cdot23CaO\cdot12SiO_2$ (abbreviated as $KC_{23}S_{12}$) can be formed by alkalis and silicate minerals, $Na_2O\cdot8CaO\cdot3Al_2O_3$ (NC_8A_3) can be formed by alkalis and C_3A, while the CaO is precipitated. Thus, C_3S is decreased and the NC_8A_3 appears. Alkalis may also affect the content of C_3S. Other minor oxides such as TiO_2, MgO and P_2O_5 also affect the mineral composition of the clinker.

Although there are certain differences between the calculated mineral composition and the measured value, it can basically explain the effect on clinker calcination and performance. It is also the only feasible method to calculate the mineral composition of a cement clinker or raw materials composition. Therefore, it is widely used in cement industry.

2.3 Raw materials and raw materials proportioning of Portland cement

2.3.1 Raw materials of cement

The main elements in cement clinker are calcium, silicon, aluminum and iron. In principle, the minerals containing these elements in nature can be used as raw materials for cement production, which include the sedimentary rocks, igneous rocks and metamorphic rocks that make up the earth. However, cement products are large in quantity, their quality is homogeneous and are of low price. These features limit their orientation. Raw materials of cement are usually made up of three types of materials:

(1) the main raw materials: calcareous materials (limestone);
(2) auxiliary raw materials: siliceous and aluminous materials (clay);
(3) regulating raw materials: siliceous, aluminous and irony.

2.3.1.1 Calcareous materials

Calcareous raw material is the main raw material for cement production. It mainly provides CaO. Production of 1 t clinker requires about 1.5 t of raw materials, of which about 1.1 t is calcareous raw material. Calcareous raw materials can be divided into two kinds: natural calcareous raw materials and industrial solid residues (also named industrial waste). Limestone, which is the most abundant in nature, is currently used.

(1) Natural calcareous raw materials

The commonly used calcareous raw materials are lime rock, griotte, marlite and chalk. The lime rock is widely used in most of the Chinese cement factories. Griotte and marlite are also used in a few factories. There is a small amount of chalk in Henan, Shanxi and Sichuan provinces.

Lime rock is deposited by chemical and biochemical processes. The purest limestone is calcite and aragonite. Generally, limestone contains impurities like dolomite [$CaMg(CO_3)_2$], siliceous materials (quartz or chert) and clay. Calcite belongs to hexagonal system, and the relative density is 2.7. Aragonite belongs to orthorhombic system and the relative density is 2.95. Griotte is a variety of coarse-grained calcite.

Limestone is a natural calcareous material of lime rock. Generally, it is a dense rock with fine grain crystal structure. Due to the presence of various impurities, the color is often gray, black, brown and so on, and the purest limestone is white. The hardness of limestone depends on the geological age, usually the older the geological age is, the harder the limestone becomes. The hardness of limestone in the Mohs standard is 1.8–3.0, and the relative density is 2.6–2.8. In a knife it can be easily carved white traces with a shell-like fracture.

Marlite is a rock between lime rock and clay, which is a homogeneous mixture composed of calcareous and clayey materials. In geology, marlite is a sedimentary rock obtained by the deposition of calcium carbonate and clayey materials.

Chalk is a sedimentary rock by the accretion of marine organism (shell, foraminifera and so on). The main component is fine-grained calcilutite, with a small amount of biological debris, and the content is not high in general. The color is usually white, light yellow and light green. The structure is loose. It can be crushed by the finger, and some are like soil but easy to grind; the content of $CaCO_3$ can reach up to 90%.

The kinds and genetic types of natural calcareous materials have a direct influence on the phases of digging, crushing, grinding and calcination in cement manufacturing. For instance, the mineral morphology, crystallinity, particle size, cementation medium, compaction rate, impurity presence type and mineral morphology of limestone have obvious effects on its behaviors during cement production (e.g. grinding property, chemical reactivity and sintering property).

(2) Industrial solid residues

Comprehensive utilization of industrial solid residues has become a major task of the cement industry. Carbide slag, filtered mud, caustic dross and so on can be used as calcareous raw materials.

The carbide slag is the slaked lime slurry containing 90% water generated by hydrolyzing calcium carbide in acetylene production workshop in the chemical plant. One ton calcium carbide can produce about 1.15 t carbide slag. Carbide slag consists of more than 80% fine particles with size ranges of 10–50 μm.

The main content of filtered mud by carbonic acid method in sugar refinery, and the caustic dross by chloralkali process in soda plant is calcium carbonate, which can be used as the calcareous raw material for the production of Portland cement.

The chemical composition of partial calcium raw materials is given in Table 2.4.

Table 2.4: The chemical composition of calcareous raw materials (mass fraction, %).

Ore lithology	LOI[a]	SiO$_2$	Al$_2$O$_3$	Fe$_2$O$_3$	CaO	MgO	K$_2$O	Na$_2$O	SO$_3$	Cl$^-$	Total
Limestone	39.15	7.88	1.92	1.00	48.35	1.20	0.71	0.03	0.02	0.001	99.06
Dolomitic limestone	43.80	2.27	0.61	1.08	39.62	11.96	0.24	0.02	0.00	0.017	99.62
Siliceous zebra limestone	26.92	22.92	8.08	2.86	32.85	3.66	1.46	0.13	1.08	0.001	99.96
Argillaceous limestone	41.40	4.02	1.41	0.59	47.97	3.69	0.74	0.03	0.12	0.017	99.99
Silty limestone	37.88	12.50	0.96	0.85	46.00	0.88	0.20	0.10	0.28	0.010	99.66
Sandy griotte	25.34	35.49	3.28	1.48	32.59	0.82	0.49	0.43	0.00	0.002	99.88
Griotte	43.48	0.49	0.09	0.04	55.55	0.12	0.05	0.02	0.01	0.003	99.85
Marlite	26.08	29.05	5.21	2.88	30.60	1.79	1.09	0.57	0.50	0.240	98.01
Chalk	36.37	12.22	3.26	1.40	45.84	0.81					99.90

a LOI, loss on ignition.

2.3.1.2 Siliceous and aluminous materials

The siliceous and aluminous raw materials are the main source of acid oxides (SiO_2 and Al_2O_3) in cement clinker. Siliceous and aluminous materials are divided into natural clay and industrial solid residue.

(1) Natural clay

Natural clayey materials mainly include loess, clay, shale, mudstone, siltstone, river mud, phyllite and extrusive rocks.

Loess and clay are migrated and deposited after the weathering and decomposition of granite and basalt, and the difference is only the degree of weathering. Loess is mainly distributed in north and northwest China. The primary minerals in loess are mainly quartz, followed by feldspar, muscovite, calcite and gypsum; clayey minerals are mainly illite, followed by montmorillonite and beillite; the alkali mainly comes from muscovite and feldspar. The main chemical composition is SiO_2 and Al_2O_3. The loess contains more fine sand relatively, and the silicon ratio is higher; in general, the SM = 3.5–4 and the IM = 2.5–2.8. In northern China, due to the characteristics of seasonal and perennial drought, the weathering and leaching effect is weak and the alkali content ($[w(K_2O)+w(Na_2O)]$) is high, which is around 3.5–4% in general, and sometimes it is even higher. The content of clay particles is mostly between 20% and 40%.

The clay is widely distributed in China. These are the black soil and brunisolic soil in the northeast, red earth and yellow earth in the south, and laterite in the north and the northwest. Because of the different natural conditions in different areas, the mineral composition is also different.

The clayey minerals contained in black soil and brunisolic soil are hydromica and montmorillonite. The nonclayey minerals are mainly finely dispersed quartz, followed by feldspar, calcite, mica and so on. The content of SiO_2 is relatively high; in general, SM = 2.7–3.1, IM = 2.6–2.9. The alkaline content of black soil is generally 4–5%, and it is below 3.5% in brunisolic soil.

The clay minerals contained in red earth and yellow earth are mainly kaolinite, followed by illite, hydrargillite and so on. The nonclayey minerals are mainly quartz, feldspar, hematite and so on. The alkaline content is low; the SM = 2.5–3.3 and IM = 2–3 in general.

The clayey minerals contained in laterite are mainly illite and kaolinite, followed by some nonclayey minerals, such as feldspar, quartz, calcite and muscovite. The content of SiO_2 is lower and the content of Al_2O_3 and Fe_2O_3 is relatively high, and their SM = 1.4–2.6 and IM = 2–5.

The phyllite rock is still formed by the metamorphism of the clayey, silty materials and partial intermediate-basic volcanic rocks and pyroclastic rocks in low-temperature areas. As a result of low temperature and strong stress, the phyllite is fine. It is usually constructed in a fine direction with phyllitic structure. It is a kind of rock that has deeper metamorphism than the clintheriform rock.

Silky luster and minuteness wrinkle can be commonly seen in thin schistosity. Although almost all of the original rock recrystallize, new minerals are much more and the particles are still very fine with an average particle size below 0.1 mm; hence, it is difficult to distinguish simply by the human eyes. The typical minerals in phyllite are sericite, quartz, albite, chlorite and so on, and the latter two are low in content. The auxiliary minerals are magnetite, rutile and tourmaline. Phyllite is widely distributed such as Liaohe Group in Liaodong areas, Kuangping group of Qinling Mountains, Banxi Group in the south, Hutuo Group in Shanxi province and Kunyang Group.

Shale is a clayey rock cemented and formed by crystal pressure. Its stratification is clear and its color is indefinite. It is usually gray, brown, black or green. Its chemical composition is similar to that of clay. The dominant minerals are quartz, feldspar, mica, calcite, and other rock fragments. The SM of argillaceous shale is low, and the SM in fine-grained sandy shale and the deposit that sandstone interbedded with shale are higher.

(2) Industrial solid residues
Industrial solid residues include fly ash from thermal power plants, calcium-added slag and gangue discharged from coal mines, red mud in aluminum refinery, dross and slag from steel factory. These materials are generally with low SM, which should be used as regulating raw materials.

The chemical composition of partial clayey materials is shown in Table 2.5.

2.3.1.3 Regulating raw materials
When the calcareous raw materials, siliceous and aluminous materials cannot meet the requirements of raw materials, the corresponding regulating materials should be added according to the missing components. The regulating materials are mainly silico-regulating materials, aluminum-regulating materials and irony-regulating materials.

(1) Silico-regulating materials
Silico-regulating materials are mainly quartzite, sandstone, siltstone, river sand and sandy limestone. In general, the content of SiO_2 in silico-regulating raw materials is 70–90%, or SM ≥ 4. The main direction of prospecting is higher degree of weathering, loose siliceous siltstone, sandstone, sand and so on. But the silica, quartz sandstone, quartzite sandstone and so on with high hardness are often encountered. They are highly corrosive and poorly grindable. It will bring difficulty in grinding the raw materials and in clinker calcination.

The minerals in sandstone are mainly quartz and feldspar. Crystalline SiO_2 has an adverse effect on grinding and calcination, so it should be used as less as possible. The quartz crystal of river sand is coarse and should only be used when there is no

Table 2.5: The chemical composition of partial clayey materials (mass fraction, %).

Ore lithology	LOI	SiO$_2$	Al$_2$O$_3$	Fe$_2$O$_3$	CaO	MgO	K$_2$O	Na$_2$O	SO$_3$	Cl$^-$	Total
Loess	10.38	55.86	12.23	4.67	9.73	2.29	2.27	1.61	0.04	0.040	99.12
Black soil	6.41	62.68	17.55	6.11	0.93	1.06	2.70	1.11	0.05	0.024	98.62
Laterite	8.42	49.83	23.29	12.71	0.28	0.73	1.69	0.07	0.05	0.002	97.07
Clay	5.12	69.79	14.01	5.92	0.51	0.95	1.90	0.66	0.01	0.005	98.87
Sandy loam	7.28	63.38	13.35	5.12	4.71	0.63	2.37	1.60	0.15	0.003	98.59
Phyllite	5.14	63.68	17.79	7.76	0.14	0.61	3.49	0.20	0.03	0.003	98.84
Sandy shale	6.08	65.81	14.97	7.41	1.49	1.14	2.00	0.46	0.63	0.009	99.99
Siltstone	4.39	69.65	15.76	5.33	0.36	0.71	2.65	0.26	0.04	0.001	99.15
Argillaceous siltstone	3.08	74.50	12.92	5.06	0.12	0.69	2.97	0.29	0.014	0.004	99.64
Black shale	7.50	60.27	14.45	5.48	4.21	2.52	2.37	1.58	0.36	0.030	98.77
Shale	5.34	64.19	17.00	6.91	0.82	1.42	4.13	0.09	0.01	0.004	99.91
Pyroclastic	-0.27	47.55	16.74	13.00	7.58	6.80	1.98	3.56	0.00	0.029	96.97

sandstone. It is better to use weathered sandstone or siltstone, and their content of SiO_2 is not too low, but it is easy to be ground and has little effect on calcination.

(2) Aluminum-regulating materials

When the content of Al_2O_3 in raw meal is insufficient, aluminum-regulating material must be added. Commonly used aluminum-regulating materials are bauxite, fly ash, argil and other siliceous and aluminous raw materials containing higher aluminum and less iron.

(3) Ferrous-regulating materials

Ferrous-regulating materials are mainly ironstone, sulfate slag and copper slag.

The chemical composition of partial silico-regulating materials, aluminum- and Ferrous-regulating materials is shown in Table 2.6.

2.3.1.4 Composition of requirements in the raw material

(1) CaO

The content of CaO in cement raw material is about 44%. In principle, all the raw material prepared for qualified raw material can be used for cement production. Limestone and clay are both independent and mutual penetration in nature. For the modern cement industry, a certain reserves and different grades of limestone can be put on standby in enterprises, as long as there is a certain amount of limestone whose CaO content is no less than 48% as the high calcareous-regulating material for cement production.

(2) SiO$_2$

The content of SiO_2 in cement raw material is about 14%. Generally, the siliceous and aluminous materials cannot meet the requirements of ingredients, and silico-regulating materials are needed. There are many kinds of associated minerals in limestone, and silicon is one of them, which is beneficial to a certain extent and can supplement the silicon in raw materials. But if the content of SiO_2 is too high, it should be used with high-quality limestone.

The form of silicon has a significant effect on the reactivity of raw materials. The diffusion rate of CaO in the SiO_2 lattice is 4–5 times higher than that of SiO_2 in the CaO lattice. Therefore, it is thought that the formation of mineral in clinker is the main result of diffusion that calcium ion to SiO_2 particles, so the silicon-contained phase is often referred to as a factor influencing the raw materials reaction ability. The mineral forms of SiO_2 in limestone are mainly clay, microquartz, coarse-grained quartz, chert concretion, opal and chalcedony. The rock is formed by the deposition of microquartz with uniform distribution and calcareous materials in the same environment. The molecules are closely related to each other, because the particles are small. They

Table 2.6: The chemical composition of partial silico-regulating materials, aluminum- and irony-regulating materials (mass fraction, %).

Raw materials	LOI	SiO$_2$	Al$_2$O$_3$	Fe$_2$O$_3$	CaO	MgO	K$_2$O	Na$_2$O	SO$_3$	Cl$^-$	Total	Notes
Quartz sandstone	0.21	95.06	2.02	0.11	0.52	0.02	1.17	0.08	0.01	0.001	99.20	Silico-regulating materials
Sandy griotte	25.34	35.49	3.28	1.48	32.59	0.82	0.49	0.43	0.00	0.002	99.88	
River sand	0.40	92.09	3.25	1.51	0.87	0.20	1.10	0.38			99.80	
Silica	0.34	97.76	0.20	0.35	0.00	0.55	0.07	0.08	0.02	0.010	99.38	
Sandstone	1.58	82.37	8.72	2.39	0.84	0.70	2.14	0.33	0.10	0.020	99.19	
Vein quartz	1.01	91.60	0.65	1.75	2.30	0.35	0.19	0.04			97.89	
Bauxite	13.54	36.07	39.79	9.38	0.20	0.32	0.08	0.07	0.00	0.002	99.45	Aluminum-regulating materials
Argil	7.63	62.68	21.69	3.20	0.49	1.13	2.62	0.12	0.00	0.004	99.56	
Fly ash	4.16	57.12	22.83	7.90	3.33	1.64	1.90	0.22	0.38	0.002	99.48	
Sulfate slag	0.62	23.50	4.03	59.80	4.52	2.29	1.10	0.03	2.70	0.001	98.95	Irony-regulating materials
Copper slag	−6.07	33.64	4.33	50.00	10.06	1.91	0.64	0.34	2.29	0.009	97.14	
Ironstone	5.38	23.82	5.38	58.91	3.39	0.92	1.38	0.22	0.05	0.004	99.45	

have little impact on the grindability and are beneficial to calcination. Coarse-grained quartz, chert, opal and chalcedony are generally in strip or lumps in the limestone. They have great effect on the abrasion and grindability and often influence the type selection of the crusher and mill. As for the effect on burnability, the ability of different kinds of SiO_2 to react with CaO increases in the following order:

Chalcedony $\rightarrow$ opal $\rightarrow$ α-cristobalite and tridymite $\rightarrow$ SiO_2 in feldspar $\rightarrow$ SiO_2 in mica, amphibole and clay mineral $\rightarrow$ SiO_2 in granulated blast furnace slag.

(3) Al_2O_3 and Fe_2O_3

If Al_2O_3 and Fe_2O_3 are not sufficient in the raw ingredients, regulating materials can be added. In the production of medium heat, low heat, sulfate-resistant and oil well cement, the content of Al_2O_3 in limestone should be limited. The production of white cement generally requires the content of Fe_2O_3 in limestone not more than 0.05%.

(4) MgO

When the content of MgO in limestone is between 2.5% and 3.5%, the content of MgO in clinker can be as high as 3.5–4.5%. These phenomena appear in the pre-calciner kiln when calcinated, such as ring formation, great pellet formation and the decrease in clinker quality. As for pre-calciner kiln, the high content of MgO is harmful; therefore, the content of MgO should be limited.

In addition, if the content of MgO in cement clinker is too high, it can cause destructive magnesium expansion in cement and concrete. But the standards of limitation on MgO content in cement clinker around the world are various, and some detailed key parameters are presented as follows: 3% for Belgium; 4% for France, Italy, Britain, Mexico and New Zealand; 4.5% for Australia; 5% for Germany, Austria, Canada, Finland, Norway, Holland, Spain, Greece, the United States, Russia, Japan, Argentina, Chile and most of the countries; 6% for some states such as the United States, China, India and Brazil; 7% for Columbia. It is stipulated in China that when the content of MgO in clinker is more than 5%, the cement autoclaved test should be done every day and the expansion rate of test pieces should be less than 0.5%.

In China, based on some standards, it can be noticed that the limitation on MgO content (by mass) should be lower than 5.0%. If the cement can smoothly pass the autoclave test, the limited MgO content can be increased to 6.0%. Normally, the critical MgO amount in limestone is about 3%. When MgO content is low in other raw materials, or there is limestone with low content of MgO used in combination, the limited MgO content can be appropriately enhanced.

(5) SO_3

During the process of new dry method production, the organic sulfur, sulfide and sulfate in raw materials generate SO_3 when heated, which escapes with exhaust gas. Partial SO_3 functions with $CaCO_3$ in the upper preheater, raw material mill and dust collector, and $CaSO_4$ is generated. Some SO_3 are discharged into the atmosphere with exhaust gases. The absorbed $CaSO_4$ comes back to the preheater system again, while the content of SO_3 in the exhaust gas discharged to the atmosphere must comply with the limitation of *The emission standards of air pollution in cement industries* (GB 4915–2013), which is promulgated by the State Environmental Protection Administration.

In order to reduce sulfate ring formation, crust and blocking, the content of SO_3 must be limited in every raw material. It is stipulated in the *Technical specifications for engineering investigation of limestone mines* (GB 50955–2013) that the content of SO_3 must be less than 1.0%, but for the specific mines, it should be determined with the content of SO_3, K_2O and Na_2O in other raw materials and fuels to ensure that the content of SO_3 is less than 1.5% and there is a suitable sulfur alkali ratio. As for the enterprises setting bypass system, the limitation of SO_3 content in limestone is flexible.

(6) K_2O and Na_2O

The cement with high content of K_2O and Na_2O easily reacts with alkali-active aggregate, which can cause destructive expansion. When the content of K_2O and Na_2O is high, the later strength (28 days) decreases, the setting time becomes short and the soundness is below the standard; in the operation, there is ring formation in kiln, crust and blocking in preheater at the end of kiln. So, the content of alkali must be limited. However, this limitation is different because of the production method, the adaptability of the equipment itself and the variety of cement products. The understanding of this problem is various around the world. According to the practices in China, when the equivalent alkali content ($[w(K_2O)+w(Na_2O)]$) is ≥1.5% in the clinker, the physical properties of the clinker began to deteriorate significantly, and the ring formation and crust appear in operation; many of the world's cement workers argue that the maximum range of alkali content in the clinker should be less than 1.0–1.2%. When the alkali content in the clinker fluctuates at the limitation boundary, special attention should be paid to adjust the sulfur alkali ratio to control volatilization of volatile substances or to improve the clinker performance.

It is stipulated in the *Technical specifications for engineering investigation of limestone mines* (GB 50955–2013) that $[w(K_2O)+w(Na_2O)]$ should be no higher than 0.6%, but this limitation is only as a reference, because when the alkali content in limestone and clay reaches the corresponding specifications (0.6% and 4%, respectively), the content of alkali in the clinker prepared by the limestone, clay and iron – three-component ingredients must be more than 1.5%. As for the enterprises producing low alkali cement, the limitation can be decreased to 0.4% or lower. As for the enterprises setting bypass system, the limitation of alkali content in limestone is flexible.

(7) Cl^-

Cl^- in limestone is present in the form of KCl, NaCl and so on. If the chloride content in cement is high, it will accelerate the corrosion of reinforcement in concrete, especially for prestressed concrete. In addition, chlorine in the raw materials (including fuel) is volatilized in the kiln and condensed at the temperature of 800 °C at the end of the kiln, resulting in crust and blocking of the preheater and the kiln tail affecting the production. It is required that the content of Cl^- in raw materials should be less than 0.015 wt% (sometimes 0.02 wt%). As for the enterprises setting bypass system, the limitation of alkali content in limestone is flexible.

(8) Others

P_2O_5, TiO_2, Mn_2O_3 and other oxides are also limited in the cement clinker. Sometimes, there is a small amount of apatite in limestone, but as long as the content of P_2O_5 in the raw materials is within 2–2.5%, it will not cause considerable impact. Other oxides are rare in limestone.

2.3.1.5 Fuel

The so-called fuel usually refers to a reasonable material in the economy that can have intense oxidation reaction with oxygen, releasing heat. The coal is used as the primary fuel in cement industries in China. Combustible garbage as an alternative solid fuel is also developed and utilized in recent years.

Coal is able to produce the required high temperature for clinker production and the ash content in coal is a component of clinker. It would fall into clinker and involve in the formation of clinker. Thus, the quality of coal had a direct impact on the clinker quality. In order to ensure that the flame temperature of the rotary kiln is higher and the coal ash falling into the clinker is relatively uniform, the lump coal is usually pulverized and then poured into the kiln because the pulverized coal burns faster than the lump coal and can reach higher temperature and is easy to adjust.

In order to ensure the complete combustion of pulverized coal, and avoid reducing gas that has harmful effect on clinker quality in the process of the formation clinker, rotary kiln is usually operated in excessive air state. The level of volatile matter in pulverized coal will directly affect the combustion characteristics of pulverized coal, such as ignition temperature, combustion stability and burning time. In order to achieve the required temperature and flame shape of calcination, the requirements of particle size of the pulverized coal will change with the volatile content. The pulverized coal with low volatiles must be ground finer than pulverized coal with high volatiles.

If the conditions permit, it is ideal to use bituminous coal in the rotary kiln, which is conducive to the control of the flame and kiln operation and improve the quality of clinker; while it is better to use anthracite in vertical kiln, which is conducive to reducing the heat losses that volatiles volatilize on the surface of kiln, so it can reduce heat consumption of the clinker. The distribution of coal storage in China is largely nonuniform. There is no or rare bituminous coal in many places, but rich in anthracite resources, such as Guangdong and Fujian provinces. In addition, poor-quality (high ash content and low calorific value) coal rejected by the cement industry in the past is now widely used with technological advances due to the energy shortage. The quality of coal for cement kilns should generally be in accordance with the general quality requirements of *Standard of design for cement plant* (GB 50295–2016), as shown in Table 2.7. Under the premise of satisfying the quality of clinker, the coal can be of low quality and low grade or as an alternative fuel.

Table 2.7: The quality requirements of coal in cement production.

Sequence	Name	Symbol	Value
1	Ash content	$w(A_{ad})$	≤28.00%
2	Volatile matter content	$w(V_{ad})$	≤35.00%
3	Sulfur content	$w(S_{ad})$	≤2.00%
4	Net calorific power	$Q_{net,ad}$	≤23,000 kJ/kg
5	Moisture content	$w(M_{ad})$	≤15.00%

2.3.2 Clinker composition design

The selection of cement clinker's mineral composition should be considered comprehensively according to the varieties of the cements and strength grades, raw material and fuel quality, the preparation of raw materials and clinker calcination process, so that it can achieve high quality, high yield, low consumption and long-term safe operation.

2.3.2.1 Cement varieties and strength grades

When producing ordinary Portland cement, in terms of ensuring the cement strength grades, the normal setting time and ideal stability, the chemical composition can fluctuate within a certain range. High iron, low iron, low silicon, high silicon, high saturation coefficient, low saturation coefficient and the other ingredients program can be used. The three modulus values should be coordinated with each other properly. A certain modulus value cannot be individually emphasized. KH or SM should be higher in clinker that requires high strength grades. On the contrary, the requirement can be lower. In general, when high KH is adopted, it is appropriate to reduce the SM; when high SM is adopted, it is appropriate to reduce KH; when both high KH and high SM are adopted, due to the demand for higher calcination temperature, C_3S is often difficult to react completely in actual production, so there will be more f-CaO, affecting the soundness of cement, or the long-term safe production of kilns.

The appropriate mineral composition should be selected in the production of special cement or characteristic cement according to its special requirements. In the production of fast-hardening Portland cement, it is required that C_3S and C_3A content should be high, and KH and IM should be increased. If increasing the C_3A content is difficult, the C_3S content can appropriately increase again. In the production of sulfate-resistant cement, C_3A and C_3S content should be reduced, which means the KH and IM should be low. In the production of moderate or low heat cement, first, the C_3A content in the clinker should be reduced, and the C_3S content should be appropriately reduced; usually C_3S content in the cement clinker is 40–55% and C_3A content is not more than 6.0%.

2.3.2.2 Raw material quality

The chemical composition, burnability and process performance of the cement raw materials have remarkable influences on the selection of clinker's mineral composition. If the raw materials present good burnability, the KH or SM can be increased appropriately, which is beneficial to improve the clinker strength. If the burnability is poor, KH or SM should not be too high; otherwise, f-CaO content in the clinker will be too high, which will further affect the soundness of the cement. In general, the ingredients program with two or three kinds of raw materials should be adopted. Unless the ingredients program cannot guarantee normal production, it is considered that the raw materials should be replaced or other regulating raw materials should be added. For example, when the content of silicon in the clay is low, the SM of clinker is difficult to increase. At this point if the use of silico-regulating raw materials can increase the SM of clinker, the addition of a raw material will bring difficulties to the production process; therefore, sometimes it is preferable to use ingredients program with low silicon and high saturation coefficient rather than silico-regulating raw materials.

In addition, when the flint content in the limestone and coarse sand content in clay is high, because it is difficult to grind raw materials and calcination of clinker, KH should not be too high. If the alkali content is too high in raw materials, KH should be reduced appropriately. If MgO content of raw materials is too high, as liquid phase increases, the liquid viscosity decreased, but for the pre-decomposition kiln, SM and IM should be increased appropriately.

2.3.2.3 Fuel quality

Fuel quality not only affects the calcination process, but also affects the quality of clinker. In general, as for the high-quality fuel with high calorific value, the flame temperature is high, and KH can be higher. If the fuel quality is poor, in addition to the low flame temperature, the coal ash sinks inhomogeneously, clinker quality is poor, so KH can be lower.

Usually anthracite is suitable for shaft kiln and bituminous coal is suitable for rotary kiln. If the bituminous coal is used in shaft kiln, volatiles in the coal are easily volatilized from the surface of the kiln resulting in losses; hence, the heat consumption of raw material should be improved. If anthracite is used in rotary kiln, due to the high ignition temperature of anthracite, the fineness of coal should be improved and the four-channel burner adapting to anthracite should be adopted. For pre-decomposition kiln, it is advisable to increase the volume of the decomposition furnace and prolong the residence time of the pulverized coal in the furnace.

In addition to the shaft kiln of the black raw material, when coal ash is added into the clinker, the distribution is often inhomogeneous, which has a great impact on the quality of clinker. According to statistics, due to inhomogeneous incorporation of coal ash, KH will be reduced by 0.04–0.16; SM will be decreased by 0.05–0.20; IM will be increased by 0.05–0.30. When the content of coal ash increases, the clinker strength decreases, the fineness of coal and the performance of multichannel burner should be improved, or KH could be reduced appropriately to facilitate the normal production.

When the coal quality changes, clinker composition should be adjusted accordingly. For the rotary kiln, the heat of coal is high and the volatiles are low; hence, the flame is long, the burning part is short, the heat concentration is concentrated, the clinker is easy to form chunk, f-CaO increases and the service life of the refractory brick is shortened. At this point in addition to attempting to extend the flame burning part, the KH could be reduced and IM could be increased appropriately. When low-quality coal is used due to the shortage of coal resources, special attention should be paid to fluctuations in coal, and coal pre-homogenization is necessary. Meanwhile, KH should be reduced appropriately in order to ensure the stability of clinker component and the improvement of cement quality.

If the liquid or gas fuel is used, the flame intensity is high and it is easy to control the shape of it. There is almost no ash content, so KH can be appropriately increased.

2.3.2.4 Fineness and uniformity of raw materials
The uniformity of cement raw materials' chemical composition not only affect the stability of the kiln thermal system and the improvement of operating rate, but also affect the quality of clinker and determination of ingredients program.

In general, if the raw material uniformity is good, KH can be higher. The standard deviation of KH is usually less than 0.02, and that of SM should be less than 0.1. If there are large fluctuations in the raw material composition, as for the rotary kiln, KH should be appropriately reduced; but for the vertical kiln, the low KH easily lead to chunk in vertical kiln. In order to ensure the normal calcination in the vertical kiln, the program of high KH and low SM should be adopted. If the raw material is coarse, because the chemical reaction is difficult to react completely, KH should also be appropriately reduced.

2.3.2.5 Kiln type and specification
The materials are heated and calcined differently in different types of kilns; hence, the composition of the clinker should be different. Compared to vertical kiln and

Lepol kiln, the material rolling constantly in the rotary kiln is heated evenly and the added coal ash is uniform. The material reaction process is more consistent, so KH could be appropriately higher.

Hot air of Lepol kiln flows from the top to the bottom into the material layer of heating machine. Most of the coal ash subside in the upper material surface, and the upper material temperature is higher than the lower materials; hence, KH of the upper material is low and decomposition rate is high, and KH of the lower material is high and decomposition rate is low. Therefore, KH should be appropriately reduced.

Aeration and calcination are not uniform in vertical kiln, so KH in clinker without mineralizer should be appropriately lower. For the clinker with mineralizer, because the liquid phase appears earlier and the viscosity is lower, the sintering temperature range is widened, and the ingredients program of high KH, low SM and high IM is generally used.

The raw materials are preheated well and the decomposition rate is high in pre-decomposition kiln. In addition, due to heat loss of the unit yield kiln is small, decomposition zone of the carbonate which has the maximum heat consumption has been moved out of the kiln, so air temperature is high in kiln. In order to adhere coating conveniently and prevent crusting, clogging and forming chunk, at present, it is a tendency to the ingredients program of low liquid phase. In China, the ingredients program of high SM, high IM and moderate KH is mostly adopted in large-scale pre-decomposition kiln.

There are many factors influencing the selection of clinker composition. A reasonable ingredient program should consider both the quality of clinker and the burnability of the material. It is necessary to consider not only modulus values and the absolute value of mineral composition, but also the internal relationships among them. In principle, the three values cannot be high or low at the same time. The reference ranges for the modulus values of the different kiln-type Portland cement clinker are shown in Table 2.8.

Table 2.8: The reference ranges for the modulus values of the different kiln-type Portland cement clinker.

Type of kiln	KH	SM	IM	Heat consumption of clinker $q/(kJ/kg)$
Pre-decomposition kiln	0.88–0.92	2.4–2.8	1.4–1.9	2,920–3,200
Long kiln with wet process	0.88–0.91	1.8–2.4	1.1–1.7	5,833–6,667
Kiln with dry process	0.86–0.89	2.0–2.4	1.0–1.6	5,850–7,520
Lepol kiln	0.85–0.88	1.9–2.3	1.0–1.8	4,000–5,850
Vertical kiln (without mineralizer)	0.88–0.92	1.9–2.2	1.1–1.4	4,200–5,430
Vertical kiln (with compound mineralizer)	0.92–0.97	1.7–2.2	1.1–1.7	3,750–5,000

2.3.3 Proportioning calculation of raw materials

2.3.3.1 Basic concepts

After clinker composition is determined, according to the raw materials used, the ingredient is calculated to obtain the raw material mix proportion which meets the clinker composition requirements. Proportioning calculation on the basis of material balance, the amount of reactant should be equal to the amount of product. Before proportioning calculation is introduced, the following basic concepts should be known:

(1) Complete black raw materials, semi-black raw materials and common raw materials

In the preparation of raw materials, the raw material obtained by grinding all the coal together with the raw material is known as complete black raw material; the raw material obtained by grinding partial coal with the raw material (the rest of the coal is added to the raw material when calcination) is known as semi-black raw material; coal-free raw material is called common raw material (coal is added from the head of kiln).

(2) Dry basis

Once the material is dried, it is in a dry state. The mass of dry state is set as the calculation units, which is known as the dry basis. The dry raw material mix proportion and the chemical composition of the feedstock are usually expressed by dry basis.

(3) Burning basis

After burning and excluding the loss of ignition, the material is in the ignition state. When the calculation is based on the material's ignition state, it is called as burning basis. The cement raw material is in a burning state after removing the loss of ignition (crystallization water, carbon dioxide and volatile matter, etc.). If the loss of production is not considered, then the following relationships are obtained:

The mass of ignition of complete black raw material = the mass of the clinker;
the mass of the ignition semiblack raw material (or common raw material) + the mass of coal ash incorporated into the clinker = the mass of the clinker.

There are many methods of raw material proportioning calculation, such as algebraic method, graphic method, trial and error method (including decreasing trial and error method and accumulative trial and error method), mineral composition method, least squares method and so on. With the development of science and technology, the application of computer has been gradually spread to various fields, and developed intelligent proportioning calculation programs have been equipped in "Expert system of laboratory in cement factory" in the market.

Now, accumulative trial and error method is mainly introduced. The principle is: based on the requirements of clinker chemical composition, a variety of raw materials are added successively and the chemical composition of the added raw materials is calculated. Then, the clinker composition is calculated accumulatively; if the composition does not meet the requirements, then try to do this again until the requirements are satisfied. Here is an example.

2.3.3.2 Proportioning calculation

Example The relevant analysis data of known raw materials and fuels are shown in Tables 2.9 and 2.10. Assuming that three raw materials are used for production by the pre-decomposition kiln, it requires the three modulus values of clinker: KH = 0.90, SM = 2.6, IM = 1.6. The unit clinker heat consumption is 3,260 kJ/kg. Calculate the mix proportion.

Table 2.9: The chemical composition of raw materials and coal ash (mass fraction, %).

Name	LOI	SiO_2	Al_2O_3	Fe_2O_3	CaO	MgO	Others	Total
Limestone	42.66	2.42	0.31	0.19	53.13	0.57	0.72	100.00
Clay	5.27	70.25	14.72	5.48	1.41	0.92	1.95	100.00
Iron powder	0.00	34.42	11.53	48.27	3.53	0.09	2.16	100.00
Coal ash	0.00	61.52	27.34	4.46	4.79	1.19	0.70	100.00

Table 2.10: Proximate analysis of coal.

$w(M_{ad})$	$w(V_{ad})$	$w(A_{ad})$	$w(F_{c,ad})$	$Q_{net,ad}$
0.60%	22.42%	28.56%	49.02%	20,930 kJ/kg

Solution

(1) Calculate the amount of coal ash
The amount of coal ash in 100 kg clinker can be calculated approximately according to the following formula:

$$G_a = \frac{q \cdot w(A_{ad}) \cdot S}{100 Q_{net, ad.}} = \frac{P \cdot w(A_{ad}) \cdot S}{100} = \frac{3,260 \times 28,056 \times 100}{100 \times 20,930} = 4.45\%$$

where
G_a is the amount of coal ash in clinker, %; q is the unit clinker heat consumption, kJ/kg; $Q_{net,ad}$ is the calorific value of coal, kJ/kg; $w(A_{ad})$ is the ash content of air dry basis of coal, %; S is the sinking rate of coal, %, the value is selected as 100%; P is the coal consumption, kJ/kg.

(2) According to the modulus values, estimate the chemical composition of the clinker
When KH = 0.90, SM = 2.6 and IM = 1.6 are known, according to formulas (2.60)–(2.63), assume $\Sigma =$ 97.5%, the chemical composition of clinker can be calculated as follows:

$$w(Fe_2O_3) = \frac{\Sigma}{(2.8KH+1)\ (IM+1)\ SM+2.65IM+1.35)} = 3.32\%$$

$$w\ (Al_2O_3) = IM\cdot\ w\ (Fe_2O_3) = 5.31\%$$

$$w(SiO_2) = SM\cdot\ [w\ (Al_2O_3)\ +w\ (Fe_2O_3)] = 22.43\%$$

$$w(CaO) = \Sigma -\ [w(SiO_2)\ +w(Al_2O_3)+w(Fe_2O_3)] = 66.44\%$$

(3) Accumulative trial and error method

The calculation of the cumulative trial and error method of raw materials proportioning is shown in Table 2.11. The last line "accumulated clinker component" in the table is the chemical composition of the clinker, and the three values (KH = 0.902, SM = 2.60, IM = 1.56) and clinker heat consumption (3360.3 kJ/kg) in the column are the modulus values and the heat consumption of the calculated clinker, which is very close to the ingredients program of the required clinker. It is worth noting that the total value of the accumulated clinker does not have to be equal to 100, as long as the modulus values and heat consumption of the checked clinker meet the requirements, but at the moment components of clinker must be converted into mass fraction (%).

As a result of three kinds of raw materials (limestone, clay and iron powder) for proportioning calculation, it can only meet the requirements of the two modulus values of clinker from the mathematical principle, and it is impossible to meet the requirements of three modulus values. If the third modulus value is different from the required modulus value, the raw materials (change the raw material composition) can be replaced, or the other two modulus values can be adjusted to take into account the requirements of the third modulus value. If the third modulus value is different greatly from the required value so that it cannot meet the requirements of clinker composition design, nor by adjusting the other two modulus values to take into consideration the third value, the regulating raw materials can only be used. That is to say, four kinds of raw materials are used for proportioning calculation. Its calculation processes and the principles are the same, they are not repeated here.

(4) Calculate the materials consumption of clinker

From Table 2.11, the demand dry materials for preparation of clinker are as follows:

$$m(Limestone) = \frac{124}{99.99} \times 100\% = 124.01\ kg$$

$$m(Clay) = \frac{124 - 2.0}{99.99} \times 100\% = 22.00\ kg$$

$$m(Iron\ powder) = \frac{3.2+0.4}{99.99} \times 100\% = 3.60\ kg$$

(5) Calculate the raw material mix proportion

$$w(Limestone) = \frac{124}{124.01+22.00+3.60} \times 100\% = 82.89\%$$

$$w(Clay) = \frac{22.00}{124.01+22.00+3.60} \times 100\% = 14.70\%$$

Table 2.11: Process of accumulative trial and error method (based on 100 kg clinker).

Calculation procedures	SiO$_2$	Al$_2$O$_3$	Fe$_2$O$_3$	CaO	MgO	Others	Total	Notes
Chemical composition of calculated clinker	22.43	5.31	3.32	66.44			97.50	
Coal ash (+4.45)	2.7376	1.2166	0.1985	0.2132	0.053	0.0312		
Limestone (+124)	3.0008	0.3844	0.2356	65.8812	0.7068	0.8928		$(66.44-0.2132)/0.5313 = 124$
Clay (+24)	16.86	3.5328	1.3152	0.3384	0.2208	0.468		$(22.43-2.7376-3.0008)/0.7025 = 24$
Iron powder (+3.2)	1.1014	0.369	1.5446	0.113	0.0029	0.0691		$(3.32-0.1985-0.2356-1.3152)/0.4827 = 3.2$
Accumulated clinker component	23.6998	5.5028	3.2939	66.5458	0.9835	1.4611	101.49	KH = 0.849, SM = 2.69, IM = 1.67
Clay (−2.0)	1.405	0.2944	0.1096	0.0282	0.0184	0.039		
Accumulated clinker component	22.2948	5.2084	3.1843	66.5176	0.9651	1.4221	99.59	KH = 0.910, SM = 2.66, IM = 1.64
Iron powder (+0.4)	0.1377	0.0461	0.1931	0.0141	0.0004	0.0086		
Accumulated clinker component	22.4325	5.2546	3.3774	66.5317	0.9655	1.4307	99.99	KH = 0.902, SM = 2.60, IM = 1.56 $q = 100 \times 3{,}260/99.99 = 3{,}360.3$

$$w(\text{Ironpowder}) = \frac{3.60}{124.01 + 22.00 + 3.60} \times 100\% = 2.41\%$$

Note: The above proportion is the dry raw material mix proportion. If raw materials contain water, they can be converted according to the moisture content.

2.4 Calcination of Portland cement clinker

After ground and homogenized, the cement raw materials are put into the cement kiln system, and burned to become the cement clinker, by drying, preheating and pre-calcination (including pre-decomposition). This process is called the cement clinker calcination and is summarized as follows:

−150 °C	Physical water evaporates from cement raw materials
About 500 °C	Clay materials release bound water and begin to decompose oxides such as SiO_2 and Al_2O_3
About 900 °C	Carbonate decomposes and releases CO_2 and new ecological CaO
900–1,200 °C	The amorphous dehydration product of clay begins to crystallize, and solid-phase reaction carried out among various oxides
1,250–1,280 °C	The produced minerals fully melt and liquid phase appears
1,280–1,450 °C	As the liquid phase increases, C_2S absorbs CaO to form C_3S, until the clinker minerals are formed
1,450–1,300 °C	Clinker mineral cooling

2.4.1 Drying and dehydration

(1) Drying
Raw materials that applied into the kiln contain a certain amount of moisture. Generally, the moisture content of the new dry kiln raw materials is not more than 1%, while the moisture content of vertical kiln and lepol kiln is between 12% and 15%, and the moisture content of the slurry in the wet-process kiln is usually 30–40%. As the temperature increases to 100–150 °C, the moisture in the raw meal is completely eliminated. This process is called the drying process. As 1 kg water evaporates, the latent heat is up to 2,257 kJ (100 °C). Therefore, reducing the moisture of raw materials can reduce the heat consumption of clinker and increase the output of kiln.

(2) Dehydration of clay minerals

There are two kinds of combined water in clay minerals: one of which exists in the crystal layer structure in the ionic state (OH^-), known as the crystal coordination water; the other is adsorbed in the crystal structure in water molecules state (H_2O), called interlayer water or interlayer adsorbed water. The interlayer water can be removed at about 100 °C, while the coordination water must be removed at temperatures as high as 400–600 °C.

After the raw material is dried, it continues to be heated and the temperature rises rapidly. When the temperature rises to 500 °C, the main composition of the clay will be dehydrated and decomposed. The reaction formula is as follows:

$$Al_2O_3 \cdot 2SiO_2 \cdot 2H_2O \rightarrow Al_2O_3 \cdot 2SiO_2 + 2H_2O$$

When kaolin loses chemical bonding of water, its crystal structure is destroyed, producing amorphous metakaolin. Therefore, the reactivity of kaolin after dehydration is higher than before, and when it continues to be heated to 970–1,050 °C, the amorphous metakaolin converts to crystalline mullite and releases heat at the same time.

Montmorillonite and Illite still have crystal structure after dehydration; as a result, their reactivity is worse than that of kaolin. Illite is accompanied with volume expansion when dehydrated, while kaolin and montmorillonite are with the volume shrinkage; hence, when it comes to vertical kiln and lepol kiln production, it is not suitable to use clay dominated by Illite. Otherwise, the poor thermal stability of the ball will cause a blow after putting into the kiln and make a great effect on the ventilation of kiln.

The dehydration and decomposition reaction of clay minerals is an endothermic process. Every 1 kg kaolin absorbs 934 kJ at 450 °C, but due to the low content of raw materials in raw meal, the endothermic reaction is not significant.

2.4.2 Decomposition of carbonate

When the temperature continuously rises to about 600 °C, carbonates in the raw meal begin to decompose, mainly owing to the presence of $CaCO_3$ in limestone and $MgCO_3$ in the raw material. The reactions are as follows:

$$MgCO_3 \rightleftharpoons MgO + CO_2 \uparrow$$

$$CaCO_3 \rightleftharpoons CaO + CO_2 \uparrow$$

2.4.2.1 Characteristics of carbonate decomposition

(1) Reversible reaction

Decomposition of silicate material is the reversible reaction. Because of the great impact of the system temperature and the partial pressure of CO_2 in the surrounding medium, a high reaction temperature must be maintained to reduce CO_2 partial pressure in the surrounding medium or to decrease the concentration of CO_2 in order to proceed the decomposition smoothly.

(2) Strong endothermic reaction

Decomposition of carbonate needs to absorb a lot of heat, which is a process that consumes most of the heat in the process of clinker formation and requires heat that accounts for about one-third of the total heat consumption of wet production, while accounting for about one-half of the suspended preheating kiln or pre-decomposition kiln. Therefore, sufficient heat must be supplied to ensure that the $CaCO_3$ decomposition reaction can be carried out completely.

(3) Increasing decomposition rate with a rise of temperature

The starting temperature of the carbonate decomposition is relatively low, and the decomposition of $CaCO_3$ takes place at about 600 °C, but the reaction rate is very slow. At 894 °C, when the decomposition pressure of CO_2 is 0.1 MPa, the decomposition rate is accelerated. At the temperature of 1,100–1,200 °C, the decomposition rate is extremely high. It can be seen from the experiment that as the temperature is increased by 50 °C, the decomposition rate constantly increases by 100% and the decomposition time is shortened by 50%.

2.4.2.2 The decomposition process of calcium carbonate

The decomposition process of $CaCO_3$ can be summarized as follows:

(1) Heat transfer processes

(i)   Heat flows transfer heat to the surface of particles
(ii) Hear transfers from the material surface to the decomposition surface in a conduction manner

(2) Chemical reaction

On the decomposition surface, $CaCO_3$ is decomposed and release CO_2.

(3) Mass transfer processes

(i)   Released CO_2 gas diffuses through the decomposition layer to the surface
(ii) CO_2 of surface spreads into the gas stream

In these processes, both heat transfer and mass transfer are physical processes, and there is only one chemical reaction process. Due to the different resistance of each process, the decomposition rate of $CaCO_3$ is controlled by one of the slowest processes.

(i)   In general, the decomposition rate of $CaCO_3$ in the rotary kiln mainly depends on the heat transfer process because the material is stacked in the kiln, the heat transfer area is very small and the heat transfer coefficient is also very low.

(ii) While it is necessary to pellet the raw materials for vertical kiln and lepol kiln. The decomposition rate of $CaCO_3$ depends on the heat transfer and mass transfer process due to the large ball diameter, the low heat transfer rate and the high mass transfer resistance.

(iii) In the new dry process, the decomposition rate of $CaCO_3$ depends on the chemical reaction speed because the raw material powder can be suspended in the airflow, the heat transfer area is large, the heat transfer coefficient is high and the mass transfer resistance is small.

2.4.2.3 Factors that affect the decomposition reaction of calcium carbonate

(1) The structural and physical properties of limestone
It is difficult for limestone with dense structure, fine particle arrangement, large crystallization, little crystal and hard texture to decompose, such as marble. Cretaceous with soft texture and marlstone with other containing components have lower activation energy and are easily decomposed.

(2) Fineness of raw materials
It is easy to take the decomposition reaction because the raw material is fine, the particle is even, the coarse grain is small in case that the ratio of surface area of raw material increases and the heat transfer and mass transfer speed are accelerated.

(3) Reaction conditions
Increase in the reaction temperature could speed up the rate of decomposition reaction and diffusion of CO_2. The CO_2 reaction can be accelerated by strengthening the ventilation and discharging the reaction gas in time.

(4) Suspension dispersion degree of raw materials
In the new dry method, the suspension dispersion of raw material powder is ideal in the preheater and in the decomposition furnace, which can increase the heat transfer area, reduce mass transfer resistance and improve the decomposition speed.

(5) Properties of clay components
If the dominant mineral in the clay material is high-reactive kaolin, it can accelerate the decomposition reaction of $CaCO_3$ due to its easy solid reaction with the decomposition product CaO to produce low calcium minerals. Conversely, if the dominant minerals in clay are high-reactive montmorillonite and illite, the decomposition rate of $CaCO_3$ will be negatively affected. The reactivity of quartz sand composed of crystalline SiO_2 is the lowest.

2.4.3 Solid-phase reaction

2.4.3.1 The process of reaction
In the process of clinker formation, reactive f-CaO appears at the beginning of $CaCO_3$ decomposition, which participates in solid-phase reaction with SiO_2, Fe_2O_3 and Al_2O_3 within the raw materials through the mutual diffusion of particles to form clinker minerals. The process of solid-phase reaction is complex, and the process can be listed as follows:

–800 °C	Begins to form $CaO \cdot Al_2O_3$ (CA), $CaO \cdot Fe_2O_3$ (CF), $2CaO \cdot SiO_2$
800–900 °C	Begins to form $12CaO \cdot 7Al_2O_3$ ($C_{12}A_7$), $2CaO \cdot Fe_2O_3$ (C_2F)
900–1,100 °C	$2CaO \cdot Al_2O_3 \cdot SiO_2$ (C_2AS) formed is then broken down, and C_3A and C_4AF begin to form. All calcium carbonate is decomposed and the content of f-CaO reaches the maximum.
1,100–1,200 °C	A large number of C_3A and C_4AF are formed and the content of C_2S reached the maximum.

The formation of clinker minerals including C_3A, C_4AF and C_2S is a complex multistage reaction with a cross process. The solid-phase reaction of the clinker minerals is exothermic, and the heat of which is about 420–500 kJ/kg when the raw material is used.

Because the solid atom, molecule or ion exhibits a great force between them, the reactivity of solid-phase reaction is low, so as the reaction speed. In general, solid-phase reactions always occur on two sets of interfaces, which are heterogeneous reactions. For granular materials, reaction is first carried out through the point of contact between particles or surfaces, then reactants diffuse through the product layer. Therefore, the solid-phase reaction generally includes two processes, which are reactions on interfaces and migration of the materials.

2.4.3.2 The main factors that affect solid-phase reactions

(1) The fineness and uniformity of raw materials
As the raw material is finer, the particle size is smaller, the specific surface area is larger, the partial free energy on surface is bigger, the reaction and diffusion capacity are stronger and reaction rate is faster. However, to a certain extent, if the grinding is continued, the increase in the speed of the solid-phase reaction won't be obvious, while the milling output will be greatly reduced, and the grinding power consumption will be sharply increased. Especially for precalcining kiln, too fine raw materials will lead to the declination of cyclone tube separation efficiency, resulting in the increase in the external circulation of materials between preheater and dust catcher, and reducing the kiln production, while increasing the clinker heat loss. Therefore, it is necessary to comprehensively balance and optimize the fineness of raw materials. For the precalciner kiln, the size of the raw materials should be as homogeneous as

possible, and the sieve margin of 0.2 mm shall be reduced as much as possible, while the sieve margin of 0.08 mm may be appropriately relaxed.

The homogeneity of the raw materials is good, that is, the mixing of components in the raw material can increase the contact between the components, so it can accelerate the solid reaction.

(2) Temperature and duration

Under low temperatures, the chemical reactivity of the solid is low, and so are the diffusion and migration of the particle. Hence, the solid-phase reaction should be carried out under a high temperature, as raising the reaction temperature can accelerate the solid reaction. Due to the time that diffusion and migration of ions needs, it is necessary to ensure that a certain period of time can be carried out to complete the opposite.

(3) Properties of feedstock

When crystalline SiO_2 (such as flint and quartz sand) and crystalline calcite are present in raw materials, the speed of the solid-phase reaction will be significantly reduced due to the difficulty in breaking its lattice, especially when the raw materials contain coarse quartz sand, the influence is greater.

(4) Mineralizer

A small amount of admixture that can accelerate the formation of crystalline compounds and can make cement raw materials burned easily is called the mineralizer. Mineralizer can enhance the reaction ability by forming a solid solution with the reactants resulting in activating the lattice, or form eutectic mixture with reactant, which leads to the appearance of the liquid phase at a low temperature, accelerate the diffusion and dissolution of the solid phase, and accelerate the reaction by breaking the bond of the reactant. Therefore, the addition of mineralizer can accelerate the solid-phase reaction.

2.4.4 Calcination of clinker

When the material temperatures rise to 1,250–1,280 °C, that is, meeting the minimum eutectic temperature, liquid phases with Al_2O_3, Fe_2O_3 and CaO as the main contents and MgO and alkali included begin to appear. In the role of liquid phases under high temperatures, the materials are gradually sintered, and change from the loose state to clinker with gray and black colors and dense structure. This process is accompanied with volume shrinkage. At the same time, some C_2S and f-CaO are gradually dissolved in liquid phases, and Ca^{2+} diffuses and reacts with silicate ions, that is, C_2S absorbs CaO and forms C_3S, the main mineral of Portland cement clinker. The reaction is as follows:

$$C_2S + CaO \rightarrow C_3S$$

With the increase in temperature and the extension of time, the liquid phase increases, the liquid viscosity decreases, CaO and C_2S continue to dissolve and spread and C_3S continues to form. Small crystals grow up gradually, and ultimately forms dozens of microns of the size of well-developed Alite crystals, then complete the process of clinker firing.

2.4.4.1 Minimum eutectic temperatures

During the heating process, the temperature of the liquid phase of two or more components is called the minimum eutectic temperature. Some minimum eutectic temperatures of different systems are listed in Table 2.12. As is shown, the properties and quantities of components can affect the minimum eutectic temperature of the system. The minimum eutectic temperature of Portland cement clinker is about 1,250–1,280 °C due to some secondary oxides including magnesia, burnt potash, sodium oxide, sulfuric anhydride, titanium oxide and phosphorus oxide. The appropriate amount of mineralizer and other trace elements has a certain effect on reducing the minimum eutectic temperature of the clinker.

Table 2.12: Minimum eutectic temperatures of some systems.

System	Minimum eutectic temperature (°C)
$C_3S-C_2S-C_3A$	1450
$C_3S-C_2S-C_3A-Na_2O$	1430
$C_3S-C_2S-C_3A-MgO$	1375
$C_3S-C_2S-C_3A-Na_2O-MgO$	1365
$C_3S-C_2S-C_3A-C_4AF$	1338
$C_3S-C_2S-C_3A-Fe_2O_3$	1315
$C_3S-C_2S-C_3A-Fe_2O_3-MgO$	1300
$C_3S-C_2S-C_3A-Na_2O-MgO-Fe_2O_3$	1280

2.4.4.2 The amount of liquid phases

Some liquid phases must exist in clinker burning process, because liquid phase is the necessary condition of forming C_2S. With the increase in liquid content, the dissolved amount of CaO and C_2S is increased, and the formation of C_3S is faster. However, when the liquid content is too high, blocks are easy to be packed in the process of calcination, resulting in rotary kiln ring forming, shaft kiln refining and furnace accretion, which will affect the normal production.

The liquid phase depends not only on the composition but also on the content of the composition and the sintering temperature of the clinker. The liquid phase quantity P at different temperatures is calculated according to the following formulas:

$$1,400\,°C \qquad P = 2.95w(Al_2O_3) + 2.2w(Fe_2O_3) \tag{2.72}$$

$$1,450\,°C \qquad P = 3.0w(Al_2O_3) + 2.25w(Fe_2O_3) \tag{2.73}$$

$$1,500\,°C \qquad P = 3.3w(Al_2O_3) + 2.6w(Fe_2O_3) \tag{2.74}$$

Here, $w(Al_2O_3)$ and $w(Fe_2O_3)$ are the mass fraction of Al_2O_3 and Fe_2O_3 in the clinker.

Due to the addition of MgO, K_2O, Na_2O and other components in the clinker, all of these components can be considered to be liquid phases. Therefore, the content of MgO and alkali $[w(Na_2O)+w(K_2O)]$ will be added to the calculation. That is:

$$1,400\,°C \qquad P = 2.95w(Al_2O_3) + 2.2w(Fe_2O_3) + w(MgO) + w(K_2O) + w(Na_2O) \tag{2.75}$$

Usually, the liquid phase of clinker is about 20–30% in the calcination stage, and the white cement clinker is only about 15% as the content of Fe_2O_3 is low.

2.4.4.3 Liquid viscosity

The liquid viscosity influences the formation rate and crystal size of C_3S directly. When the liquid viscosity is small, the viscosity is small and the diffusion velocity of particles in the liquid phase is accelerated, which is beneficial to the formation of C_3S and the growth and development of crystals, and vice versa. The liquid viscosity of clinker varies with temperatures and composition (including a small amount of oxides). With high temperatures and low viscosity, the IM of clinker increases and so is the liquid viscosity.

2.4.4.4 Surface tension of liquid phase

The smaller the surface liquid tension is, the easier to damp the clinker particles or solid-phase materials, which is beneficial to the solid-phase reaction, the solid–liquid phase reaction and the formation of clinker minerals, especially C_3S. The experimental results show that the surface tension of liquid phase decreases with the increase in temperature. Also, when magnesium, alkali and sulfur exist in the clinker, the surface tension of liquid phase will be reduced and consequently promote the sintering of clinker. But when the liquid surface tension decreases, the diameter of the clinker particles will become smaller. If the liquid surface tension is too small, the sand will be produced in the rotary kiln because of the small particle size.

2.4.4.5 The rate of calcium oxide dissolved in the liquid phase of the clinker

The dissolution of CaO in the clinker liquid phase, or the rate of CaO dissolved in the clinker liquid phase, which is related to the size of the CaO particles and the calcination temperature, has a very important effect on the reaction of CaO with C_2S to generate C_3S. With small limestone particles in raw materials and high clinker calcination temperatures, the dissolution rate is fast.

2.4.4.6 The presence of reactants

The results showed that after the calcination of clinker, CaO and C_2S crystals are small in size and have many new structures with large crystal defects. The reactivity is high and the activation energy is small and it is easy to be dissolved in the liquid phase. The experiment also showed that the rapid rise in temperature (>600 °C/min) can result in the dehydration of clay minerals, decomposition of carbonate and solid–phase reaction, and solid–liquid reaction almost coincide; as a result, the reaction is in a new high reactive state. Also, liquid phase, C_2S and C_3S are engendered in a short time. The formation of clinker is always in the process of solid–liquid reaction, which can greatly accelerate the diffusion rate of particles or ions, reduce the activation energy of ion diffusion, speed up the reaction and is beneficial to the rapid formation of C_3S.

2.4.5 Clinker cooling

Generally speaking, the cooling process refers to the period after the solidification of liquid phase (<1,300 °C). But to be strict, after the highest temperature (about 1,450 °C), it is in the cooling stage. The cooling of clinker is not only the decrease of the clinker temperature, but also a series of physical and chemical changes, and the coagulation and phase transition of liquid phase. The clinker cooling has the following effect:

(1) Improve the quality of clinker

When the clinker is cooled, phase transformation occurs in the minerals formed; for instance, β-C_2S is converted to γ-C_2S at slow cooling, and at the same time the volume expansion is about 10%, which makes the clinker powder. Because γ-C_2S is almost anhydrous, the quality of clinker will decrease. Fast cooling and solid solution of some ions can avoid β-C_2S being converted to γ-C_2S, thus obtaining high hydraulicity.

C_3S is unstable below 1,250 °C and decomposes slowly to C_2S and secondary f-CaO, which decreases the hydraulicity. Therefore, increasing the cooling rate can prevent the decomposition of C_3S.

Cement stability is affected by periclase crystal size seriously, and the greater, the more serious. Rapid cooling can bring the absent crystallization of MgO in the vitreous, or fine the crystallization to make it dispersed, thereby reducing the impact on the stability of cement.

Clinker quick cooling can enhance the sulfate resistance of cement. Because clinker is cold, C_3A is mainly vitreous, so its ability to resist sulfate corrosion enhances.

(2) Improving grindability of clinker

The high content of vitreous in the clinker makes the clinker to produce internal stress and the mineral crystal of clinker is smaller; hence, the quick cooling can improve the grindability of clinker remarkably.

(3) Recovery of waste heat

Clinkers posses vast quantities of heat with temperature of 1100 - 1300 °C when entering cooler. If it is cooled to room temperature, about 837 kJ/kg heat can be recycled with secondary air, which is conducive to the combustion of fuel in the kiln and can improve the thermal efficiency of the kiln.

(4) Being favorable for conveying, storing and grinding of clinker

In order to ensure the safe operation of the conveying equipment, the clinker temperature should be lower than 100 °C. Under high temperatures, circular reinforced concrete storage pool of clinker is prone to crack. So to prevent "false coagulation" phenomenon caused by high temperature inside the mill, the grinding body producing balls and the cement packing bag breakage because of too high temperature caused in cement grinding, clinker must be cooled to a relatively low temperature.

2.4.6 The role of other components

2.4.6.1 Calcium fluoride

Calcium fluoride (CaF_2, also known as fluorite) is the most widely used mineralizing agent. In the process of clinker calcination, fluorine ion can destroy the lattice of raw material components, increase the reactivity of raw materials, promote the decomposition of carbonates and accelerate the solid-phase reaction.

When the raw materials contain feldspar and other alkali minerals (such as potassium feldspar), the addition of CaF_2 can reduce decomposition temperature and accelerate decomposition and volatilization. CaF_2 under 1,000–1,200 °C can promote the decomposition of C_3A into $C_{12}A_7$ and CaO, then precipitated CaO is combined with C_2S to generate C_3S, which increases the content of Alite. The above impact is more obvious when raw materials with high Al_2O_3 are calcined (such as the production of white cement).

CaF_2 can significantly reduce the temperature of liquid phase and the sintering temperature of clinker. With the addition of 0.6–1.2% of CaF_2, the sintering temperature can be reduced by 50–100 °C, and the range of firing temperature is enlarged, which is equal to the length of the burning zone that is prolonged, and the reaction time of the material is increased. In addition, CaF_2 can also reduce the viscosity of liquid phase, favor the diffusion of particles in liquid phase and accelerate the formation of C_3S.

Recent studies have shown that the addition of CaF_2 makes C_3S form when the temperature is below 1,200 °C, and the Portland cement clinker can be sintered at about 1,350 °C. The clinker composition contains C_3S, C_2S, $C_{11}A_7 \cdot CaF_2$, C_4AF and other minerals, and sometimes C_3A can also be generated. The quality of the clinker is good and the stability is qualified. The clinker may be fired at a temperature of 1,400 °C or higher to obtain a cement clinker which has a common mineral composition.

When the mineralizer CaF_2 is added, the clinker should be cooled down quickly, which is propitious to increase the content of C_3S.

It should be noted that the use of CaF_2 is suitable for the vertical kiln. The use of CaF_2 in rotary kiln will prolong the setting time of clinker, which is easy to circle and corrode refractory bricks and so on. Hence, CaF_2 is normally not used in the rotary kiln production process.

2.4.6.2 Sulfide

Clay raw materials or shale contains a small amount of sulfur. The sulfur brought by fuel is usually more than that of raw materials. In the oxidizing atmosphere, the sulfur-containing compounds are finally oxidized to SO_3, which is distributed in the clinker, waste gas and fly ash. Sulfur can improve the formation of clinker: SO_3 can reduce the liquid viscosity, increase the liquid phase and be conducive to the formation of C_3S. And SO_3 can form $2C_2S \cdot CaSO_4$ and anhydrous calcium sulfoaluminate $4CaO \cdot 3Al_2O_3 \cdot SO_3$ (abbreviated C_4A_3S). $2C_2S \cdot CaSO_4$ is an intermediate transition compound, which begins to form at about $1,050\,°C$ and decomposes to $\alpha'\text{-}C_2S$ and $CaSO_4$ at about $1,300\,°C$.

$C_4A_3\bar{S}$ is formed at about $950\,°C$ and remains stable at $1,350\,°C$. At near $1,400\,°C$, $C_4A_3\bar{S}$ begins to decompose into CA, CaO and SO_3, and a large amount of decomposition occurs when the temperature is above $1,400\,°C$. $C_4A_3\bar{S}$ is an early strength mineral, so it is advantageous to have an appropriate amount of $C_4A_3\bar{S}$ in the cement clinker.

SO_3 can reduce the temperature of the liquid phase and decrease the liquid viscosity and surface tension. Therefore, SO_3 can significantly promote the growth process of Alite crystal, which is beneficial to its growth into large particles of crystal. But because of the weak hydraulicity of sulfate-containing Alite crystals, this must be noted when sulfide is used as mineralizer alone. Like CaF_2, SO_3 is usually used as a mineralizer in a shaft kiln, but not in the rotary kiln generally.

2.4.6.3 Compound mineralizer of fluorite and gypsum

The compound mineralizer is usually composed of two or more mineralizers, and the most common compound mineralizer is fluorite and gypsum composite mineralizer (namely, fluorine sulfur compound mineralizer).

With the addition of fluorite and gypsum compound mineralizer, the complex process of clinker formation is influenced by many factors, including the composition of clinker (KH and IM), $w(CaF_2)/w(SO_3)$, sintering temperature and so on. The clinker minerals are not exactly the same under different producing conditions. With fluorine sulfur compound mineralizer, the $2C_2S \cdot CaSO_4 \cdot CaF_2$ is formed at $900\text{--}950\,°C$, and when the quaternary transition phase disappears about $1,150\,°C$, there is a large amount of liquid in the material at the same time. The addition of fluorine and sulfur compound mineralizer can remarkably reduce the temperature of liquid phase and reduce the viscosity of liquid phase when the clinker is burned, thus reducing the formation temperature of Alite by 150–200 °C, and promoting the formation of Alite.

The test showed that after adding fluorine and sulfur compound mineralizer, Portland cement clinker can be sintered at a low temperature of 1,300–1,350 °C, which contains high content of Alite and low content of f-CaO. It is also possible to form either high early strength minerals such as $C_4A_3\bar{S}$, $C_{11}A_7 \cdot CaF_2$ or both, so that the early strength of clinker is high. When the calcination temperature is above 1,400 °C, despite the decomposition of the early strength minerals such as $C_4A_3\bar{S}$ and $C_{11}A_7 \cdot CaF_2$, the large quantity of formed Alite and good crystal growth also can contribute to obtaining high quality of cement clinker, of which the ultimate strength is higher than that of the low-temperature sintered clinker.

The ingredient scheme of high KH, low SM and high IM is usually used in the cement clinker with fluorine and sulfur compound mineralizer, of which appropriate gypsum content should be 1.5–2.5% of the SO_3 dosage in the clinker. The advisable dosage of fluorite should be 0.6–1.2% of the dosage of CaF_2 in the clinker, and the proper ratio of fluorine to sulfur $[w(CaF_2)/w(SO_3)]$ is 0.35–0.6.

It is noteworthy that the clinker with fluorine sulfur compound mineralizer will show abnormal condensation phenomenon, including flash setting and slow setting. Flash setting often appears when the KH is low, the calcination temperature is low and the reducing atmosphere in the kiln and the IM is high. When the calcination temperature is too high, the IM is low, the KH is high and the content of MgO and CaF_2 is high. The phenomenon of slow setting will occur. In addition, attention should be paid to corrosion of furnace lining and the air pollution caused by fluorine sulfur compound mineralizer.

Fluorine sulfur composite mineralizer is widely used in shaft kiln calcination. The use of fluorine and sulfur compound mineralizer in rotary kiln will prolong the clinker setting time, make it easy to circle and corrode refractory bricks and so on. So it is usually not used.

2.4.6.4 Alkali

The alkali in cement clinker mainly refers to the two elements of potassium and sodium (represented by R), which are mainly from raw materials. Feldspar, mica and other impurities in clay and limestone are alkaline aluminates. In the use of coal for fuel, there is also a small amount of alkali. In the process of material calcination, caustic alkali and chlor-alkali first volatile, then the alkali carbonate and sulfate follow, and the alkali existed in the feldspar, mica and illite can volatile only at a high temperature. Only a small amount of volatile alkali is discharged into the atmosphere, and the rest will condense on the raw material with lower temperature when the flue gas moves into the low-temperature region of the kiln. For the preheater kiln, it is usually condensed in the lowest two-stage preheater, and then enters the kiln with the raw material. When the temperature rises, it also volatilize, which results in the alkaline circulation. As alkaline circulates and accumulates to a certain extent, it will lead to the adhesion of alkali chloride sulfate (RCl) and alkali (R_2SO_4) compounds on the lowest two-stage preheater cone part or unloading chute and the crust formation. There will be serious blockage, which can affect the normal production. Therefore, when the alkali content of raw materials is high, the kiln with cyclone preheater exhaust alkali should be bypassed.

A small amount of alkali can reduce the minimum eutectic temperature and the sintering temperature of the clinker, but increase the amount of liquid phase and act as a fluxing agent. It does not cause much harm to the clinker performance. However, when alkali content is high, calcination encounters difficulty. At the same time, alkali minerals and solid solutions $KC_{23}S_{12}$ and NC_8A_3 produced by alkali and clinker minerals reaction will make it difficult to form C_3S and will increase f-CaO content, thus affecting the strength of the clinker.

The presence of sulfur in the clinker, due to the formation of sulfides, can mitigate the detrimental effects of alkali. The high alkali content of cement, due to alkali to produce potassium gypsum ($K_2SO_4 \cdot CaSO_4 \cdot H_2O$), can help caking of cement silo and quick setting of cement. Alkali can make the concrete surface bloom (blaze). When making hydraulic concrete, the alkali energy in cement will react with the reactive aggregate to produce the alkali aggregate reaction, resulting in local expansion, which further results in deformation and cracking of the structure.

In general, the alkali content of clinker should be less than 1.3% by Na_2O. When producing low heat cement, it should be less than 0.6% in hydraulic construction. The cyclone preheater and precalciner kiln [$w(K_2O) + w(Na_2O)$] should be less than 1%.

2.4.6.5 Magnesia

Limestone often contains a certain amount of $MgCO_3$ and decomposes MgO involved in the clinker calcination process. Some of the MgO bonded with clinker minerals in the solid solution is partly soluble in the glass phase. A small amount of magnesium oxide can reduce the firing temperature of the clinker and the liquid viscosity, and increase the liquid phase. Also, it can play a role in the fluxing action, which is conducive to the formation of C_3S. However, in the precalciner kiln, it is detrimental because the firing temperature is high and also is the MgO content (>3%). Because it is contrary to the fact that the pre-decomposition kiln requires a low liquid phase and a high viscosity, MgO usually reduces the quality and yields the precalciner clinker. When the MgO content of precalciner clinker is too high, the content of Fe_2O_3 in clinker should be reduced to release the harm caused by MgO.

MgO can also modify the color of cement. A small amount of MgO and C_4AF forms a solid solution, which can change C_4AF from brown to olive green, thereby changing the color of cement into dark green. In the silicate cement clinker, the dissolution of MgO is up to 2%, and the excess MgO exists in the form of periclase and in free state. Therefore, too high content of MgO will affect the stability of cement.

2.4.6.6 Phosphorus oxide

Generally, the content of phosphorus oxide (P_2O_5) in the clinker is very low. When the raw material contains phosphorus, such as phosphorus lime or phosphorus containing compound as a mineralizer, a small amount of phosphorus can be brought into it. When the content of P_2O_5 in the clinker ranges from 0.1% to 0.3%, the clinker strength can be increased. When the content of P_2O_5 in the clinker is too high, it will lead to decomposition of C_3S. The addition of 1% of P_2O_5 will decrease C_3S by 9.9% and will increase C_2S by 10.9%. When the content of P_2O_5 is about 7%, the content of C_3S in the clinker will be reduced to zero. Therefore, when phosphorus is present in raw materials, the content of CaO in the raw materials should be reduced properly so as to avoid the high content of f-CaO. For the clinker's low ratio of $w(C_3S)/w(C_2S)$, the developmental strength is slow and the setting time is long. When apatite contains fluorine, the decomposition of C_3S can be reduced, and the temperature of the liquid phase will be decreased. Therefore, if there is phosphorus in the raw material, fluorite can be added to offset the adverse effects of partial P_2O_5.

2.4.6.7 Titania

The clay material contains a small amount of titanium oxide (TiO_2). Generally, the content of TiO_2 in the clinker is not more than 0.3%. When the clinker contains a small amount of TiO_2 (0.5–1.0%), it can form solid solution with various cement clinker minerals, especially, it has a stabilizing effect on β-C_2S, so the quality of clinker can be improved. But if the content is too high, then it is used to generate the nonhydraulic calcium titanium ore ($CaO \cdot TiO_2$) and so on, which can reduce the content of the ritter in the clinker, thus affecting the strength of the cement. Therefore, the content of TiO_2 in the clinker should be less than 1%.

2.4.7 Main equipments for clinker calcination

The new dry kiln system is a modern cement production method, which is based on suspension preheating and pre-decomposition technology, the modern science and technology and industrial production of the latest achievements, such as raw material mining computer control networked mining, raw materials pre-homogenization, raw materials homogenization, extrusion grinding, wide application of new heat resistance, fire resistance, insulation material and IT technology in cement production process; thus, the production of cement with high efficiency and high quality, resource conservation and cleaner production conform to the requirements of environmental protection and large scale, automation and scientific management features. Figure 2.6 depicts the schematic diagram of pre-decomposition kiln system production process, which mainly comprises the tube (cyclone), pipe (heat transfer pipe), furnace (decomposition furnace), kiln (rotary kiln) and machine (cooler).

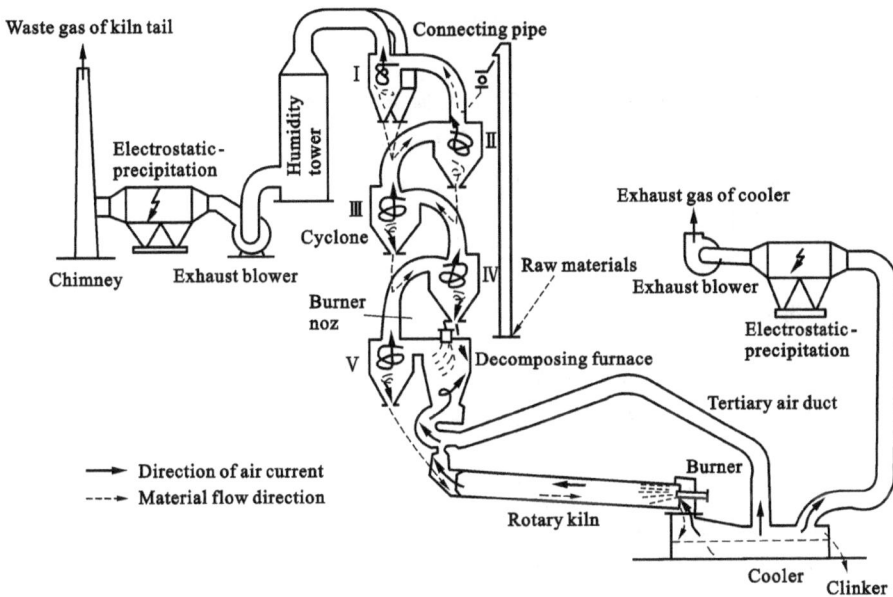

Figure 2.6: Schematic diagram of production process of precalciner kiln system.

2.4.7.1 Cyclone tube

Each stage of the cyclone preheater consists of a cyclone tube and a heat exchange tube (as shown in Figure 2.7). The main task of the cyclone is the separation of gas and solid. In this way, the raw material heated by the preheating unit in the upper stage is separated from the cyclone tube before entering the lower stage heat exchange unit to continue heating up. Therefore, the design of cyclones should mainly be focused on how to obtain higher separation efficiency and lower pressure loss.

The airflow is mainly affected by the centrifugal force and the friction of the wall when the dust-laden airflow rotates in the cyclone. The dust is mainly affected by the centrifugal force, the friction of the wall and the resistance of the airflow. In addition, both of them are simultaneously subjected to the downward thrust produced by the continuous extrusion of the dust-laden air from the upper part of the cyclone, which is the reason for the downward movement of the dusty airflow. It can be seen that the force of airflow and dust in dust-laden airflow is basically equal. However, due to the different physical properties of air flow and dust, one is a gaseous substance, which is small in mass and easy to deform, while the other is a solid substance, which is large in mass and difficult to deform. Therefore, when the dust-laden air flow is concentrated in the inner wall of the cyclone cylinder, the centrifugal force is larger than the gas. So the dust in mechanic conditions will force out air and condense on the wall of the cylinder, and the air is attached to the dust layer, so that the dust airflow finally gets separated.

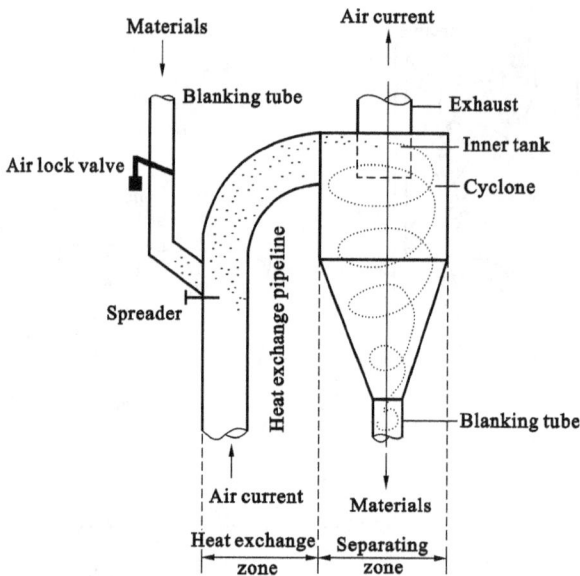

Figure 2.7: Schematic diagram of functional structure of cyclone heat exchanger unit.

2.4.7.2 Heat exchange pipelines

Heat exchange pipelines are important equipment of cyclone preheater system, which provide connection and transport channels for gas-particle flow between cyclones on the top and bottom, and also there are places for uniform distribution of raw materials, air blocking and heat exchange between gas and solid phases. As shown in Figure 2.7, apart from pipelines, heat exchange pipelines are equipped with blanking tubes, spreaders, air locks and so on. Combined with cyclones, a unit for heat exchange is established.

Because there is a remarkable temperature difference and relatively high speed between raw meal particulate and hot air flow, the raw meal powder is suspended by the airflow, from which drastic heat exchange is realized. The heat exchange of raw materials and air flow is mainly (more than 80%) conducted in connecting pipelines. When the wind speed in pipelines is too low, prolonging heat exchange is required but heat transfer efficiency will be reduced. Moreover, raw materials will accumulate and settle down because of insufficient suspension, and makes the pipe area excessively large. When the wind speed in pipelines is too fast, the resistance in system increases, which means more electric energy will be consumed and separation efficiency of cyclone preheater will be influenced. Therefore, determining suitable gas speed in pipelines is the priority step in designing dimension of heat exchange pipelines. The gas speed of heat exchange pipelines in the range of 12–18 m/s is widely selected in various categories of cyclone preheaters.

The function of a material-spreading device is to avoid blocking of the materials before they enter into heat exchange pipelines by preventing raw materials from falling too fast to go straight into the next cyclone, which is also beneficial for better spattering and dispersion of falling materials into gas flow. Despite their small size, material-spreading device plays quite an important role in maintaining sufficient heat exchange between gas and particles in heat exchange pipelines.

Air-lock revolve plate with ash valve (air lock for short) is an important appertain equipment of the preheater system, which is arranged in an appropriate position between blanking tubes in the upper cyclone and material entrance in the heat exchange pipeline locating at the exit of lower cyclone. Air lock is used to keep blanking tubes maintain long-term sealing condition, ensuring uniform feeding and sealing area in blanking tubes that cannot be filled by materials. In this way, gas blocking and leakage easily caused by the pressure difference at the exit of heat exchange pipelines between upper and lower cyclones are significantly avoided, from which the requirement that gas flow in heat exchange pipelines and materials in blacking tubes should go a separate way is achieved. On one hand, these can prevent hot gas flow in heat exchange pipelines from rising to upper cyclone through blacking tubes and collected materials from floating again, which reduces separation efficiency to some degree. On the other hand, problem that hot gas flow in heat exchange pipelines sneaks into lower cyclone from upper cyclone without heat exchange with materials is avoided. Unnecessary heat loss is reduced and the higher heat exchange efficiency is achieved in this process. Hence, air lock should possess resealable and flexible structure.

2.4.7.3 Calciner

Calciner is a kind of new thermal equipment with various types and forms, in which fuel burning, heat exchange and decomposition reactions are carried out simultaneously. Basic principles of calciner are as follows: After calciner is synchronously fed with preheated raw materials, certain amount of fuel and hot gas, raw materials reach the suspended or boiling state in the furnace. Flameless combustion is carried out, where heat transfer and decomposition of calcite are completed efficiently. It takes 2–10 s to burn fuels and decompose calcite. Decomposition rate of calcite in raw materials reaches about 80–90% and the temperature of pretreated raw materials reaches to 800–850 °C. Solid, liquid or gaseous fuels can be employed in calciner. Pulverized coal is widely used as fuel in our country and fuels consumed in calciner accounts for 55–65% of the overall.

Based on the working mechanism, calciner can be divided into spiral-flow type, spouted type, turbulence type, swirl combustion type, boiling type and many other types. However, they possess similar basic principles. The NSF (New Suspension Preheatcr Flash Calciner) decomposition furnace produced by ishikawa Island Company is illustrated in Figure 2.8. The uptake flue at rotary kiln inlet is connected to

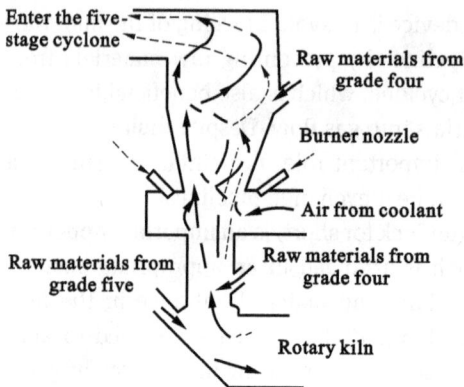

Figure 2.8: Schematic diagram of NSF calciner.

the bottom of NSF calciner, enabling high-temperature flue gas from rotary kiln to enter into lower volute through bottom of calciner and encounter hot air from cooler. Together with raw meal and pulverized coal, rising hot gas moves spirally inside the wall of reaction chamber. High-temperature flue gas rises to upper volute and then goes into the last cyclone through gas pipelines. Particles of raw materials and fuels mix with and diffuse through gas under vortex and whirlwind. Differing from rotary kiln, bright flame is not observed when fuel particles are consumed in calciner, which should be attributed to the fact that fuels are burnt in suspension state. Meanwhile, heat released by fuel combustion is transferred to raw meal particles directly through forced convection, which builds up the combustion zone for calcite decomposition in the whole furnace. Flameless combustion state with uniform temperature field locating at 800–900 °C, whose temperature is relatively lower, is maintained in the furnace. In this way, heat transfer efficiency is promoted and decomposition rate of calcite reaches 85–95%.

2.4.7.4 Rotary kiln

As shown in Figure 2.9, the rotary kiln consists of a kiln body, wheel, roller, transmission device and so on. Inside the kiln, there is a brick ring and a fireproof brick, and the cold end is also welded with a retaining ring, and it is equipped with a feeding device. The hot end of the kiln is connected to a fire cover, and the kiln head is provided with a pulverized coal burning device, and the cold end is connected with the smoke chamber. Rotary kiln is a cement clinker calcining machine which integrates combustion, heat transfer, mixing, reaction heat storage and conveying various functions.

In the precalciner kiln, the decomposition rate of $CaCO_3$ of raw materials has reached 85–95%. Therefore, the precalciner kiln can be divided into three zones:

(1) Transition zone

Transition zone is the part from the kiln end to the position where the material temperature is about 1,300 °C, in which the temperature of the material is 1,000–

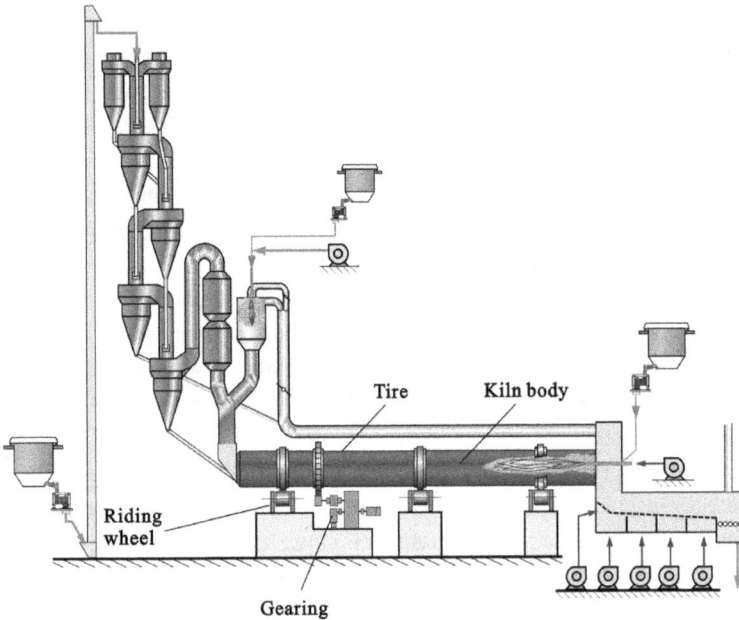

Figure 2.9: Schematic diagram of cement rotary kiln.

1,300 °C and the temperature of gas is 1,400–1,600 °C. In the transition zone, CaO produced by carbonate decomposition with Fe_2O_3, Al_2O_3 and SiO_2 undergoes further solid-phase reaction to form clinker minerals such as C_2S, C_3A and C_4AF, and releases some heat.

(2) Clinkering zone

From the beginning of the liquid phase to the solidification of liquid is called the burning zone. The material temperature of the burning zone is between 1300 and 1450 °C. In this region, the material is heated directly by the flame, and the liquid phase appears since it enters. Until 1,450 °C, the liquid phase continues to increase, which greatly promotes the solid-phase reaction. And f-CaO reacts with SiO_2 to form a large number of C_3S, until the cement clinker burning.

As the formation rate of C_3S increases rapidly with the increase in temperature, a high temperature must be ensured in the burning zone. In the case of no damage to the kiln skin, by properly increasing the temperature of the burning zone can promote the rapid formation of clinker and improve the yield and quality of the clinker. It is also necessary for the clinkering zone to have a certain length, mainly to keep the material present at the sintering temperature for a period of time to make the chemical reaction of C_3S to be as complete as possible, and the clinker is with the least amount of f-CaO. The residence time of the general material in the burning zone is about 15–20 min, and the length of the firing zone of the precalciner kiln is generally 4.5–5.5D (kiln diameter), and the average value is about 5.2D.

(3) Cooling zone

Except for the transition zone and the firing zone, the precalciner kiln is called the cooling zone. In the cooling zone, the clinker temperature begins to drop from about 1,300 °C of the burned zone, and the liquid phase solidifies into solid grayish black particles, enters the cooler and further cools down. However, there is almost no cooling zone in the large precalciner kiln, and the materials with the temperature up to 1,300 °C enter the cooling machine immediately, which can improve the quality of clinker and improve the grindability of clinker.

If the cooling zone is too long, the quenching effect of the clinker decreases, the toughness of the clinker increases, the vitreous feature becomes worse and the abrasion reduces. The magnesia stone used in high magnesium raw materials cannot be fast cooled; otherwise, the grain size will be bigger and the uncertainty of the clinker and cement products will be increased.

It should be noted that, the division of the rotary kiln is artificial, reactions in these zones are often cross react or at the same time, it cannot be completely separated. If the raw material is heated unevenly or the heat transfer is slow, it will increase the cross of the various reactions. Therefore, the division of the rotary kiln is rough.

2.4.7.5 Clinker cooler

Clinker cooler is a heat exchange device to transfer heat from high-temperature clinker to low-temperature gas. As a process equipment, it undertakes the task of quenching high-temperature clinker. And as a thermal equipment, it is responsible for the heating and warming up of the two air entering the rotary kiln and the three blast of the calciner. As a heat recovery equipment, it undertakes the task of a large number of heat recovery with the enthalpy of kiln clinker, and as a clinker-conveying equipment, it undertakes the task of conveying high-temperature clinker.

Figure 2.10 shows the fourth-generation grate-type cooling machine diagram. The grate type cooler is composed of the upper shell and the lower shell, grate bed, driving device, supporting device, clinker crusher, material leakage zipper machine, automatic lubrication system and cooling fan unit.

The hot clinker is discharged from the kiln to the grate bed and is distributed along the length of the grate bed under the push of the reciprocating pushing grate plate to form a certain thickness on the bed. The cooling air is blown into the layer from the bottom of the bed and permeates and diffuses the hot clinker. When the clinker is cooled, the cooling air becomes hot air, the hot air at the hot end is used as the combustion air into the kiln (secondary air) and the calciner (tertiary air), and part of the hot air can also be used for drying or waste heat power generation. The use of hot air can achieve heat recovery, thereby reducing the heat consumption of the system. The excess low-temperature hot air is discharged into the atmosphere after dust collection. After cooling, the small clinker passes through the grid sieve and falls into the conveyor behind the grate cooler. The large clinker is broken and cooled and

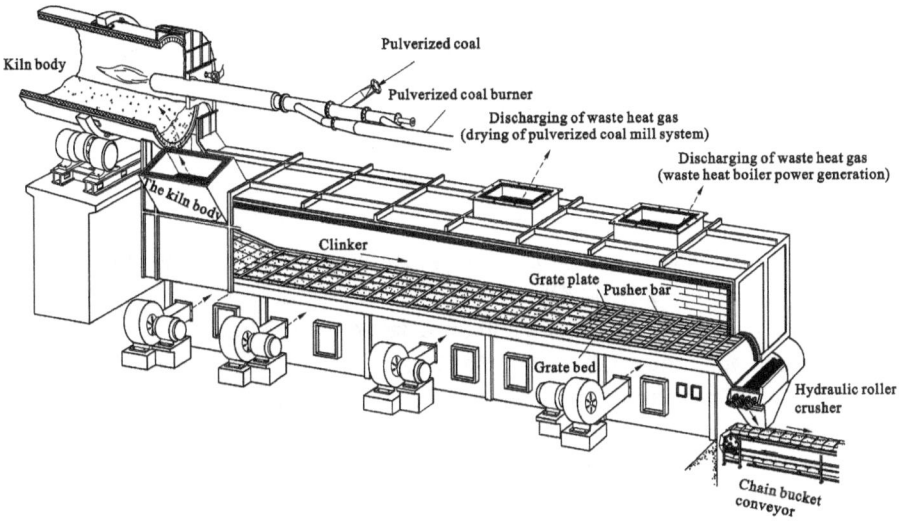

Figure 2.10: Schematic diagram of grate cooling machine of the fourth generation.

then fed into the conveyor and fine clinker, and dust enters the hopper through the grate gap of grate bed and the leakage of grate hole. When the material level in the hopper reaches a certain height, the lock valve which is controlled by the material-level sensing system is automatically opened, and the leaking fine material is transported to the machine to leak the zipper machine. When the remaining fine material in the hopper can still seal the air lock valve, the valve plate is closed so as to ensure no air leakage.

The performance requirements of modern grate cooler are high cooling efficiency, high heat recovery rate and high operation rate.

2.5 Production and standards for common Portland cement

The cement is mainly made of Portland cement clinker that is known as Portland cement.

During the cement production process, in order to improve the cement properties and strength class, some mineral admixtures are normally added, which can be called as cement admixtures. When incorporated with many admixtures, the name of Portland cement should be titled with the mixed materials such as ordinary Portland cement, slag cement and fly ash cement. The common Portland cement refers to the cement for general purpose in large quantities of civil engineering, which mainly includes Portland cement, ordinary Portland cement, slag Portland cement, pozzolanic Portland cement and fly ash Portland cement.

The production of Portland cement is a progress of achieving the quality requirements with appropriate compositions of Portland cement clinker, gypsum and admixture by grinding, shortage and homogenization.

2.5.1 Cement admixture

There are many kinds of admixtures for cement production, and the classification methods are different as well. According to different sources, the admixture can be divided into the natural admixture and artificial admixture (mainly industrial solid residue, short for industrial waste). But it is usually classified according to the properties of admixture, namely the role in the progress of cement hydration, which are active admixture and inactive admixture.

The active admixture refers to the pozzolanic activity, potential hydraulicity and the combination of pozzolanic activity and potential hydraulicity. The main varieties contain various industrial slag (blast furnace slag, steel slag, cupola slag, phosphorus slag, etc.), pozzolanic admixture and fly ash, whose indicators should meet the relevant national standards or industrial standards.

The so-called pozzolanic activity refers to a kind of powder material by grinding without hydraulicity separately, but it shows hydraulicity properties at normal temperature when contacting with lime and water. While the potential hydraulicity is referred to some materials without hydraulicity separately, however, it exhibits hydraulicity under some activators. There are two types of commonly used activators, which are alkali-activator (cement clinker and lime) and sulfate activator (all kinds of natural gypsum or $CaSO_4$ as the main composition of chemical by-products, such as fluorine gypsum and phosphor gypsum).

The inactive admixture is referred to as mineral materials which plays a role as fill staff in cement without influencing the properties of cement. That is to say that the active indicator cannot meet the requirements of active admixture such as slag, pozzolanic materials, fly ash, limestone, sandstone and raw shale. Generally, the requirements of inactive blended material is that they are harmless to the cement.

Incorporating admixture into cement can increase the output of cement; besides, it can reduce the production cost of cement and improve and adjust certain properties as well. On the other hand, the comprehensive utilization of industrial waste residue can reduce the environmental impact.

2.5.1.1 Granulated blast furnace slag

During smelting iron in blast furnace, the obtained melt mainly containing calcium silicate and calcium aluminate is granulated blast furnace slag or slag and water granulated slag after quenching. Nowadays, it is the largest consumed and best qualified admixture in the domestic cement industry. However, the by-product presents bulk or powdery without active and belongs to the inactive admixture.

During smelting of ferromanganese in blast furnace, the acquired residue is named as ferromanganese slag, which is suggested that other compositions and properties of ferromanganese slag are similar to the granulated blast furnace slag in the general smelting progress except for its higher MnO content. Therefore, the ferromanganese slags are usually incorporated in the granulated blast furnace slag.

The main chemical compositions of granulated blast furnace slag are silicon dioxide (SiO_2), aluminum oxide (Al_2O_3), calcium oxide (CaO), magnesium oxide (MgO), manganese oxide (MnO), ferrous oxide (FeO) and sulfur trioxide (SO_3). In addition, some slags also contain minute amount of TiO_2, V_2O_5, Na_2O, BaO, P_2O_5, Cr_2O_3 and so on. The amount of CaO, SiO_2 and Al_2O_3 of granulated blast furnace slag exceeds 90%. The content ratio of alkaline oxides (CaO and MgO, wt%) to acidic oxides (SiO_2 and Al_2O_3) is called alkalinity coefficient (M_0), as shown in formula (2.76):

$$M_0 = \frac{w(CaO) + w(MgO)}{w(SiO_2) + w(Al_2O_3)} \qquad (2.76)$$

According to the value of M_0, the slag can be divided into three types:
(1) When $M_0 > 1$, the slag is called alkaline slag
(2) When $M_0 = 1$, the slag is called neutral slag
(3) When $M_0 < 1$, the slag is called acidity slag

Based on the types of iron smelting, the slag can be divided into another three types:
(1) Cast iron slag
(2) Steel pig slag
(3) Special iron slag (ferromanganese slag and magnesium iron slag)

Moreover, based on cooling method, physical properties and appearances, the slag can be divided into another two types:
(1) Slow cooling slag (blocky and powdery)
(2) Quenching slag (granular, fibrous, porous, pumice)

The chemical compositions of several blast furnace slags are shown in Table 2.13.

The oxide constituents of blast furnace slag exist in varied forms as silicate minerals or glass phases.

(1) Alkaline blast furnace slag
The most common minerals of alkaline blast furnace slag are melilite, dicalcium silicate, olivine, rankinite, aedelforsite and spinel.

Table 2.13: Chemical compositions of blast furnace slag (wt%).

Types	CaO	SiO_2	Al_2O_3	MgO	MnO	Fe_2O_3	S	TiO_2	V_2O_5
Steelmaking, casting blast furnace slag	32–49	32–41	6–17	2–13	0.1–4	0.2–4	0.2–2	–	–
Ferromanganese slag	25–47	21–37	7–23	1–9	3–24	0.1–1.7	0.2–2	–	–
Vanadium–titanium slag	20–31	19–32	13–17	7–9	0.3–1.2	0.2–1.9	0.2–1	6–25	0.06–1

(2) Acidity blast furnace slag

The acidity slag also has different mineral formation due to its different cooling speed. During the rapid cooling period, the whole acidity slag coagulates vitreous, while the acidity slag especially the weak acid slag tends to appear as crystal mineral phases such as melilite, wollastonite, pyroxene and plagioclase by slow cooling.

(3) Titanium blast furnace slag

The mineral compositions of high titanium blast furnace slag mostly contain titanium.

(4) Magnesium iron blast furnace slag

Manganese olivine ($2MnO \cdot SiO_2$) and rhodonite ($MnO \cdot SiO_2$) exist in magnesium iron slag.

(5) Aluminum blast furnace slag

High aluminum slag contains large amount of calcium aluminate ($CaO \cdot Al_2O_3$), calcium trialuminate ($5CaO \cdot 3Al_2O_3$), calcium dialuminate ($CaO \cdot 2Al_2O_3$) and so on.

The activity of slag mainly depends on its chemical compositions and granulation quality. Its activity can be expressed by the ratios that the amount of CaO, MgO and Al_2O_3 when compared with the amount of SiO_2, MnO and TiO_2 in terms of chemical compositions, which is called the coefficient of quality (K) as shown in eq. (2.77). The larger the value of K, the higher the reactivity of slag:

$$K = \frac{w(CaO) + w(MgO) + w(Al_2O_3)}{w(SiO_2) + w(MnO) + w(TiO_2)} \tag{2.77}$$

The quality of slag can also be verified by the strength test method except the evaluation of K. The slag cement and Portland cement with different content(wt.%) of the identical clinker are prepared, of which the 28d compressive are tested under the strict control of specific surface area and gypsum content. Then the quality of slag cement can be evaluated by the following formula:

$$R = \frac{\text{The 28-day compressive strength of slag cement}}{\text{The 28-day compressive strength of Portland cement} \times (1 - \text{slag cement})\%} \tag{2.78}$$

When the quality of slag is poor, R is less than 1, while the larger R indicates a better quality.

The slag can be divided into qualified product and superior quality according to the quality coefficient, chemical composition, bulk density and particle size. The quality coefficient and chemical composition of slag should meet the requirements as presented in Table 2.14. The radioactivity of slag should comply with the regulation based on *Building materials radionuclide limited* (GB 6566-2010), and the specific value is determined with the additional content of slag by cement plant. Besides, the

Table 2.14: Requirements of quality coefficient and chemical composition of slag.

Technical index	Qualified product	Superior quality
Quality coefficient	≥1.20	≥1.60
Titanium dioxide (TiO_2) content (%)	≤10.0	≤2.0
Manganese oxide (MnO) content (%)	≤4.0 ≤15.0 (smelting ferromanganese)	≤2.0
Fluoride content (F %)	≤2.0	≤2.0
Sulfide content (S %)	≤3.0	≤2.0

Table 2.15: Requirements of bulk density and particle size of slag.

Technical index	Qualified product	Superior quality
Loose bulk density/(kg·L^{-1})	≤1.20	≤1.00
Maximum particle size/mm	≤100	≤50
More than 10 mm particles content (wt%)	≤8	≤3

bulk density and particle size of slag should conform to the requirements as shown in Table 2.15. The slag must not be mixed with extrinsic contaminants such as iron-bearing sludge and inadequate quenching slag.

2.5.1.2 Pozzolanic admixture

No matter the main composition of raw mineral contains natural or artificial SiO_2 and Al_2O_3, it will not harden when mixed with water. However, it can harden in the air, and also continuously hardens in water when mixed with lime, which is called pozzolanic admixture. The pozzolanic admixture can be divided into natural pozzolanic admixture and artificial pozzolanic admixture based on its origin.

(1) Natural pozzolanic admixture
(i) Pozzolan

Pozzolan represents fine-grained loose deposits of vulcanian eruption.
(ii) Tuff

Tuff is a compacted rock formed by volcanic ash deposits.

(iii) Zeolite rock

Tufa formed a rock containing hydrous aluminum silicate minerals of an alkali or alkaline-earth metal as the main component by the environmental medium.

(iv) Pumice

Pumice represents porous vitreous rocks of vulcanian eruption.

(v) Diatomite and diatom stone

Diatomite and diatom stone is a rock formed by very detailed diatom shell aggregation and deposition.

(2) Artificial pozzolanic admixture

(i) Coal gangue

Coal gangue is a product of carbonaceous shale by the spontaneous combustion or after calcination in coal stream.

(i) Calcined shale

Calcined shale is a product of shale or oil shale by calcining or spontaneous combustion.

(ii) Burnt clay

Burnt clay is a product of clay by calcining.

(iii) Cinder

Cinder is a residue after burning coal.

(iv) Silica residue

Silica residue is a residue.

(v) Silica fume

Silica fume is a by-product obtained from the progress of smelting silica or ferrosilicon. The SiO_2 mainly contains glassy state with above 90% and the average particle size is about 0.1 mm, which has very high pozzolanic activity.

The chemical compositions of pozzolanic admixture are SiO_2 and Al_2O_3, and their contents are about 70%. However, the mineral compositions are largely affected by their causes due to the lower CaO content.

The activity of pozzolanic admixture is pozzolanicity, and there are usually two methods to evaluate its activity, one is chemical method and the other is physical method.

(1) Chemical method

Chemical method, namely the pozzolan test, is based on GB/T 2847-2005 "Pozzolan admixture for cement":

(i) The 30% replacement level weighing 20 ± 0.01 g of pozzolanic admixture cement is mixed with 100 mL of distilled water made from turbid fluid, and then kept at a temperature of (40 ± 1) °C for 8 days, afterwards filter the solution.

(ii) The total alkalinity (mmol/L) is determined by the selected filtrate.

(iii) The CaO content of filtrate is measured.

(iv) The test results are plotted on a graph of pozzolanicity (Figure 2.11) based on the total alkalinity (OH⁻ concentration) as abscissa and the CaO content (CaO concentration) as ordinate.

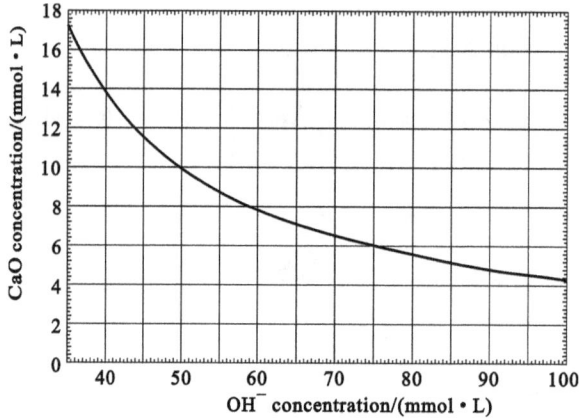

Figure 2.11: Curve graph of pozzolanic active evaluation based on $c(OH^-)$ and $c(CaO)$.

(i) The evaluation of results
   (a) If the test point is under the curve of solubility of calcium hydroxide at 40 °C, the admixture of pozzolanicity is considered as qualified.
   (b) If the test point is above or on the curve, the experiment should be reworked for 15 days. If the test point is below the curve, it can still be considered that the ash is qualified; otherwise, it will not be qualified.

(2) Physical method

Physical method, namely comparison of compressive strength of mortar after 28 days, is to measure the strength of the 30% of pozzolanic admixture compared to cement and it should be higher than 42.5 MPa without any pozzolanic admixture in accordance with the *Strength test for cement standard specimen* (GSB 14-1510). The specific test method is carried out according to the *Industrial waste residue test method for cement mixing* (GB/T 12957-2005).

The pozzolanic admixture used in cement must be in accordance with the specifications of GB/T 2847-2005:

(i) Loss on ignition. The loss on ignition of artificial pozzolanic admixture is not higher than 10%.
(ii) SO_3 content should not exceed 3.5%.
(iii) Pozzolanic test. According to the test method of GB/T 2847-2005 appendix A, it must be qualified.
(iv) The strength ratio of cement mortar of 28 days is not less than 65%.
(v) Radioactive substances should comply with the regulations of GB 6566-2010.

The pozzolanic admixtures conforming to the above quality requirements are active admixtures. The pozzolanic admixtures required by subsections (i),(ii) and (v) are inactive mixtures. While the pozzolanic admixtures without any of conforming above (i),(ii) and (v) should not be used as cement mixtures.

2.5.1.3 Fly ash

The powder collected from the flue gas of pulverized coal furnace is called fly ash (excluding the following conditions: (i) when burning municipal refuse or other wastes with coal; (ii) to calcinate industry or municipal waste in incinerator; and (iii) the use of circulating fluidized bed boiler combustion). The fly ash can be divided into class F and class C according to the coal type. Class F fly ash is collected from anthracite or bituminous coal burning. Class C fly ash is collected from lignite or secondary bituminous coal calcining, in which the CaO content is generally higher than 10%. In the thermal power plant, the pulverized coal is burned in the boiler at a high temperature of 1,100–1,500 °C, then general 70–80% of powdery ash is exhausted with flue gas and collected by dust collector, namely the fly ash, while 20–30% of sintered ash falls into the bottom of furnace, called furnace bottom ash or slag. Fly ash is one of the most common and the largest industrial waste residues.

The chemical compositions of fly ash fluctuate within a wide range depending on the coal type, combustion condition and dust collection. However, the main compositions are SiO_2 and Al_2O_3, with a small amount of Fe_2O_3 and CaO. The activity of fly ash depends on the soluble SiO_2, Al_2O_3 and vitreous phase as well as its fineness. Additionally, the loss on ignition indicates the degree of carbon content, namely, complete degree of combustion also affects its quality. The particle size of fly ash generally ranges from 0.5 to 200 μm, and the main particle size is between 1 and 50 μm. The allowance through sieve of 80 mesh is 3–40%, while the mass density and volume density are 2.0–2.3 g/cm^3 and 0.6–1.0 g/cm^3, respectively.

As shown in Table 2.16, the use of fly ash for premixed concrete and mortar should be conformed to the specification of *Fly ash for cement and concrete* (GB/T 1596-2017). The cement active admixture for fly ash should meet the technical requirements as shown in Table 2.17. The radioactivity of fly ash should meet the requirements of the construction subject materials stipulated in the national standard of *Radioactive nuclear element in building materials* (GB 6566-2010). The alkali content is calculated according to the Na_2O equivalent [$w(Na_2O) + 0.658w(K_2O)$]. When the alkali content is required in the application of fly ash, it is determined by the two parties through consultation. When using dry or semidry desulfurization process, the content of calcium sulfite should be detected, which means the w ($CaSO_3 \cdot 0.5H_2O$) should be less than 3.0%. The homogeneity of fly ash is characterized by fineness, and the fineness of the single sample should not exceed the maximum deviation of 10 previous samples average fineness (if the number of samples is less

Table 2.16: Specifications of the used fly ash for premixed concrete and mortar.

Item		I Grade I	Grade II	Grade III
		Specifications		
Fineness (allowance through sieve of 45 μm, %)	Class F fly ash Class C fly ash	≤12.0	≤30.0	≤45.0
Ratio of water demand (%)	Class F fly ash Class C fly ash	≤95	≤105	≤115
Loss on ignition (%)	Class F fly ash Class C fly ash	≤5.0	≤8.0	≤10.0
Water content (%)	Class F fly ash Class C fly ash	≤1.0		
$w(SO_3)$ (%)	Class F fly ash Class C fly ash	≤3.0		
w(f-CaO) (%)	Class F fly ash Class C fly ash	≤1.0		
Total content of SiO_2, Al_2O_3 and Fe_2O_3 (%)	Class F fly ash Class C fly ash	≥70.0 ≥50.0		
Density $\rho/(g/cm^3)$	Class F fly ash Class C fly ash	≤2.6		
Soundness/mm	Class C fly ash	≤5.0		
Intensity activity index (%)	Class F fly ash Class C fly ash	≥70.0		

than 10, then chose the average of all the previous samples). The maximum deviation range shall be determined between the buyer and the seller.

Determination of water demand is adapted by the fluidity of mortar with a cement/sand ratio of 1:3. The contrasted cement mortar with strength grade is higher than 42.5 MPa without any admixture based on GSB 14-1510 standard, and 30 wt% fly ash is added with a certain amount of water, making the fluidity of all mortar reach 130–140 mm and the water added is the ratio of water demand.

Strength active index is measured through tested mortar and contrasted mortar in accordance with the *Test method of strength for cement mortar* (GB/T 17671-2005), then the activity index of the tested mortar is determined by the strength ratio. The mix proportion of tested mortar is with cement of 315 g, fly ash of 135 g, standard sand

Table 2.17: Specifications of the used fly ash for active cement.

Item		Specifications
Loss on ignition (%)	Class F fly ash	
	Class C fly ash	≤8.0
Water content (%)	Class F fly ash	
	Class C fly ash	≤1.0
$w(SO_3)$ (%)	Class F fly ash	
	Class C fly ash	≤3.5
$w(f\text{-}CaO)$ (%)	Class F fly ash	
	Class C fly ash	≤1.0
Total content of SiO_2, Al_2O_3	Class F fly ash	≥70.0
and Fe_2O_3 (%)	Class C fly ash	≥50.0
Density $\rho/(g/cm)$	Class F fly ash	
	Class C fly ash	≤2.6
Soundness/mm	Class C fly ash	≤5.0
Strength active index (%)	Class F fly ash	
	Class C fly ash	≥70.0

of 1,350 g and water of 225 mL; while the mix proportion of contrasted mortar is with cement of 450 g, standard sand of 1,350 g and water of 225 mL.

2.5.1.4 Other admixture

Other admixtures indicate all kinds of industrial slags that can be used as admixtures except the specifications of *Ground granulated blast furnace slag used for cement* (GB/T 203-2008), *Pozzolanic admixture used for cement* (GB/T 2847-2005) and *Pulverized-fuel ash used for cement* (GB/T 1596-2005). Also, the activity of other admixtures is divided into active and inactive. The 28 days compressive strength ratio of mortar not lower than 75% is active admixture, while it is inactive admixture when the 28 days compressive strength ratio of cement mortar is lower than 75%.

(1) Iron-melting furnace slag

Iron-melting furnace slag is a discharged waste residue in iron and steel plant, which becomes granulated iron slag by water quenching and rapid cooling in a melting state. The mineral composition of iron-melting furnace slag contains C_2AS, CAS_2 and CS and a small quantity of $C_2S_3 \cdot CaF$ and CaF_2 that are similar to that of ground granulated blast furnace slag.

The iron-melting furnace slag can be used as cement admixture, manufacturing of clinker-free cement and low-clinker cement, or some special cement is the same as blast furnace slag.

(2) Granulated electric furnace phosphorus slag

Granulated electric furnace phosphorus slag is a waste residue obtained from the production of yellow phosphorus by using phosphate rock, silica and coke with electrical sublimation method in an electric furnace. It is prepared by water quenching and rapid cooling in a melting state.

The chemical compositions of phosphorus slag are similar to that of blast furnace slag, and the difference is that the content of CaO and SiO_2 is slightly higher while the content of Al_2O_3 is slightly lower. In addition, phosphorus slag contains small quantity of P_2O_5 and CaF_2 and the activity of phosphorus slag is slightly lower than the blast furnace slag. When phosphorus slag is used for cement admixture, its quality should comply with the requirements of *Granulated electric furnace phosphorus slag for cement* (GB/T 6645-2008). The characteristic of manufacturing cement is high early compressive strength; however, the growth rate of later strength is large with a long setting time.

(3) Granulated blast furnace titanium slag

When using vanadium-bearing titanomagnetite as raw material in pig-iron smelting, the obtained molten slag that contains titanium silicate and perovskite becomes granulated blast furnace titanium slag after quenching and granulating.

The content of TiO_2 in granulated blast furnace titanium slag is different from the normal ground granulated blast furnace slag, and is usually more than 20%, resulting in a great reduction of its activity. The granulated blast furnace titanium slag is black brown in color and mixes up with a small amount of iron with strong crystallization ability, and a few glassy phases are usually used as inactive admixtures. The quality should comply with the requirements of *Granulated blast furnace titanium slag used for cement* (JC/T 418-2009).

(4) Calcium-enriched cyclone furnace slag

The manufacturing of coal dust with incorporating appropriate limestone under coal fired in power plant is discharged from furnace in smelting state, and becomes calcium-enriched cyclone furnace slag by quenching.

Compared to ground granulated blast furnace slag, the content of CaO in calcium-enriched cyclone furnace slag is less while the Al_2O_3 content is higher. When the content of CaO is more than 25%, its activity is next only to that of ground granulated blast furnace slag and much higher than that of fly ash. Calcium-enriched cyclone furnace slag by quenching is a kind of latent hydraulic material which contains more than 95% glassy phase minerals. The range of its mass density is 2.7–3.0 g/cm³, while the bulk density is 1.2–1.4 g/cm³. The quality should comply with the requirements of *Calcium-enriched cyclone-furnace slag used for cement* [JC/T 454-1992(1996)].

(5) Steel slag

Steel slag mainly refers to the open-hearth steelmaking slag, vessel slag and electric arc furnace restored slag. Among them, the main chemical compositions of open-hearth steelmaking slag and vessel slag are approached to that of clinker. However, the content of Fe_2O_3 and MgO is slightly higher while the content of CaO and Al_2O_3 is slightly lower. The chemical compositions of electric arc furnace restored slag with high content of Al_2O_3 and less content of FeO and Fe_2O_3 are similar to ground granulated blast furnace slag. They all contain a certain amount of hydraulic minerals such as C_3S, C_2S and C_4AF that cause hydraulicity.

The utilization of steel slag for admixture together with other materials usually generates activated reaction and ensures the soundness of such composite cement qualified that can obtain excellent strength. The quality should comply with the requirements of *Steel slag used for cement* (YB/T 022-2008).

(6) Fluidized furnace slag

Fluidized furnace slag is the slag discharged from the ebullition boiler or fluidized bed combustion boiler in the combustion process of low heat coal gangue. As most of the boilers are used for coal gangue or for poor quality coal, the chemical composition, mineral compositions and basic properties of fluidized furnace slag belong to pozzolanic admixture materials that are similar to coal gangue and fly ash.

(7) Cement kiln dust

Cement kiln dust is collected from the exhaust fumes of kiln during the production of cement clinker in cement rotary kiln. Cement kiln ash is divided into two categories: one is a kiln ash that is discharged from general air dry, wet process and semidry rotary kiln; and the other is the kiln ash discharged from the by-pass of precalcining kiln cement. The latter has very high content of f-CaO as well as R_2O, SO_3 and Cl^-, which is difficult to make full use in cement industry. Nowadays, one composition of the cement admixture is the former kiln dust.

The chemical compositions of cement kiln ash are basically between raw material and clinker; however, the compositions possess great difference with the different raw materials, fuel, calcining equipment and collection systems. The loss on ignition of cement kiln ash is 10–25%, and f-CaO is around 10% by mass, while the content of SO_3 mainly depends on the sulfur content of used coal.

Cement kiln dust mainly consists of $CaCO_3$, K_2SO_4, Na_2SO_4, $CaSO_4$, calcined clay materials, clinker minerals and coal ash glassy ball.

Cement kiln dust is neither considered as active admixture nor inactive admixture in the composition of cement. However, cement kiln dust is usually used as admixture in cement for the following reasons:

(i)   Cement kiln dust contains a certain amount of clinker minerals and calcined clay with pozzolanic activity. These minerals will be hydrated with cement and have a certain effect on the strength of cement.

(ii) The fine-powder $CaCO_3$ in cement kiln dust can accelerate the hydration of C_3S during the hydration process, forming calcium carboaluminate of needle crystallization with aluminate; in addition, it can contribute to the early strength due to its microfilling effect.

(iii) The ingredients of K_2SO_4 and Na_2SO_4 can be used as early strength agent while $CaSO_4$ can be used as a retarding effect of gypsum.

Although the content of f-CaO is much higher than that of clinker, it is mostly soft burnt lime in a finely dispersed state, which can hydrate fast and dose not damage the soundness of cement; therefore, the national standard stipulates that the cement can be mixed into a certain amount of cement kiln ash. The quality should comply with the requirements of *Mixing cement kiln ash for cement* (JC/T 742-2009).

2.5.2 Definition, classification and specifications for common Portland cement

The national standard has strict rules on the composition materials, strength grade, specifications and test analysis methods of common Portland cement. The definition, classification and specifications of common Portland cement are described below according to the *Common Portland cement* (GB 175-2007).

2.5.2.1 Definition and classification of common Portland cement
The hydraulicity cementitious material made up of Portland cement clinker, a moderate amount of gypsum and the specified admixture is the common Portland cement.

Common Portland cement can be divided into Portland cement, ordinary Portland cement, Portland blast furnace cement, Portland pozzolana cement, Portland fly ash cement and composite Portland cement based on the varieties and dosage of admixture material. The components and product codes of each species should conform to the specifications of Table 2.18.

(1) Portland cement
Portland cement is the hydraulicity cementitious material made up of Portland cement clinker, less than 5% limestone or granulated blast furnace slag and a certain amount of gypsum by grinding. There are two types of Portland cement, one is P·I Portland cement without admixture materials, the other is P·II Portland cement with the addition of not more than 5% limestone or granulated blast furnace slag during the cement clinker grinding.

Table 2.18: Variety and components of common Portland cement.

Variety	Product code	Composition (wt%)				
		Clinker + gypsum	Granulated blast furnace slag	Pozzolana admixture	Fly ash	Limestone
Portland cement	P·I	100	–	–	–	–
	P· II	≥95	≤5	–	–	–
		≥95	–	–	–	≤5
Ordinary Portland cement	P·O	≥80 and <95	>5 and ≤20[a]			–
Portland blast furnace slag cement	P·S·A	≥50 and <80	>20 and ≤50[b]	–	–	–
	P·S·B	≥30 and <50	>50 and ≤70[b]	–	–	–
Portland pozzolana cement	P·P	≥60 and <80	–	>20 and ≤40	–	–
Portland fly ash cement	P·F	≥60 and <80	–	–	>20 and ≤40	–
Composite Portland cement	P·C	≥50 and <80	>20 and ≤50[c]			

[a] The materials in this group are active blending materials qualified with standards, where they can be replaced by inactive blending materials qualified with standards in an amount of less than 8% to cement mass or qualified kiln ash with an amount of less than 5% to cement mass.

[b] The materials in this group are slag or slag powders qualified with standards, where they can be replaced by any of active blending materials, inactive blending materials or kiln ash qualified with standards in an amount of less than 8% to cement mass.

[c] The materials in this group are composed of two or more qualified active and/or inactive blending materials, where they can be replaced by qualified kiln ash with an amount of less than 8% to cement mass. When mixed with slag, the amount of blending materials must not be repeatedly calculated with the amount of slag in Portland cement.

(2) Ordinary Portland cement

Ordinary Portland cement is the hydraulicity cementitious material made up of Portland cement clinker, greater than 5% and less than 20% admixture material and a certain amount of gypsum by grinding, which is coded as P·O.

When mixed with admixture material, the maximum amount should not exceed 20%, which can be replaced by kiln ash with not more than 5% or inactive admixture materials with not more than 8% of cement.

(3) Portland blast furnace slag cement

The hydraulicity cementitious material which is made of Portland cement clinker, granulated blast furnace slag and a certain amount of gypsum by grinding is called Portland blast furnace slag cement. The Portland blast furnace slag cement is divided into type A and type B. The content of cement blast furnace slag in type A is greater than 20% and less than 50% according to the quality percentage, which is coded as P·S·A, while the content of cement blast furnace slag in type A is greater than 50% and less than 70%, which is coded as P·S·B.

The slag in cement blast furnace slag should conform to the specifications of GB/T 203-2008 *Granulated blast furnace slag used for cement* or GB/T 18046-2017 *Granulated blast furnace slag for cement and concrete*, which is allowed to replace any other active admixtures, inactive admixtures or kiln ash that are not more than 8% of the cement.

(4) Portland pozzolana cement

The hydraulicity cementitious material made of Portland cement clinker, pozzolana admixture and a certain amount of gypsum by grinding is Portland pozzolana cement, which is coded as P·P. The pozzolana admixture in cement should conform to the specification of GB/T 2847-2005, and the content of pozzolana admixture is higher than 20% and less than 40% according to the quality percentage.

(5) Portland fly ash cement

The hydraulicity cementitious material made of Portland cement clinker, fly ash and a certain amount of gypsum by grinding is Portland fly ash cement, which is coded as P·F. The fly ash in cement should conform to the specification of GB/T 15962017, and the content of fly ash must be higher than 20% and less than 40% according to the quality percentage.

(6) Composite Portland cement

The hydraulicity cementitious material made of Portland cement clinker, two or more kinds of specified admixture materials and a certain amount of gypsum by grinding is composite Portland cement. The total content of admixture in cement is higher than 20% and less than 50% according to the quality percentage. No more than 8% kiln ash is allowed to replace part of admixture materials according to JC/T 742-2009, while the content of admixture should not be duplicated with Portland blast furnace slag cement when incorporating slag.

2.5.2.2 Components of common Portland cement

(1) Clinker
Clinker, also known as cement clinker, is short for Portland cement clinker. It is a hydraulic material made from grinding and calcining raw materials mainly containing CaO, SiO_2, Al_2O_3 and Fe_2O_3. In the clinker, the calcium silicate mineral should be not less than 66%, and the quality ratio of calcium oxide and silica must not be less than 2.0.

(2) Gypsum
Gypsum should be better than the G-class of M-class gypsum or compounded gypsum specified by the national standard *Natural gypsum* (GB/T 5483-2008).

Industrial secondary gypsum is an industrial by-product with calcium sulfate as its main component. Before using, it should be proved that this industrial secondary gypsum has no negative effect on cement.

(3) Active admixture
Granulated blast furnace slag, ground granulated blast furnace slag, fly ash and pozzolanic admixture should meet the requirements of GB/T 203-2008, GB/T 18046-2008, GB/T 1596-2005 and GB/T 2847-2005, respectively.

(4) Inactive admixture
Inactive admixture means the active index of granulated blast furnace slag, ground granulated blast furnace slag, fly ash and pozzolanic admixture which should be lower than the requirements of GB/T 203-2008, GB/T 18046-2008, GB/T 1596-2005 and GB/T 2847-2005, respectively. Besides, the Al_2O_3 content in limestone should not be higher than 2.5%.

(5) Kiln dust
Cement kiln dust is collected from the exhaust fumes of kiln during the production of cement clinker in cement rotary kiln, which comply with the requirements of JC/T 742-2009.

(6) Grinding aids
The grinding aids are allowed to be used in the cement and the mass fraction should be not more than 0.5%. Besides, the quality of grinding aid should comply with the requirements of GB/T 26748-2011.

2.5.2.3 Main performance of common Portland cement
Setting time, soundness and strength are three main performances of common Portland cement.

(1) Setting time

The setting time of cement refers to the time when the cement starts from the addition of water to the loss of fluidity that is required to develop from the state of plasticity to the solid state, which includes the initial setting time and the final setting time. The initial setting time refers to the time when the cement starts to the initial loss of plastic state from adding water, while the final setting time refers to the time when the cement is mixed with the water and starts to completely lose plasticity. In order to ensure that the used cement in mortar or concrete has sufficient time to stir, transport and cast, it must be a certain initial setting time. After casting is finished, in order to improve construction speed and increase the turnover rate of formwork, concrete is expected to harden quickly. Therefore, it is also required that the cement has a relatively short final setting time. Mineral composition of clinker, type and dosage of gypsum, alkali content, fineness and so on affect the setting time of cement. Appropriate amount of gypsum in Portland cement can adjust the setting time and improve strength. However, overdose of gypsum will not only decrease cement strength but also lead to poor soundness of cement. Hence, maximum SO_3 content is limited to the national standard. Cement has an optimum amount of gypsum, and it is usually determined by the tests such as the comprehensive consideration of strength, setting time and so on.

(2) Soundness

The homogeneity of volume changes in cement after the hardening is called soundness, which refers to the cement slurry that can maintain a certain shape and properties with no cracking, no deformation and no breakdown after hydrating and hardening. Generally, most cement presents slight volume shrinkage during the hardening process, except expensive cement which has slight expansion. However, these expansion and shrinkage are completed before it hardened; therefore, the volume of hardened cement paste, mortar and concrete changes uniformly indicating good soundness. If a chemical reaction of some components in cement accompanied by a volume change occurs after hardening rather than before hardening, harmful internal stress within the hardened cement matrix will take place. If this internal stress is large enough to reduce the strength of cement and even rupture, it has a poor soundness.

The poor soundness of cement is caused by f-CaO, crystallization of MgO or excessive mixing of gypsum to cement. Among them, free lime is the most common and serious factor. The lime hydrates very slowly in dead roasting state, and continuously forms hexagonal platelet morphology of $Ca(OH)_2$ crystal with water, which causes volume changes by 100% and generates expansive stress, resulting in damage to the hardened cement matrix. Second, this is the crystallization of MgO, namely periclase, which hydrates more slowly and expends 148% in volume when $Mg(OH)_2$ is generated during hydration. However, the crystallization of magnesia in the clinker by quenching is small and has little effect on soundness. Finally, the content of SO_3 is overtopping in cement, that is to say the excessive incorporating of gypsum and the excess amount of gypsum will continue to form AFt with water and C_3A after cement hardening, which result in expansion and generating expansive stress, affecting the soundness of cement.

The poor soundness caused by different reasons should be tested by using different methods. Due to comparatively fast hydration speed of f-CaO, simply heated to 100 °C can determine whether poor soundness of cement can be caused in a short time, so the boiling method is used to test. Due to the slow hydration of periclase, even if heated to 100 °C, it cannot be judged and must use a higher temperature and higher pressure (215.7 °C and 2.0 MPa) to draw the conclusion in a short time; hence, the autoclave method is needed. Because the MgO in cement is not all magnesia, and the damaging degree of magnesia is related to its crystallite size. Numerous experiments demonstrated that as long as the MgO content in cement conforms to the requirements listed in Table 2.19, no harmful damage will occur without autoclave expansive test. As to the poor soundness caused by SO_3, a mass of experiments demonstrated that as long as the SO_3 content in cement is less than 3.5% (for slag cement is less than 4.0%), no problem in soundness will be caused by SO_3. Due to the decomposition of AFt at high

Table 2.19: Chemical requirements of common Portland cement.

Variety	Code	Insolubles	LOI	SO_3	Magnesium oxide	Chloride ion
Portland cement	P·I	≤0.75	≤3.0			
	P·II	≤1.50	≤3.5	≤3.5	≤5.0ᵃ	
Ordinary Portland cement	P·O	—	≤5.0			
Portland blast furnace slag cement	P·S·A	—	—	≤4.0	≤6.0ᵇ	≤0.06ᶜ
	P·S·B	—	—		—	
Portland pozzolana cement	P·P	—	—			
Portland fly ash cement	P·F	—	—	≤3.5	≤6.0ᵇ	
Composite Portland cement	P·C	—	—			

ᵃ The content of magnesia in cement (mass fraction, %) is allowed to be broadened to 6% if the autoclaving test of cement is qualified.
ᵇ It is necessary to carry out the test of cement steaming stability and it must be qualified when the content of magnesia in cement (mass fraction, %) is higher than 6%.
ᶜ The index is determined by both of producers and buyers when there are more stringent requirements.

temperature, the unsoundness caused by excessive SO_3 must be tested using water immersion method (28 days immersion in water under 20 °C).

(3) Strength

Strength is a key index of cement as well as an important parameter of mix design in concrete. The strength should increase gradually during the hardening process of qualified cement. The poor soundness of cement may cause strength decrement and even collapse without strength after 3 days, which is called strength decreasing. Three-day strength is usually called as early strength, while 28-day strength as later strength. The 3-month strength is called long-term strength. Since the strength of cement reaches the maximum value nearly after 28 days, the strength grade of cement is divided according to the 28-day strength. Every type of strength-grade cement must meet the requirements about compressive and flexural strength listed in Table 2.20. If any parameter cannot achieve the limited value, the strength grade should be determined according to the lowest strength of any certain age.

Table 2.20: Strength requirement of all strength grades for common Portland cement.

Variety	Strength grade	Compressive strength (MPa)		Flexural strength (MPa)	
		3 days	28 days	3 days	28 days
Portland cement	42.5	≥17.0	≥42.5	≥3.5	≥6.5
	42.5R	≥22.0		≥4.0	
	52.5	≥23.0	≥52.5	≥4.0	≥7.0
	52.5R	≥27.0		≥5.0	
	62.5	≥28.0	≥62.5	≥5.0	≥8.0
	62.5R	≥32.0		≥5.5	
Ordinary Portland cement	42.5	≥17.0	≥42.5	≥3.5	≥6.5
	42.5R	≥22.0		≥4.0	
	52.5	≥23.0	≥52.5	≥4.0	≥7.0
	52.5R	≥27.0		≥5.0	
Composited Portland cement	32.5R	≥15.0	≥32.5	≥3.5	≥5.5
	42.5	≥15.0	≥42.5	≥3.5	≥6.5
	42.5R	≥19.0		≥4.0	
	52.5	≥21.0	≥52.5	≥4.0	≥7.0
	52.5R	≥23.0		≥4.5	
Portland blast furnace slag cement	32.5	≥10.0	≥32.5	≥2.5	≥5.5
	32.5R	≥15.0		≥3.5	

Table 2.20 (continued)

Variety	Strength grade	Compressive strength (MPa)		Flexural strength (MPa)	
		3 days	28 days	3 days	28 days
Portland pozzolana cement	42.5	≥15.0	≥42.5	≥3.5	≥6.5
	42.5R	≥19.0		≥4.0	
Portland fly ash cement	52.5	≥21.0	≥52.5	≥4.0	≥7.0
	52.5R	≥23.0		≥4.5	

2.5.2.4 Strength grade of common Portland cement

Six strength grades of Portland cement are 42.5 grade, 42.5R grade, 52.5 grade, 52.5R grade, 62.5 grade and 62.5R grade.

Four strength grades of ordinary Portland cement are 42.5 grade, 42.5R grade, 52.5 grade and 52.5R grade.

Five strength grades of composite Portland cement are 32.5R grade, 42.5 grade, 42.5R grade, 52.5 grade and 52.5R grade.

Six strength grades of Portland blast furnace slag cement, Portland pozzolana cement and Portland fly ash cement are 32.5 grade, 32.5R grade, 42.5 grade, 42.5R grade, 52.5 grade and 52.5R grade.

Among them, R indicates the grade with high early strength, which has higher requirements for 3-day strength.

2.5.2.5 Technical requirements

(1) Chemical requirements

The chemical requirement of common Portland cement should conform to the requirements presented in Table 2.19.

(2) Alkali content (selection index)

Alkali content of cement is expressed as alkali equivalent $w_{eq}(Na_2O)$:

$$w_{eq}(Na_2O) = w(Na_2O) + 0.658w(K_2O)$$

When the reactive aggregate is used and the customer asks for low alkali cement, the $w_{eq}(Na_2O)$ should be not more than 0.60% or agreed by the purchaser and manufacturer.

(3) Physical requirements

(i) Setting time

For Portland cement, the initial setting time should be not less than 45 min and the final setting time is not more than 390 min.

For ordinary Portland cement, Portland blast furnace slag cement, Portland pozzolana cement, Portland fly ash cement and composite Portland cement, the initial setting time should be not less than 45 min and the final setting time is not more than 600 min.

(ii) Soundness
The soundness should be qualified with the boiling test.

(iii) Strength
The compressive strength and flexural strength of all types of strength-grade cements at 3, 7 and 28 days should conform to the requirements as shown in Table 2.20.

(iv) Fineness (selection index)
The fineness expressed as specific surface area in Portland cement and ordinary Portland cement should be at least 300 m²/kg. The fineness of Portland blast furnace slag cement, Portland pozzolana cement, Portland fly ash cement and composite Portland cement is expressed as sieve residue, of which 80 µm sieve residue should be not more than 10% or 45 µm sieve residue should be not more than 30%

Before cement delivery, a series of tests including insoluble substance, loss on ignition, content of MgO, SO_3, Cl^-, setting time, soundness and strength should be conducted. The cement that meets the technical requirements of these projects is a qualified product, and only if all these parameters meet the requirements, the delivery of cement can be qualified

2.5.3 Portland limestone cement

Portland limestone cement is a hydraulicity cementitious material made of Portland cement clinker, limestone and appropriate amount of gypsum, where the content of limestone (mass fraction) is 10–25%. Portland limestone cement is coded as P·L. The content of $CaCO_3$ in the used limestone should be at least 75.0% by mass, while the content of Al_2O_3 content should be not more than 2.0% by mass.

Although Portland limestone cement belongs to a kind or variety of Portland cement family, it is also generally used as cement; therefore, it has not been classified as a kind of common Portland cement due to its less scale of production and consumption.

Portland limestone cement can be classified into four strength grades, including 32.5, 32.5R, 42.5 and 42.5R. The technical requirements in JC/T 600-2002 for Portland limestone cement are as follows:

(1) MgO
The content of MgO in clinker should be not more than 5.0%. If the results of autoclave method for soundness are qualified, the limit content can be extended to 6.0%.

(2) SO$_3$
The content of SO$_3$ in cement should be not more than 3.5%.

(3) Chloride ion
The content of chloride ion in cement should be not more than 0.06%.

(4) Fineness
The specific surface area of cement should be not less than 350 m^2/kg.

(5) Setting time
The initial setting time should be not less than 45 min, while the final setting time should be not more than 600 min.

(6) Soundness
The result of boiling test should meet the requirement.

(7) Compressive strength
The compressive strength and flexural strength of all strength-grade cements with different ages should be not less than the limit values in Table 2.21.

Table 2.21: Strength norm of different strength grade on different ages.

Strength grade	Compressive strength (MPa)		Flexural strength (MPa)	
	3 days	28 days	3 days	28 days
32.5	11.0	32.5	2.5	5.5
32.5R	16.0	32.5	3.5	5.5
42.5	16.0	42.5	3.5	6.5
42.5R	21.0	42.5	4.0	6.5

(8) Alkali content

The alkali content in cement is calculated by the Na$_2$O equivalent [w(Na$_2$O) + 0.658w (K$_2$O)]. To determine whether there is a limitation for this value, the buyer and seller should have a discussion. If the active aggregate is used, the alkali content in cement should be as follows: w(Na$_2$O) + 0.658w(K$_2$O) is less than or equal to 0.60%, or it should be discussed by the purchaser and manufacturer.

2.5.4 Grinding of cement

2.5.4.1 Technology progress of grinding cement
Cement grinding system contains open circuit and closed circuit. During the grinding period, the new pulverized grinding process is being popularized in cement grinding system with the development of vertical roller mill and compressed grinding technology.

(1) Grinding system of open circuit

The flow of raw material: Based on production requirements, the cement composition materials are well mixed by batching plant, and are then fed to grinding by the feed equipment, and the cement that reaches the requirement is transported to storage in cement silo by the conveying equipment.

The flow of airflow: The airflow accompanying materials pass into the grinding, which are discharged from the grinding tail and entered into the dust collecting system, then are emitted into the atmosphere by the exhaust fan.

(2) Grinding system of closed circuit

The flow of raw materials: Based on production requirements, the cement composition materials are well prepared by batching plant before fed into grinding by the feeding equipment. The materials qualified are send to classifying equipment through the conveying equipment. The raw materials are returned to the flour mill and then pulverized again. The cement that reaches the fineness requirement is transported to storage in cement silo by the conveying equipment.

The flow of airflow: The flow of airflow is the same as that of open circuit.

2.5.4.2 Influencing factor of quality for grinding cement

(1) The particle size of raw materials in the mill

The size of raw materials in the feed equipment is one of the main factors that affect the output of mill. The small size of raw materials in the grinding mill increases the production and reduces the power consumption per unit.

(2) Moisture content of raw material in the mill

The moisture content of raw materials has a greater influence on efficiency in mill. When the average moisture content of raw materials is higher, the produced fine powder will stick to the grinding medium and the under-boarding that blocks the fine-draw of bulkhead, which result in a reduction of the efficiency in mill. However, the excessively dry raw materials are unnecessary because the moisture of raw materials can reduce the temperature in a mill, which is beneficial to reduce the electrostatic effect and improves the efficiency of grinding. Therefore, the average moisture content of raw material in a mill is controlled at 0.5–1.0%.

(3) Grindability of raw materials in the mill

The grindability (or fragility) of raw materials indicates the complexity of itself during crushing. The distinction in grindability of clinker with different composition is larger. The grindability is good and easy to grinding when the content of C_3S increases in clinker, while the grindability is poor when the content of C_2S and C_4AF increases. The grinding of cement clinker is also related to calcination conditions such as super-burning material or yellow heart clinker that has poor grindability, while the clinker from rapid cooling is easy to grinding. The incorporated admixture with different kinds and contents will lead to different grindability when grinding cement.

(4) Temperature of raw materials in the mill
The temperature of raw materials in the mill has an influence on the output of mill and the quality of cement. The high temperature of raw materials increases the internal temperature, and easily causes adhesion to the surface of grinding by cement powder. This is the so-called bag ball phenomenon, which affects the grinding efficiency in the mill.

Besides, the high temperature of mill also cause that the gypsum is dehydrated into semihydrated gypsum or even small amount of anhydrite, which induce the false set and affect the quality of cement.

(5) Fineness of product
When the required fineness of raw materials is smaller, raw materials stay in the mill for longer time. In order to perform sufficient grinding and make the grinding materials in the mill to meet the requirement of fineness, it is necessary to reduce the amount of feeding materials and decrease the flow rate of feeding material in the mill. On the other hand, if the required fineness is finer, the finer powder will result in larger adhesion effect and the more serious adhesion phenomenon will produce in the mill, all of which will reduce the production efficiency of the mill. Therefore, in order to meet the requirements of cement variety and strength grade, it is important to determine the fine grinding index of grinding.

(6) Mill ventilating
The ventilation in mill is one of the important factors affecting the efficiency of grinding mill. The strengthening of mill ventilating can discharge the micropowder timely, reduce the overcrushing phenomenon and the buffer effect, thus increasing the grinding efficiency. In addition, the strengthening of ventilating can discharge water vapor and heat in the mill timely, reduce the adhesion phenomenon and prevent the blockage of hole in the compartment. It can also reduce the temperature in the mill, increase the materials flow speed in the mill, prevent the graying of mill head, improve the environment sanitation and reduce the wear of equipment.

(7) Grading efficiency and rate of circulating load
The height of grading efficiency in powder concentrator with closed-circuit grinding system has remarkable influence on mill output, because the separator can separate the qualified fine powder from raw material, which improves the grinding condition and increases the efficiency of grinding mill. However, if the grading efficiency is high, the mill output is not necessary as the mill itself is not pulverized, nor does it increase the surface area of raw materials. Therefore, the function of the separator must be

matched with the grinding capacity of the mill and rate of circulating load, and improve the output of mill.

The rate of circulating load refers to the coarse powder (back grinding powder) in the closed-circuit grinding system that is divided into qualified products and needs to be returned to the grinding mill after stabilizing operation, and the amount of grinding powder remains stable. The ratio of cycle to product quality is called the rate of circulating load. The rate of circulating load is higher and the total amount of raw materials entering the mill is larger. The movement speed of materials from the feed end to discharge end can also be improved. The grinding time of materials in grinding is shortened, and the grinding phenomenon is greatly reduced, which is beneficial for improving mill production. But if the circulation rate is too large, not only to increase the mill load lifting equipment of material but also significantly reduces the classification efficiency of classifier, so the rate of circulating load and grading efficiency should be controlled within a certain range.

(8) Ball material ratio and the speed of raw material flow in the mill

The ball material ratio refers to the ratio of quality of grinding body to the quality of raw materials in the mill. According to the production experience, the grinding machine of normal production should be stopped and inspected, and the steel ball should be exposed to half of the ball in the first cabin and the grinding body of second cabin should be buried 1–2 cm under the raw material.

The flow rate of raw materials in the mill is an important factor affecting the quality of production and energy consumption. When the flow rate of raw materials in the mill is too fast, it runs coarse material easily and it is difficult to guarantee the product fineness, while if the flow rate is too slow, it can produce over-pulverization that increases pulverization resistance and reduces grinding efficiency. Therefore, the flow rate of raw materials must be controlled properly during the production. The flow rate of raw materials can be adjusted by means of the ball material ratio in the mill, the form of compartment plate, the size of hole, the grading of grinding body and the loading capacity.

2.6 Hydration and hardening of general-purpose Portland cement

When cement is mixed with water, it will transform into a plastic paste that binds sand and gravel together. This synthetic rock gradually gains strength through a setting and hardening process, accompanied by the potential heat liberation and volumetric deformation. This demonstrates that the binary cement–water system produces long-lasting complex physical and chemical changes, which enables a sustained strength increase. Because cement is a multiphase mixture with complex hydration processes, the constituent phases are normally investigated separately before elucidating its hydration chemistry.

2.6.1 Hydration of cement clinkers

2.6.1.1 Tricalcium silicate (C₃S)

In Portland cement clinker, C_3S is the main constituent that accounts for 50 wt% and sometimes more than 60 wt%. As a result, the properties of hardened cement paste largely depend on the hydration characteristics of C_3S.

The reaction of C_3S hydration at ambient temperature can be approximated by the following equation with the major hydration products as calcium silicate hydrates (C–S–H) and calcium hydroxide (CH).

$$3CaO \cdot SiO_2 + nH_2O = xCaO \cdot SiO_2 \cdot H_2O + (3-x)Ca(OH)_2$$

Or

$$C_3S + nH = C\text{–}S\text{–}H + (3-x)CH$$

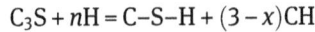

The hydration heat liberation rate of C_3S and Ca^{2+} concentration with respect to the hydration time is shown in Figure 2.12.

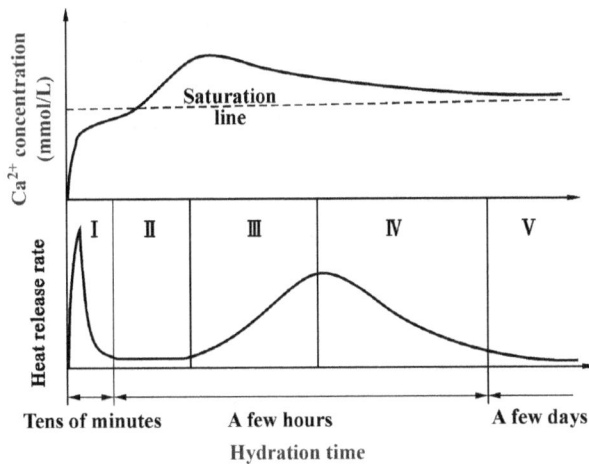

Figure 2.12: Heat liberation rate and Ca^{2+} concentration curves with the elapse of hydration time. I, preinduction period; II, induction (dormant) period; III, acceleration period; IV, deceleration period; V, stable period.

C_3S hydrates fast and its hydration process can be classified into five distinct stages based on the heat liberation rate curve.

(1) Preinduction period

An intensive but short-lived reaction upon the contact of cement with water. The duration of this period is typically not more than 15 min.

(2) Induction period (or dormant period)
This period has a significantly low rate of reaction and is known as the dormant period lasting for 1–4 h.

(3) Acceleration period
The reaction accelerates again, and the reaction rate increases with time. Then a second exothermic peak appears and reaches the peak at the end of this period (4–8 h). At this point the final setting is over and the paste begins to harden.

(4) Deceleration period
The reaction rate decreases with time and lasts for about 12–24 h. Hydration is gradually switches to the diffusion-controlled mode.

(5) Stable period
The reaction rate is low at a substantially steady stage, and hydration is completely controlled by the irons diffusion rate. Some literature will collectively term stages 4 and 5 as the diffusion control stage.

Thus, in the early period of water addition, the reaction is very fast, but soon transforms into the induction period with a much slower reaction rate. At the end of this period, hydration reaccelerates to generate more hydration products. The rate of hydration then decreases gradually over time. In addition, C_3S hydration process can be more generally divided into three stages, namely, the early hydration stage (including the preinduction and induction stages), the mid-hydration stage (including the acceleration and deceleration stages) and the stable stage as the later hydration stage.

(i) Early hydration of C_3S
The performance of the hardened paste is largely determined at the early hydration stage. The termination of the induction period is related to the initial setting, whereas the final setting occurs roughly in the middle of the acceleration period. Researchers have done a lot of researches on this, and there are different opinions about the essence of the induction period, which is the beginning and the end of the induction period.

H. N. Stein and others believe that the induction period is due to the formation of a protective film generated from hydration. When the protective film is damaged, the induction period is over. This is the so-called protective film theory. They assume that C_3S is uniformly soluble in water. The first hydrate C_3SH_n, which was first formed, soon forms a dense protective film around C_3S particles, hindering its further hydration, slowing the exothermic reaction and also the dissolution of Ca^{2+}. This corresponds to the beginning of the induction period. Hydration reaccelerates when the first hydrate is converted into the second hydrate [n $(CaO)/n(SiO_2) = 0.8–1.5$]. When hydration is reaccelerated, more Ca^{2+} and OH^-

enter the liquid to achieve supersaturation, and accelerates the exothermic reaction. The induction period is over.

Another group of scholars put forward "the nucleus formation delay theory," in which they claim that it takes some time for the formation and growth of $Ca(OH)_2$ or C–S–H nucleus, so that the hydration is delayed, leading to the induction period. For example, M. E. Tadros et al. thought that C_3S was initially discordant, which is mainly due to the dissolution of Ca^{2+} and OH^-, and the ratio of $n(CaO)/n(SiO_2)$ in the liquid phase is much higher than 3, leading to the formation of calcium-deficient but silicon-enriched layer. Then, Ca^{2+} is adsorbed to the silicon-rich surface to make it positively charged to form the electric double layer. Thus, the dissolution of Ca^{2+} from C_3S is slowed down, leading to the induction period. Until the Ca^{2+} and OH^- in the liquid phase slowly grow to reach a sufficient degree of supersaturation (1.5–2.0 times saturation), a stable $Ca(OH)_2$ nucleus is formed. When it grows to a considerable size and the number is large enough, the Ca^{2+} and OH^- in the liquid phase rapidly precipitates $Ca(OH)_2$ crystals, which prompts the C_3S to dissolve quickly and the hydration is reaccelerated. The maximum Ca^{2+} concentration in the liquid phase as shown in Figure 2.12 appears at the end of the induction period, which can be used as an argument for the above hypothesis. At the same time, Ca $(OH)_2$ also combines with silicate ions and thus also serves as the nucleus of C–S–H. More recent studies show that the induction period is more likely to be due to the delayed nucleation and growth of C–S–H.

Based on the experimental results of the induction length and the lattice defect of C_3S, some researchers claim that the early C_3S hydration reaction does not cover the entire surface, but only in the lattice defects where reaction is most likely to occur with water. Nonuniform distribution of hydration products will not form the protective film. They then proposed that the type and number of lattice defects was the main factors to determine the length of induction period.

J. Skalng and J. F. Young put forward the following comprehensive viewpoints: When the C_3S is in contact with water, hydrolysis occurs on the activation site (Figure 2.13) of its surface with lattice defects. Ca^{2+} and OH^- enter the solution, forming a calcium-deficient but silicon-rich layer on the C_3S surface. Then, Ca^{2+} is adsorbed to the "silicon-rich layer" surface to form an electric double layer, so that C_3S dissolution is hindered, leading to the induction period. On the other hand, due to the zeta potential formed by the electric double layer, agglomeration does not occur until the zeta potential drops close to zero even if the particles remain dispersed in the solution. Since C_3S still hydrates slowly, $Ca(OH)_2$ crystallizes when its concentration in the solution reaches a certain degree of supersaturation, leading to the end of the induction period. At the same time, precipitation of CSH will also occur because the silicate ions are more difficult to migrate than Ca^{2+}, so the growth of C–S–H is confined to the surface; the crystals of $Ca(OH)_2$ may also begin to grow on the surface of C_3S, but some crystals will stay away from the particles or form in in the water-filled pores. The surface morphology of C_3S approaching the end of induction period is shown in Figure 2.13.

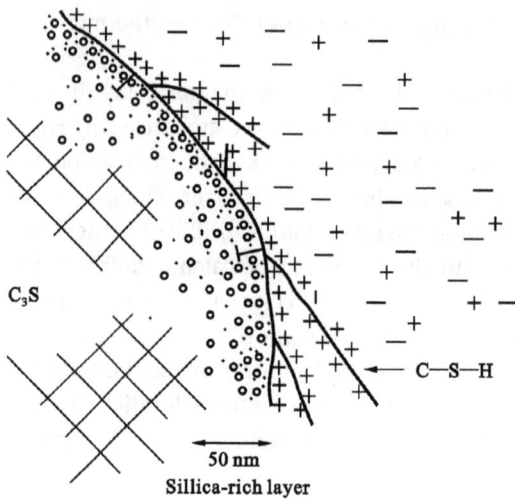

Figure 2.13: Surface morphology of C_3S toward the end of induction period. o, $H_3SiO_4^-$ or $H_2SiO_4^{2-}$; ·, H^+ or H_2O; +, Ca^{2+}; −, OH^-; ⊥, activation sites.

(ii) Midstage hydration of C_3S

With the formation and growth of $Ca(OH)_2$ and C–S–H during the acceleration period of C_3S hydration, the supersaturation of $Ca(OH)_2$ and C–S–H in the liquid phase decreases and accordingly $Ca(OH)_2$ and C–S–H grow slower. With the formation of hydration products around the particles, the hydration of C_3S is also hindered. As a result, the acceleration stage gradually transforms into deceleration phase. Most of the initial products grow in the water-filled space beyond the original perimeter of the grain, whereas later growth takes place within the original perimeter of the grain. The C–S–Hs of these two parts are called external and internal products, respectively. With the formation and development of internal products, the hydration of C_3S changes from the deceleration phase to the steady phase and gradually enters the later stage of hydration.

(iii) C_3S late hydration

Regarding the posthydration of C_3S, F.H.W. Taylor presented a schematic as shown in Figure 2.14. He believes that there is an interface zone in the hydration process and progresses toward the interior of the particle. The H^+ from the dissociation of water transfers from an oxygen atom (or water molecule) to another oxygen atom in the internal product until it reaches and interacts with C_3S, which is almost the same as the case of being in direct contact with water. While some Ca^{2+} and Si^{4+} in the interfacial region migrate outwardly through the internal product to $Ca(OH)_2$ and the exterior of C–S–H. Therefore, H^+ is obtained in the interface region while Ca^{2+} and Si^{4+} are lost, and the ions recombine to transform C_3S into internal C–S–H. As a result, hydration continues as the interface zone advances inward. Due to space constraints and changes in ion concentration, C–S–H, which is an internal product and is usually denser, will vary in morphology and composition compared to external C–S–H. Hydration of the C_3S stages can be illustrated in Figure 2.15.

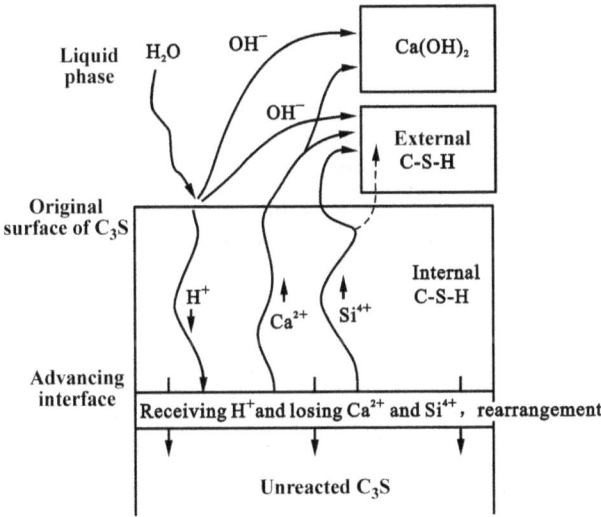

Figure 2.14: Diagram of late-stage hydration of C_3S

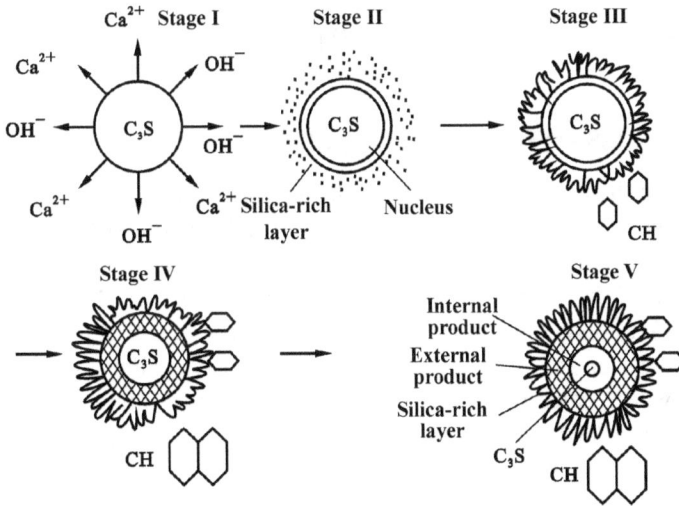

Figure 2.15: Schematic for C_3S hydration.
I, preinduction period; II, induction period; III, acceleration period; IV, deceleration period; V, stable period

The chemical and dynamics processes of the C_3S stages are given in Table 2.22.

Table 2.22: Summary on the chemical processes and kinetic behavior of each stage of C_3S hydration.

Stage	Reaction	Chemical process	Kinetic behavior
Early	Preinduction period	Initial hydrolysis and the entrance of ions into solution	Rapid
	Induction period	Continued dissolution and formation of early C–S–H	Slow
Mid	Acceleration period	Stable hydration products begin to grow	Rapid
	Deceleration period	Continuous growth of hydration products and development of microstructure	Slows down
Later	Stable period	Gradual densification of microstructure	Very slow

2.6.1.2 Dicalcium silicate

Usually C_2S clinker is mostly β-type. The hydration process of β-C_2S is very similar to that of C_3S, which includes the induction and acceleration periods as well. However, its hydration rate is very slow, about 1/20 of C_3S. It has been estimated that β-C_2S takes about a few tens of hours to accelerate the square wave, even after only a few weeks covering a thin layer of amorphous hydrated calcium silicate on the surface and increasing the thickness of the hydration product layer slow. The hydration reaction of β-C_2S can be expressed as follows:

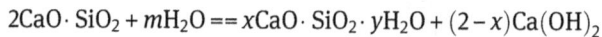

$$2CaO \cdot SiO_2 + mH_2O == xCaO \cdot SiO_2 \cdot yH_2O + (2-x)Ca(OH)_2$$

That is

$$C_2S + mH == C - S - H + (2-x)CH$$

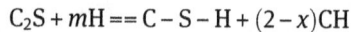

Due to the low hydration heat, it is difficult to use the exothermic rate to study the β-C_2S hydration. However, the first exothermic peak is comparable to that of C_3S, and the second peak is rather faint and even difficult to measure. Some observations indicate that some parts of the β-C_2S hydration starts earlier and the surface quickly becomes asymptotic after contact with water, much like the case of C_3S. Hydration products can be detected even at 15 s in contact with water, but the later development is extremely slow. However, calcium silicate hydrate formed by β-C_2S and C_3S does not differ too much in the $n(CaO)/n(SiO_2)$ ratio and morphology; thus, it is collectively referred to as C–S–H gel. Some test results show that although the rate of nucleation and crystal growth of hydration products during the hydration of β-C_2S is similar to C_3S; however, the diffusion rate through the hydration product layer is about eight times lower than C_3S and the surface dissolution rate is several times lower than C_3S. This indicates that the hydration reaction rate of β-C_2S is mainly controlled by the surface dissolution rate. Therefore, it is possible to accelerate its hydration rate by increasing the structural activity of C_2S, selecting the appropriate hydration medium and improving the hydration conditions. In particular, research on active C_2S is progressing due to the potential energy-saving benefits.

2.6.1.3 Tricalcium aluminate

Hydration reaction of C_3A is rapid, and the composition and structure of its hydration products are greatly influenced by the concentrations of CaO and Al_2O_3 and temperature in the solution. At normal temperature, C_3A hydration is

$$2(3CaO \cdot Al_2O_3) + 27H_2O == 4CaO \cdot Al_2O_3 \cdot 19H_2O + 2CaO \cdot Al_2O_3 \cdot 8H_2O$$

That is

$$2C_3S + 27H == C_4AH_{19} + C_2AH_8$$

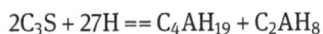

C_4AH_{19} with less than 85% relative humidity will lose 6 mol of crystal water to become C_4AH_{13}. C_4AH_{19}, C_4AH_{13} and C_2AH_8 are hexagonal platelet crystals (Figure 2.16), which are metastable at room temperature and have a tendency to transform into C_3AH_6 equiaxed crystals (Figure 2.17):

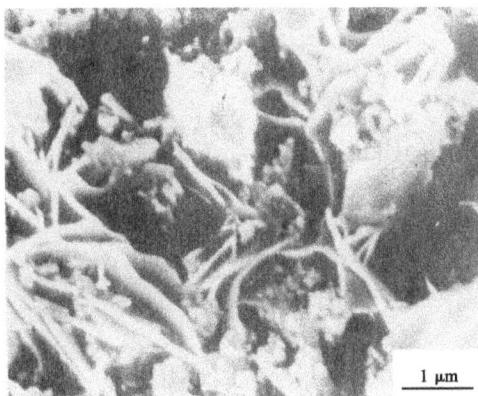

1 μm

Figure 2.16: Hexagonal platelet crystals of C_2AH_8 and C_4AH_{19}.

2 μm

Figure 2.17: Octahedron C_3AH_6.

$$4CaO \cdot Al_2O_3 \cdot 13H_2O + 2CaO \cdot Al_2O_3 \cdot 8H_2O == 2(3CaO \cdot Al_2O_3 \cdot 6H_2O) + 9H_2O$$

That is

$$C_4AH_{13} + C_2AH_8 == 2C_3AH_6 + 9H$$

The above process will be accelerated with an increasing temperature, and the hydration heat of C_3A by itself is very high, facilitating the conversion shown in the above formula. At a higher temperature (above 35 °C), C_3AH_6 crystals can be generated directly:

$$3CaO \cdot Al_2O_3 + 6H_2O == 3CaO \cdot Al_2O_3 \cdot 6H_2O$$

That is

$$C_3A + 6H == C_3AH_6$$

When the concentration of CaO in the liquid reaches saturation, C_3A may also be hydrated according to the following formula:

$$3CaO \cdot Al_2O_3 + Ca(OH)_2 + 12H_2O == 4CaO \cdot Al_2O_3 \cdot 13H_2O$$

That is

$$C_3A + CH + 12H == C_4AH_{13}$$

This reaction is most likely to occur in the alkaline liquid phase of the Portland cement paste, while C_4AH_{13} in an alkaline medium is stable at room temperature, leading to rapid increase in number and retarding the relative movement of the particles. This is considered to be a major cause of instantaneous coagulation of the paste. Therefore, cement is usually mixed with gypsum. Because of the simultaneous presence of gypsum and CaO, C_3A initially hydrates quickly to form C_4AH_{13}, but then reacts with gypsum as follows:

$$4CaO \cdot Al_2O_3 \cdot 13H_2O + 3(CaSO_4 \cdot 2H_2O) + 14H_2O == 3CaO \cdot Al_2O_3 \cdot 3CaSO_4 \cdot 32H_2O + Ca(OH)_2$$

That is

$$C_4AH_{13} + 3C\bar{S}H_2 + 14H == C_3A \cdot 3C\bar{S}H_{32} + CH$$

The formation of 6-calcium aluminate trisulfate-32-hydrate is also known as ettringite. As aluminum can be replaced by iron to form iron tri-sulfate phase, it is often expressed as AFt. When C_3A has not yet completely hydrated but the gypsum has been depleted, the C_4AH_{13} from hydration of C_3A reacts with previously formed ettringite to form tetracalcium aluminate monosulfate-12-hydrate (AFm):

$$3CaO \cdot Al_2O_3 \cdot 3CaSO_4 \cdot 32H_2O + 2(4CaO \cdot Al_2O_3 \cdot 13H_2O) ==$$

$$3(3CaO \cdot Al_2O_3 \cdot CaSO_4 \cdot 12H_2O) + 2Ca(OH)_2 + 20H_2O$$

That is

$$C_3A \cdot 3C\overline{S} \cdot H_{32} + 2C_4AH_{13} == 3(C_3A \cdot C\overline{S} \cdot H_2) + 2CH + 20H$$

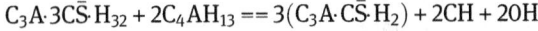

When the amount of gypsum is extremely low, unhydrated C_3A may remain after all ettringite is converted to monosulfoaluminate hydrate. In this case, a solid solution of $C_4A\overline{S}H_{12}$ and C_4AH_{13} is formed by the following formula:

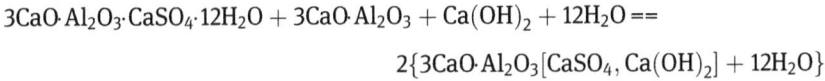

$$3CaO \cdot Al_2O_3 \cdot CaSO_4 \cdot 12H_2O + 3CaO \cdot Al_2O_3 + Ca(OH)_2 + 12H_2O ==$$

$$2\{3CaO \cdot Al_2O_3[CaSO_4, Ca(OH)_2] + 12H_2O\}$$

That is

$$C_4A\overline{S}H_{12} + {}_3A + CH + 12H == 2C_3A(C\overline{S} \cdot CH)H_{12}$$

Therefore, hydration of C_3A may produce a variety of hydration products, depending on the actual amount of gypsum present in the reaction, as shown in Table 2.23.

Table 2.23: Hydration products of C_3A.

Actual $n(C\overline{S}H_2)/n(C_3A)$	Hydration products
3.0	Ettringite (AFt)
1.0–3.0	Ettringite (AFt) + calcium monosulfoaluminate (AFm)
1.0	Calcium monosulfoaluminate (AFm)
<1.0	Monosulfo-solid solution [$C_3A(C\overline{S} \cdot CH)H_{12}$]
0	Hydrogarnet (C_3AH_6)

When C_3A is mixed with water alone, rapid reaction occurs within a few minutes and hydration can be completed within a few hours. Figure 2.18 shows the hydration exotherm. When mixed with gypsum, the reaction will continue after a few hours to accelerate hydration.

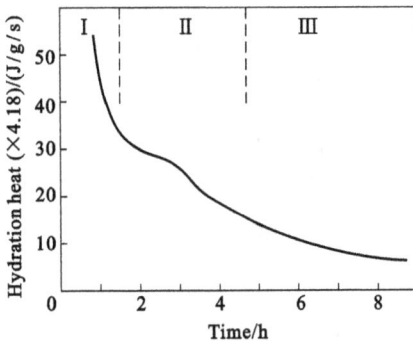

Figure 2.18: Exothermic hydration curve of C_3A in pure water.

The combination of gypsum and $Ca(OH)_2$ results in more pronounced retardation. Figure 2.19 shows an example of the hydration exotherm (exothermic rate in units of C_3A). It can be seen that the formation of calcium monosulfoaluminate and the second exothermic peak will appear a long time after the formation of the first exothermic peak of ettringite. Therefore, the presence of gypsum, its dosage and dissolution are the determining factors for the C_3A hydration rate and hydration products.

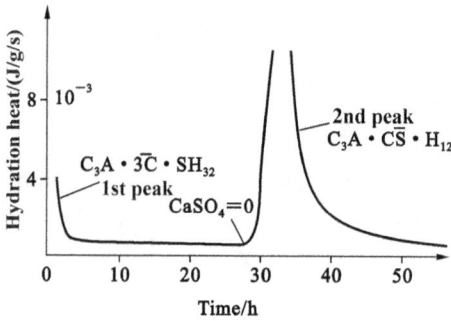

Figure 2.19: Exothermic hydration curve of the system of $C_3A\text{-}CaSO_4\cdot2H_2O\text{-}Ca(OH)_2\text{-}H_2O$. In this system, the components are 3 g of C_3A, 0.4 g of $CaSO_4\cdot2H_2O$ and 0.0196 of mol/L Ca $(OH)_2$ solution with 2 mL.

According to the general gypsum content in Portland cement, the final aluminate hydrate is often ettringite and calcium monosulfoaluminate. However, the migration of ions is limited to a certain extent in the conventional cement paste with a common w/c ratio. It is difficult to sufficiently conduct the above-mentioned various reactions. Therefore, ettringite and several other aluminate products are concentrated in the local area.

2.6.1.4 Solid solution of ferrite phase

In addition to C_4AF as a representative of a series of ferrite-based solid solution in cement clinker, Fss can also be used to represent this phase. The hydration rate of C_4AF is slightly slower than that of C_3A, and the hydration heat is low, which will not cause flash setting even if hydrated alone.

The hydration reaction and its product of C_4AF are very similar to C_3A. Fe_2O_3 plays essentially the same role as Al_2O_3 in that iron replaces part of the aluminum in the hydration product to form a solid solution of calcium sulfoaluminate hydrate and calcium sulfoferrite hydrated, or hydrated calcium aluminate and hydrated iron calcium solid solution.

In the absence of gypsum, C_4AF reacts with $Ca(OH)_2$ and water to form C_4AH_{13} in which some of the aluminum is replaced by iron, that is, $C_4(A,F)H_{13}$:

$$4CaO\cdot Al_2O_3\cdot Fe_2O_3 + 4Ca(OH)_2 + 22H_2O == 2[4CaO\cdot(Al_2O_3, Fe_2O_3)\cdot13H_2O]$$

That is

$$C_4AF + 4CH + 22H == 2C_4(A, F) + H_{13}$$

The resultant $C_4(A,F)H_{13}$ is also stable at low temperatures, but will be converted to $C_3(A,F)H_6$ at around 20 °C. However, the conversion process is slower than that of the C_3A hydration. It may be due to the low hydration heat of C_4AF, which is not easy to increase the temperature of the paste. Similar to C_3A, the presence of $Ca(OH)_2$ will also slow its conversion to the cubic crystal form $C_3(A,F)H_6$. C_4AF will be converted to $C_3(A,F)H_6$ directly at higher temperatures (>50 °C).

If gypsum exists, hydration will proceed as follows:

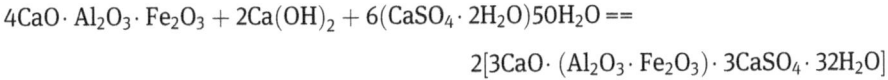

$$4CaO \cdot Al_2O_3 \cdot Fe_2O_3 + 2Ca(OH)_2 + 6(CaSO_4 \cdot 2H_2O)50H_2O ==$$

$$2[3CaO \cdot (Al_2O_3 \cdot Fe_2O_3) \cdot 3CaSO_4 \cdot 32H_2O]$$

That is

$$C_4AF + 2CH + 6C\bar{S}H_2 + 50H == 2C_3(A, F) \cdot 3C\bar{S} \cdot H_{32}$$

When the gypsum is consumed and $C_4(A,F)H_{13}$ is still present, the resultant trisulfide phase is converted to the monosulfide phase by the following formula:

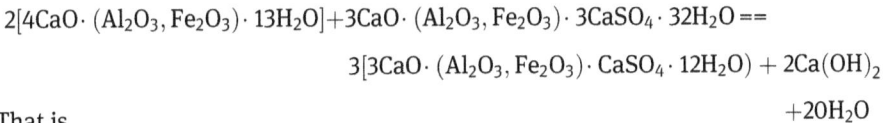

$$2[4CaO \cdot (Al_2O_3, Fe_2O_3) \cdot 13H_2O] + 3CaO \cdot (Al_2O_3, Fe_2O_3) \cdot 3CaSO_4 \cdot 32H_2O ==$$

$$3[3CaO \cdot (Al_2O_3, Fe_2O_3) \cdot CaSO_4 \cdot 12H_2O] + 2Ca(OH)_2$$

$$+20H_2O$$

That is

$$2C_4(A, F)H_{13} + C_3(A, F) \cdot 3C\bar{S} \cdot H_{32} == 3C_3(A, F) \cdot C\bar{S} \cdot H_{12} + 2CH + 20H$$

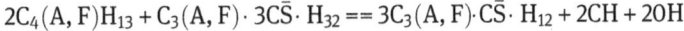

The hydration exotherm of C_4AF is also similar to that of C_3A, but the early hydration delayed more obviously by gypsum. In $Ca(OH)_2$ saturated solution, gypsum can make its exothermic rate extremely slow (Figure 2.20).

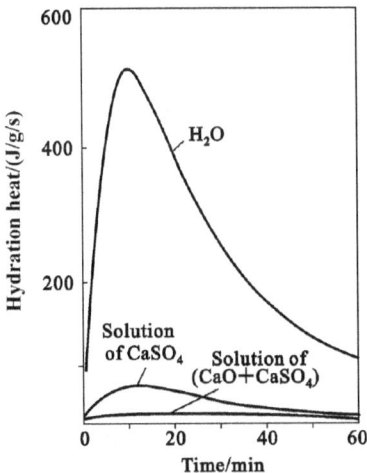

Figure 2.20: Exothermic hydration curve of C_4AF.

In a series of studies on solid solution in the C_2F–C_6A_2F range, A. Negro et al. found out that the hydration activity of the solid solution increases with an increasing $n(Al_2O_3)/n(Fe_2O_3)$ ratio. Conversely, if the Fe_2O_3 content is increased, the hydration rate is decreased. However, contradictory results exist, namely that the hydration rate of C_6A_2F is greater than other ferrous phase. In addition, the $n(Al_2O_3)/n(Fe_2O_3)$ ratio of the hydration product is generally higher than that of the prehydration solid solution so that amorphous hydrous $Fe(OH)_3$ is formed in addition to the above hydration products.

2.6.2 Hydration of Portland cement

When cement is mixed with water, a chemical reaction occurs immediately, and its various components begin to dissolve. So only after a moment, the liquid phase filled between the particles is no longer pure water, but a solution containing a variety of ions. The main changes are shown as follows:

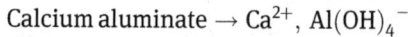

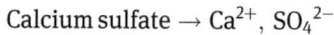

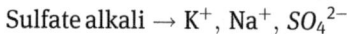

$$\text{Calcium silicate} \rightarrow Ca^{2+}, OH^-$$

$$\text{Calcium aluminate} \rightarrow Ca^{2+}, Al(OH)_4^-$$

$$\text{Calcium sulfate} \rightarrow Ca^{2+}, SO_4^{2-}$$

$$\text{Sulfate alkali} \rightarrow K^+, Na^+, SO_4^{2-}$$

Due to the rapid dissolution of C_3S to generate $Ca(OH)_2$, the incorporated gypsum will quickly dissolve in water. In particular, some of the dihydrate gypsum may dehydrate into hemihydrate gypsum or soluble anhydrite with a greater rate of dissolution during cement grinding. The alkali contained in the clinker dissolves quickly, and even 70–80% of the K_2SO_4 dissolves within a few minutes. Therefore, the hydration of cement begins essentially with a saturated solution of $Ca(OH)_2$, $CaSO_4$.

The amount of time Ca^{2+} and SO_4^{2-} are held in solution at high concentrations depends on the composition of the cement. Fujii Kinjiro et al. have determined that the supersaturation of a highly supersaturated $Ca(OH)_2$ solution sharply decreases after reaching a maximum value in the first 10 min. Afterwards, the solution becomes saturated or just weakly supersaturated. However, there are also data indicating that the high degree of supersaturation of $Ca(OH)_2$ can be maintained for 4 h or 1–3 days. The concentration of SO_4^{2-} begins to decrease after reaching the maximum value, similar to the change of Ca^{2+} concentration. This is mainly due to the depletion of sulfate by calcium aluminate to form ettringite or monosulfoaluminate hydrate, so that the SO_4^{2-} concentration in the solution declines and it gradually becomes substantially $Ca(OH)_2$, KOH and NaOH solution.

However, the solubility of calcium is reduced in the presence of potassium and sodium (Figure 2.21), which accelerates the crystallization of $Ca(OH)_2$ and finally makes the liquid phase become K^+-, Na^+- and OH^--based solutions.

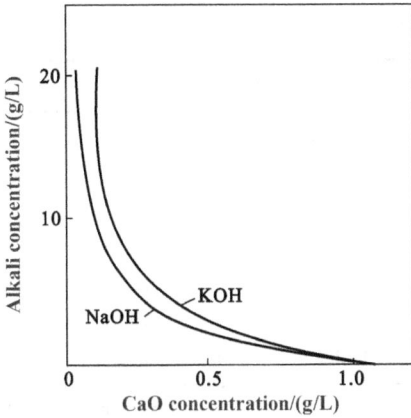

Figure 2.21: Relationship between the solubility of CaO and alkali concentration of potassium and sodium (20 °C).

Thus, the composition of the liquid phase depends on the solubility of various components in the cement, but the liquid phase composition will in turn have a profound impact on the hydration rate of each clinker mineral, and solid and liquid phases change in time with a dynamic balance. Figure 2.22 shows an example of the change in ion concentration in the liquid phase of the cement paste. Table 2.24 shows a set of results for the composition of the pore solution in the filter test.

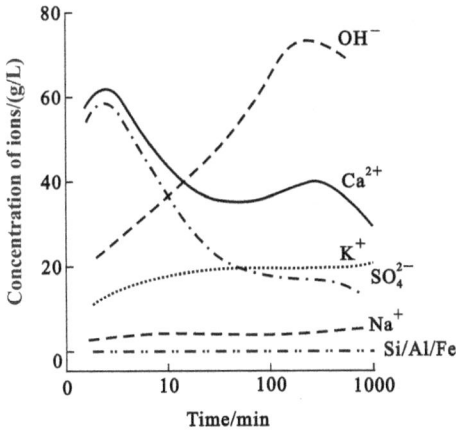

Figure 2.22: Changes in ion concentration of the cement paste ($w/c = 2$).

According to the current understandings, the hydration process of Portland cement can be summarized as shown in Figure 2.23. C_3A reacts immediately with the addition of water to cement. The hydration of C_3S and C_4AF is also fast, while C_2S hydrates slowly. When observed in the electron microscope after a few minutes of reaction, it can be observed that ettringite crystals, amorphous calcium silicate hydrate and $Ca(OH)_2$ or calcium aluminate hydrate hexagonal plate crystal are generated on the

Table 2.24: Chemical compositions in the pore solution of Portland cement (mmol/L).

Age	pH	CaO	OH⁻	SO₃	K₂O	Na₂O	SiO₂	Al₂O₃
5 h	12.99	26.1	83	24.9	39.9	6.8	0.06	—
1 d	13.24	7.2	147	1.2	59.6	12.6	0.03	0.03
3 d	13.31	5.0	195	0.5	69.0	19.2	0.04	0.04
7 d	13.38	4.2	229	—	76.1	25.4	0.13	0.01
28 d	13.48	3.6	261	1.5	100.5	27.4	0.08	—
90 d	13.52	2.4	283	—	113.8	32.3	0.11	0.03
180 d	13.52	2.0	288	—	120.0	40.3	0.13	0.02

Note: The contents of K_2O and Na_2O in Portland cement are 0.43% and 0.16%, respectively; w/c ratio is 0.5.

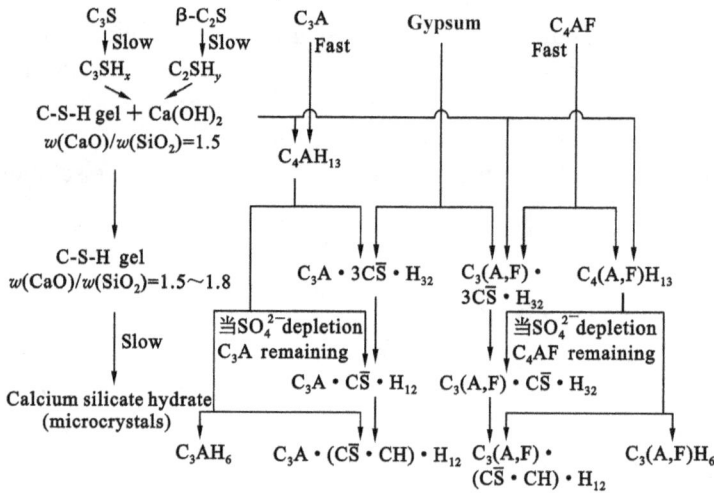

Figure 2.23: Hydration process of Portland cement.

surface of cement particles. Due to the continuous growth of ettringite, SO_4^{2-} in the liquid phase is gradually reduced. After its exhaustion, there is the presence of calcium monosulfoaluminate (iron). In the case of the lack of gypsum and the presence of remaining C_3A and C_4AF, monosulfide hydrates and solid solutions of $C_4(C,F)H_{13}$ or even $C_4(A,F)H_{13}$ will be formed, which in turn gradually transforms into equiaxed crystals of $C_3(A,F)H_6$.

Figure 2.24 shows the change in the proportion of each phase in the clinker and the final hydration products in the solid phase before and after hydration of Portland cement.

Figure 2.25 shows the exothermic curve of Portland cement in the hydration process. It can be seen that its pattern is basically the same as that of C_3S. Accordingly, the cement hydration process can be simply divided into the following three stages:

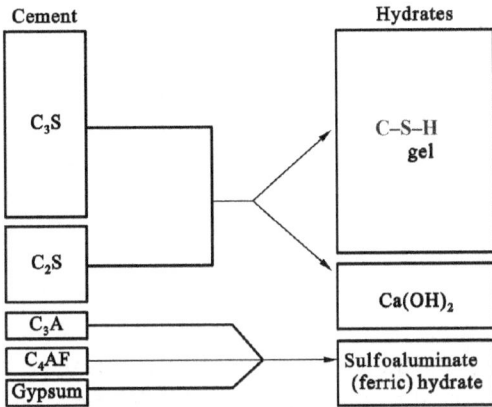

Figure 2.24: Changes of solid phase and its volume ratio before and after hydration.

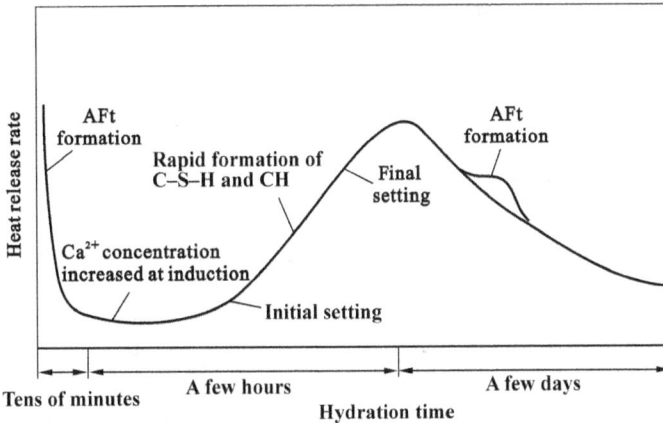

Figure 2.25: Hydration exotherm of Portland cement.

(1) Ettringite formation period

C_3A is the first one to hydrate, forming Ettringite rapidly in the presence of gypsum, which is the main factor leading to the first exothermic peak.

(2) C_3S hydration period

C_3S begins its rapid hydration generating a lot of heat and the formation of the second exothermic peak. Sometimes there is a third exothermic peak or a "shoulder" on the second exothermic peak, which is generally thought to be due to the conversion of ettringite to calcium monosulfoaluminate (iron). Of course, C_2S and ferrite phases also participate in the reaction of these two stages to some extent, and generate the corresponding hydration products.

(3) Structure formation and development period

The rate of heat release is very low and tends to be stable. With the increase of various hydration products, the space originally occupied by water is filled up, and then gradually connected, intertwined, and developed into a hardened paste structure.

However, it is worth noting that since cement is a multimineral, multicomponent system, clinker minerals cannot be hydrated by themselves and the interaction between them necessarily has an impact on the hydration process of cement. For example, the rapid hydration of C_3S increases the concentration of Ca^{2+} in the liquid phase and promotes the crystallization of $Ca(OH)_2$, thereby shortening the induction period of β-C_2S and accelerating the hydration. In another example, both C_3A and C_4AF should be combined with SO_4^{2-}, but the reaction of C_3A is faster and its increasing consumption of gypsum makes it impossible for C_4AF to form enough calcium sulfoaluminate stoichiometrically, which may delay the hydration rate. On the other hand, if the amount of gypsum is too small, the initial reaction of C_3A is not well controlled and the formation of monosulfate-type solid solution precedes the reacceleration of C_3S hydration will consume Ca^{2+}. This will affect the nucleation of $Ca(OH)_2$ and C–S–H, delaying the hydration process. However, the amount of gypsum can make the hydration of silicate slightly accelerated. At the same time, a considerable amount of SO_4^{2-}, Al^{3+}, Fe^{3+} and so on are incorporated into C–S–H, which, together with C_3A and C_4AF, C_3S, consumes SO_4^{2-}. It has been determined that the amount of sulfoaluminate salt formed in the cement paste is less than half of the theoretical value. It is also to be mentioned that the presence of alkali also affects the initial hydration of cement, especially C_3A. H.N. Stein et al. pointed out that when the concentration of Na_2O is low, the exothermic heat of initial hydration of C_3A decreases with the increase of Na_2O content. However, when the concentration of Na_2O is above 0.4 mol/L, the exothermic heat is accelerated. The former effect is due to a decrease in solubility of Ca^{2+} in NaOH solution, whereas the latter is due to the destruction of Al–O bonds in C_3A when the OH^- concentration is high.

In addition, the use of general hydration formula, in fact, is difficult to truly represent the cement hydration process. As hydration continues to progress, the C–S–H gel layer around the cement particles is continuously thickened and the rate of diffusion of water within the C–S–H layer becomes a decisive factor. Under such conditions, each clinker mineral cannot be hydrated according to its inherent characteristics. Therefore, although the degree of hydration varies greatly in the early stage, it is close to the later stage. At the same time, the actual amount of water used for mixing of paste is usually small and diminishes during the hydration process, with hydration occurring at varying concentrations, and the hydration exotherm of clinker minerals in turn causes the temperature of the hydration system to be not in constant state. Therefore, the hydration process and the general chemical reaction in solution are different, especially the migration of ions is more difficult. It is impossible to react in a very short period of time completely, but rather starting from the surface, it slowly penetrates to the center by diffusion under conditions of constant concentration and temperature. More importantly, even in fully hardened paste, it is not in equilibrium. In the center of clinker particles, hydration has often been temporarily halted. Later, when the temperature and humidity conditions are appropriate, the hydration of paste continues after replenishment of water from the outside or the redistribution of water. Therefore, the hydration process must not be treated as a general chemical reaction, and its long-term imbalance and its relationship with the surrounding environment must also be given full attention.

2.6.3 Hydration and hardening of Portland cement with supplementary cementitious materials

The hydration process of Portland cement with supplementary cementitious materials (SCM) is more complex than the Portland cement. After the addition with water, Portland cement hydrates first. This creates $Ca(OH)_2$, which reacts with the components of the SCM to form a series of hydration products.

The density of the blended cement is smaller than that of Portland cement, normally 2.7–2.9 g/cm^3 for pozzolanic cement and 2.8–3.0 g/cm^3 for slag cement. The color is lighter and the setting time is generally longer than that of Portland cement. The initial and final setting times for slag cement are generally 2–5 h and 5–9 h, respectively. Standard consistency of water consumption depends on the type of mixture, which is similar between slag cement and ordinary cement, and is higher in pozzolan cement, but lower in fly ash cement and limestone Portland cement. Strength development of slag cement, pozzolan cement and fly ash cement is sensitive to temperature. The setting and hardening process is retarded at low temperatures. Thus open-air construction in winter should be avoided. The hydration heat of the blended cement is lower than Portland cement while the water resistance is slightly better than or similar to Portland cement. Heat resistance is good and the bonding with steel is improved as well. Sulfate resistance is also superior to Portland cement. Except for limestone Portland cement with a similar frost resistance to Portland cement, the other blended cements exhibit worse frost resistance and antiatmospheric stability than the Portland cement. Premature drying and drying–wetting alternating exposure is unfavorable for their strength development. Slag cement has a higher bleeding ratio, while limestone Portland cement has good workability and small bleeding. Pozzolanic cement has a higher standard consistency of water consumption, and thus a higher dry shrinkage rate.

2.6.3.1 Hydration and hardening of slag cement

Hydration and hardening of slag cement depends on the activity and content of slag and the interaction between slag and clinker hydration products and gypsum, in addition to the hydration and hardening of Portland cement clinker.

(1) Hydration process and characteristics of slag cement
With the addition of water to slag cement, the clinker hydrates first. C_3A in clinker rapidly interacts with gypsum to form needle-like crystals of ettringite. C_3S hydrates to form C–S–H, Ca $(OH)_2$, and other products such as calcium ferrite. These hydrates have the same nature as that of Portland cement.

Due to the formation of $Ca(OH)_2$ and the presence of gypsum, the potential hydraulicity of the slag is stimulated. $Ca(OH)_2$, as an alkaline activator, plays a role of dissociating the vitreous structure of the slag and causes Ca^{2+}, AlO_4^{5-}, Al^{3+} and SiO_4^{4-} in the vitreous to enter the solution, resulting in the dispersion and dissolution of the slag; meanwhile, $Ca(OH)_2$ and active SiO_2, Al_2O_3 in the slag react to produce calcium silicate hydrate and calcium aluminate hydrate. Under the combined effect of $Ca(OH)_2$ and gypsum, the active Al_2O_3 in slag reacts to form calcium sulfoaluminate hydrate by the following process:

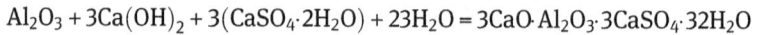

$$Al_2O_3 + 3Ca(OH)_2 + 3(CaSO_4 \cdot 2H_2O) + 23H_2O = 3CaO \cdot Al_2O_3 \cdot 3CaSO_4 \cdot 32H_2O$$

In addition, calcium ferrite hydrate, calcium aluminate silicate hydrate (C_2ASH_8), hydrated garnet and so on can be produced as well.

As the relative content of clinker in the slag cement decreases, coupled by the effect of the considerable amount of $Ca(OH)_2$ and active slag components, the basicity of hydration products is generally lower than that of Portland cement. The resultant C–S–H has a ratio of $n(CaO)/n(SiO_2)$ of 1.4–1.7 with less $Ca(OH)_2$ content and greater ettringite content.

(2) Hardening characteristics of slag cement
Like Portland cement, the hardening process of slag cement is also accompanied by its hydration. Upon contacting with water, the hydration products of clinker minerals gradually fill the space occupied by water, and cement particles gradually come closer. The interweaving network is formed by the needle-like ettringite, rod-shaped crystal overlapping each other; in particular, a large number of foil-shaped, fibrous C–S–H cross-climbing attachment, so that the previously dispersed cement particles and hydration products come to form a solid three-dimensional space. However, since the clinker mineral in slag cement is relatively reduced (compared with Portland cement), and the potential activity of slag has not yet been fully stimulated, this results in relatively few hydration products, so the early hardening of slag cement is slower, as demonstrated by the lower compressive strength at 3 and 7 days.

With the continuous hydration, the potential activity of slag is stimulated. Although the content of $Ca(OH)_2$ is decreasing, a large amount of calcium silicate hydrate, calcium aluminate hydrate and ettringite are formed, which enhances the bond and compactness between hydration products and improves the stability of three-dimensional space. The porosity of hardened paste is gradually decreased and the average pore diameter is smaller leading to an increasing strength. The 28-day strength can catch up with or even exceed Portland cement.

Slag has long been widely used as a cementitious material, but the traditional process is to grind slag and clinker together. Due to the large differences in hardness and grindability of slag and clinker, the average particle size of the slag components in the mix is large and thus its potential activity cannot be fully exerted. Slag and clinker are separately ground and then mixed, and the relative increase of slag fineness (specific surface area 350–450 m^2/kg) can significantly increase its strength or increase the content of slag. In addition, the finely ground slag powder (specific surface area 600–800 m^2/kg) can be directly incorporated into the concrete as a concrete admixture to prepare high-performance concrete, which can improve not only the strength but also a series of performance of concrete.

2.6.3.2 Hydration and hardening of pozzolans

Hydration and hardening process of pozzolans is shown as follows: after its mixing with water, cement clinker first starts to hydrate to produce calcium silicate hydrate, calcium sulfoaluminum hydrate, $Ca(OH)_2$ and so on, followed by the pozzolanic reaction of $Ca(OH)_2$ released by hydration of clinker minerals with the active component of pozzolans, that is, the reaction of $Ca(OH)_2$ on the vitreous bodies in pozzolans containing silicon and aluminum oxide, making it collapse, dissolve and react with Ca^{2+} to generate water-insoluble secondary hydration products – calcium silicate hydrate, calcium aluminate hydrate and so on. The above pozzolanic reaction reduces the content of $Ca(OH)_2$ in the clinker liquid phase, which in turn can accelerate the hydration of clinker minerals C_3S and C_3A.

Hydration products of pozzolans are generally the same as Portland cement, mainly low-lime calcium silicate hydrate gel or C–S–H (I), followed by calcium aluminum hydrate, calcium aluminum (Iron) hydrate and solid solution. If the hydration temperature is increased, hydrated garnets may also be generated. Due to the pozzolanic reaction, the ratio of $n(CaO)/n(SiO_2)$ in the C–S–H of hydration products is relatively low, generally 1.0–1.6, while the amount of $Ca(OH)_2$ is less than that of Portland cement and is gradually reduced with the extension of curing time. Even in some pozzolans, there is no $Ca(OH)_2$ in the hydration product as it is being completely absorbed by the admixture.

2.6.3.3 Hydration and hardening of fly ash cement

The hydration and hardening process and its hydration products of fly ash cement are very similar to that of pozzolanic cement. However, due to the chemical composition of fly ash, the structural state is different from that of pozzolanic material.

The hydration and hardening of fly ash cement and its hardened structure have their own characteristics.

After the fly ash cement is mixed with water, the hydration of the cement clinker occurs first, and then the active components SiO_2 and Al_2O_3 in the fly ash react with hydration products such as $Ca(OH)_2$ released by clinker mineral hydration. Since the vitreous structure of fly ash is relatively stable and the surface is quite dense, the erosion and destruction of its vitreous structure by the clinker mineral hydration products $Ca(OH)_2$ is very slow, which means the pozzolanic reaction is slow. After 7 days of cement hydration, there is almost no change in the surface of the fly ash particles. The initial hydration of the surface is not observed until 28 days with the appearance of a slightly gelatinized hydration product. After 90 days of hydration, the particle surface begins to generate a large number of cross-linked calcium silicate hydrate gel to form excellent bond strength. Therefore, the reactivity of fly ash is evaluated based on the 3-month compressive strength value.

The hydration products of fly ash cement are basically the same as those of Portland cement, including calcium silicate hydrate, calcium sulfoaluminate hydrate, calcium aluminum hydrate, $Ca(OH)_2$ and sometimes there may also be a small amount of hydrated garnet. However, the content of $Ca(OH)_2$ is lower, and the majority of the hydration products is gel phase, while the ratio of $n(CaO)/n(SiO_2)$ in C–S–H gel is lower.

2.6.3.4 Hydration and hardening of limestone Portland cement

After limestone Portland cement is mixed with water, the hydration of Portland cement clinker is first carried out to generate hydration products such as C–S–H and $Ca(OH)_2$. At this time, the finely dispersed $CaCO_3$ particles act as numerous nuclei crystals for the growth of C–S–H and $Ca(OH)_2$ on its surface, resulting in the decrease of Ca^{2+} concentration in the liquid phase and accelerating the migration of particles on the surface of C_3S particles to the solution, thereby accelerating hydration. In addition, $CaCO_3$ reacts with C_3A in the clinker to form needle-like calcium carbon aluminate ($C_3A \cdot 3CaCO_3 \cdot 32H_2O$ and $C_3A \cdot CaCO_3 \cdot 12H_2O$), promoting the continuous generation of crystals and the development of cement strength.

At the same time, the large amount of hydration products accumulated on the surface of limestone particles improves the surface state, which is in favor of the bonding with hydrated C_3S particles and the compactness of paste structure. Therefore, the early strength enhancement rate is higher. However, in the later stage of hydration, $CaCO_3$ itself does not rehydrate or reactivate its potential activity as in the case of slag, so it does not contribute to the later age strength, especially in the case of a large amount of limestone.

2.6.4 Hydration rate

The hydration rate of clinker minerals or cements is often expressed as the degree of hydration per unit time or as the depth of hydration. The degree of hydration refers to the ratio of the amount of hydration within a certain period of time to the amount of complete hydration. The hydration depth refers to the thickness of the hydrated layer. Only if the conditions of cement particle size, water–cement ratio and hydration temperature are basically the same, the hydration rate can be compared. Figure 2.26 shows a hydration depth diagram of a spherical particle (mean diameter is d_m), in which the shading indicates the hydrated fraction.

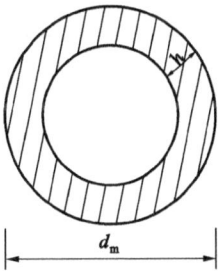

Figure 2.26: Hydration depth diagram of a cement particle.

According to the above definition on the degree of hydration, and assuming that the particle can always maintain a spherical shape, and the density is constant, the relationship between hydration depth h and degree of hydration a can be deduced:

$$h = \frac{d_m}{2}\left(1 - \sqrt[3]{1-a}\right) \tag{2.79}$$

The factors that determine the hydration rate of cement are mainly the compositions and structure of clinker minerals. The fineness of cement, the amount of water added, the temperature of curing, the properties of blending materials and admixtures of cement, all have an impact on the hydration rate.

2.6.4.1 Hydration rate of clinker minerals

There are two methods to determine the hydration rate, that is, direct method and indirect method. Direct method is to use petrographic analysis, X-ray analysis or thermal analysis to quantitatively determine the contents of hydrated and unhydrated parts. Indirect method is to determine the amounts of combined water, hydration heat or produced $Ca(OH)_2$ amount. The method by determining the amount of combined water is more convenient. The degree of hydration at different ages can be calculated by comparing the chemically combined water at a certain age with the chemically combined water when fully hydrated.

Table 2.25 shows the degree of hydration of various clinker minerals measured at different stages.

Table 2.25: Degree of hydration of various clinker minerals measured at different ages (%).

Mineral	Hydration time				
	3 days	7 days	28 days	90 days	180 days
C_3S	33.2	42.3	65.5	92.2	93.1
C_2S	6.7	9.6	10.3	27.0	27.4
C_3A	78.1	76.4	79.7	88.3	90.8
C_4AF	64.3	66.0	68.8	86.5	89.4

Table 2.26 shows the results calculated from the data listed in Table 2.25 according to the above formula.

Table 2.26: Depth of hydration of clinker minerals at different ages [μm, (d_m = 50 μm)].

Mineral	3 days	7 days	28 days	90 days	180 days
C_3S	3.1	4.2	7.5	14.3	14.7
C_2S	0.6	0.8	0.9	2.5	2.8
C_3A	9.9	9.6	10.3	12.8	13.7
C_4AF	7.3	7.6	8.0	12.2	13.2

It can be seen that the hydration depth of C_3S, C_3A and C_4AF particles with a diameter of 50 μm has reached more than one-half of the radius, while the hydration of C_2S has not yet reached one-fifth of its radius. After hydration for 28 days, the hydration depth of C_3S was three-tenth of its radius, C_3A was about two-fifth of its radius, and the hydration depth of C_4AF was slightly larger than that of C_3S, while the hydration depth of C_2S was less than one-twenty-fifth of its radius.

It is noteworthy that due to different measurement methods, the measured hydration rate is not the same. And because of differences between various processes, the same clinker minerals from different sources cannot hydrate at the same rate. In addition, the type and amount of impurities also have a considerable effect, for example, the hydrations of Alite and Belite are much faster than those of C_3S and C_2S, respectively. However, for hydration rates of these minerals, generally C_3A is the fastest, followed by C_3S and C_4AF, and C_2S is the slowest. There are different views on the hydration rate of C_4AF, and there are big differences in experimental results in all aspects. Some studies show that C_4AF even hydrates quickly, but not necessarily has a good cementitious ability. Other studies show that the hydration of iron-based solid solutions with various compositions is slow or particularly slow. However, it is given in Table 2.25 that the hydration rate of C_4AF is not low, and before 28 days its hydration rate is only second to C_3A, but much higher than that of C_3S. The results of study in China also confirmed that the early hydration rate of C_4AF was very high, but the hydration at later stage could be prevented by the formation of $Fe(OH)_3$ gel.

Figure 2.27 shows the X-ray diffraction analysis results by Yamaguchi Goro et al., which shows that the hydration degree of four clinker minerals in Portland cement is not much different from that of the four minerals alone.

The rate of reaction between clinker minerals and water varies depending on the inherent properties of the individual phases, which in turn is governed by the crystal structure and crystalline form, foreign ions and crystal defects.

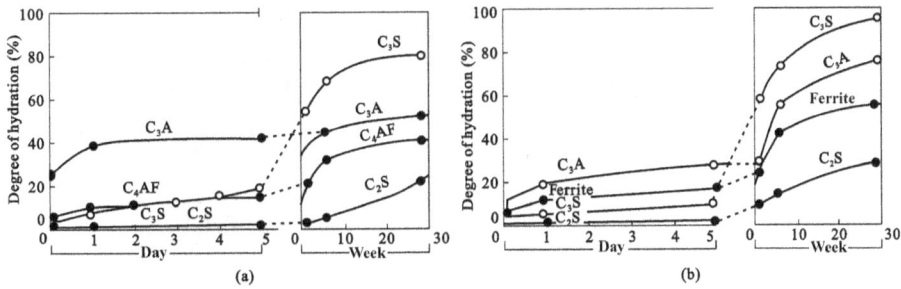

Figure 2.27: Hydration degree of cement clinker minerals at different ages: (a) hydrating alone and (b) hydrating in cement system.

For example, C_3S and β-C_2S, γ-C_2S belong to the same island silicate structure in which the vertices of $[SiO_4]^{4-}$ tetrahedra are not connected to each other and exist in an isolated state.

The coordination number of Ca^{2+} in C_3S is 6, its structure can be divided into two units, one is $[SiO_4]^{4-}$ tetrahedron and the other is $[CaO_6]^{10-}$ octahedron, which are connected through Ca^{2+}. The distribution of six O^{2-} around Ca^{2+} in the $[CaO_6]^{10-}$ octahedron is irregular and concentrates on the side of each Ca^{2+} leaving the other side a cavity that is large enough to contain a Ca^{2+}, so the water tends to infiltrate and react.

In β-C_2S, half of the coordination number of Ca^{2+} is 6, and half is 8. The distance between each O^{2-} and Ca^{2+} is not equal, the coordination is irregular and therefore unstable. But the cavity in β-C_2S crystal is smaller than the cavity in C_3S, so its hydration rate is slower.

As the coordination number of Ca^{2+} in γ-C_2S is 6 with regular distribution, the lattice does not have a cavity, the structure is quite stable and it barely hydrates at room temperature.

Similarly, in the crystal structure of C_3A, the coordination numbers of Al^{3+} are 4 and 6, and the coordination numbers of Ca^{2+} are 6 and 9. Since the O^{2-} around Ca^{2+} with coordination number 9 is extremely irregular and the distances are not the same, a large cavity is formed, and water can easily enter. As Al^{3+} with coordination number 4 is not saturated due to the valence bond, it is easy to accept two H_2O or OH^- to become a more stable coordination, so that C_3A can rapidly hydrate with water.

Many researches have been done on the relationship between the crystal form and hydration activity of clinker minerals. For example, the order of hydraulic activity of C_3S polymorphism has been reported as monoclinic < triclinic < rhombus, but the difference gradually decreases with the hydration age. At the same time, many foreign ions have a significant impact on the hydraulic activity of C_3S. When the incorporation of impurities (mass fraction) is 1%, the order of the hydration rate within 24 h is Alite > Magnesium-Alite > Iron-Alite > Aluminate-Alite.

It has also been reported that when the content of P_2O_5 and Cr_2O_3 is low and does not exceed the solubility, the hydraulic activity of C_3S increases. For β-C_2S, according to the results of Nurse, the phosphate-stabilized β-C_2S has the highest hydraulicity, followed by B_2O_3 and (Fe_2O_3 + Na_2O) stabilized β-C_2S. However, it has also been reported that pure C_2S has the highest hydraulic activity. In the β-C_2S sample doped with impurities, the order of its hydraulic activity is V > Cr > B > S. Among the various crystalline forms of C_3A, cubic crystal C_3A has the greatest hydraulic properties.

The effect of crystal form of C_3A on hydration is shown in Figure 2.28.

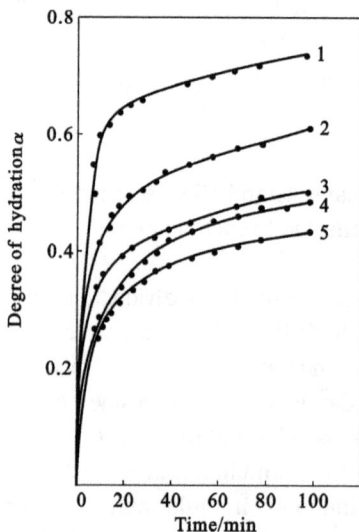

Figure 2.28: Effect of the crystal form of C_3A on hydration. 1, cubic (Na_2O-free); 2, cubic crystal form (Na_2O content of 2.42%); 3, rhomboidal (Na_2O content of 3.80%); 4, tetragonal (Na_2O content of 4.83%); 5, monoclinic (Na_2O content of 5.70%).

As shown in Figure 2.28, the degree of hydration of C_3A at the same time decreases with the doping amount of Na_2O, and the presence of C_3A is in rhombic, tetragonal or monoclinic form. Other dopants such as Cr_2O_3, TiO_2, MgO, Fe_2O_3 or SiO_2 also have similar effects.

It is noteworthy that, as hydraulic activity is usually a combination of many factors, it is difficult to carry out an explicit comparison of one specific effect, and sometimes even the completely different conclusions might be obtained. However, it can be generally considered that the hydraulic activity of clinker minerals is mainly related to irregularities in crystal structure due to irregular coordination and high coordination. Especially, the introduction of foreign ions may result in a variety of crystal defects and volume distortion, and obviously improve the hydraulic activity. In particular, the study of complex defects caused by several elements makes it more active, although there are still many problems, but it is a very meaningful work.

2.6.4.2 Effect of fineness and water–cement ratio

According to the general principles of chemical reaction kinetics, when other conditions are the same, the larger the surface area of the reactants participated in the reaction, the faster the reaction rate will be. Increasing fineness of cement leads to an increase in the surface area and a shortened induction period; therefore, the second exothermic peak appears earlier. On the contrary, the coarser the particles, the slower the reaction at all stages.

Water-to-cement ratio (w/c) between 0.25 and 1.0 has no significant effect on the early hydration rate of cement. However, when w/c is too low, the hydration reaction at the later stage may be retarded due to the insufficient amount of water required for hydration and insufficient space to accommodate the hydration products. Because the amount of water added should not only meet the needs of hydration reaction, but also fill pores inside C–S–H gel. At the same time, this part of water into gel pores is very difficult to flow and to be removed from C–S–H to make unhydrated minerals hydrate next. Therefore, for the purpose of full hydration, the mixing water should be roughly double the amount of water needed by chemical reactions. That is, during the hydration in a closed container, the w/c should be more than 0.4. The results of C_3S exothermic rate also showed that the effect of w/c on the early reaction rate was small, but when w/c was low, the hydration rate will become low at the later stage. However, it has also been reported that for a cement paste with a higher w/c, the early reaction is slightly slower, but the degree of hydration after 10 days is larger.

2.6.4.3 Relationship between temperature and hydration rate

Cement hydration reaction process also follows the general rules of chemical reaction. As the temperature increases, the hydration is accelerated, which shortens the induction period of C_3S. The second exothermic peak appears earlier, and the acceleration and deceleration periods also appear earlier. However, β-C_2S is more heavily affected by temperature and C_3A hydrates rapidly at room temperature, and has more exotherm; therefore, the effect of temperature on its hydration rate is not so obvious.

The hydration heat release curves of C_3S at different temperatures are shown in Figure 2.29.

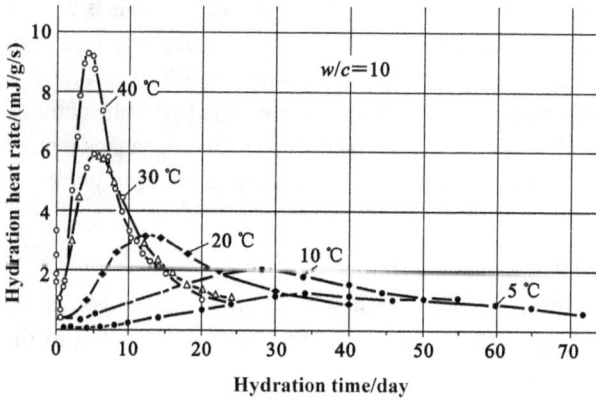

Figure 2.29: Hydration exothermic rate of C_3S at different temperatures.

The effect of temperature on hydration rate of Portland cement is shown in Figure 2.30.

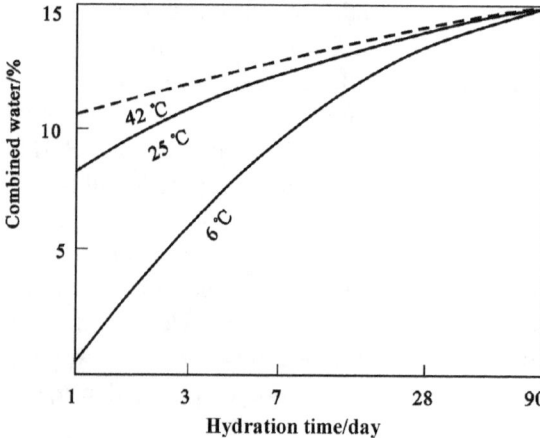

Figure 2.30: Effect of temperature on the hydration rate of Portland cement.

It can be seen from the figure that the higher the temperature, the more amount of bound water will be generated, and also the faster the hydration will be. The effect of temperature on the early hydration rate of cement is more significant, and the difference in hydration degree is gradually narrowed at the later stage.

At low temperatures, the hydration mechanism of Portland cement and its mineral composition have no obvious difference with that at room temperature. Despite prolongation of the induction period of C_3S, there is still considerable hydration rate afterwards, while β-C_2S is mostly affected. Tests show that Portland cement in the −5 °C ambient temperature can continue to hydrate, but below −10 °C, hydration is about to stop. When hydrated at a higher temperature below 100 °C, the induction period is shortened and the increased second exothermic peak is steep, indicating that the hydration is extremely rapid at the end of the induction period. However, the later hydration rate decreases, probably due to the over-crowded C–S–H gel forming a coating, with the same hydrated products as at room temperature.

2.6.5 Hardened cement paste

Cement paste mixed by cement and water initially has plasticity and mobility. With hydration the slurry gradually loses the ability of flow, and transforms into a solid with a certain strength, that is, the setting and hardening of cement. Hydration is the prerequisite for setting and hardening of cement. Setting and hardening are the result of cement hydration. The hardened cement paste is a heterogeneous system composed of various hydration products, residual clinker, water and air present in pores and it is therefore a solid–liquid–gas three-phase porous body. It has a certain mechanical strength and porosity, and other properties with an appearance similar to natural stone, so it is also often referred to as cement stone.

According to Taylor's assay, the volume fractions of the various compositions in the hydrated Portland cement paste for 90 days (w/c = 0.5) were C–S–H gel 40%, $Ca(OH)_2$ 12% (including 1% $CaCO_3$), MON 16%, pore 24% and unreacted cement clinker 8%. It has also been reported that hydrogarnet (with an average composition of nearly $Ca_3Al_{1.2}Fe_{0.8}SiO_{12}H_8$) is the predominant aluminate phase in the cement slurry that has been hydrated for 23 years.

2.6.5.1 The formation and development of cement paste structure
A lot of researches have been done on the cement setting hardening process, the formation and development of hardened paste structure.

In 1887, H. Lechatelier proposed the theory of crystallization. He believed that the reason why cement has gelling ability is that the crystals produced by hydration crossed each other and linked into a whole. According to this theory, the hydration and hardening processes of cement are as follows: the clinker minerals in cement dissolve first in water and react with water, and the resulting hydration products precipitate out as the solubility is less than that of the reactants. Clinker minerals then continue to dissolve, hydration products continue to precipitate, so dissolved precipitation continues. That is to say cement hydration is carried out in liquid phase as ordinary chemical reactions, the so-called dissolution–precipitation process, and then by crystallization and cross-linking of hydration products to set and harden.

In 1892, W. Michaelis proposed the theory of colloids. He believed that hydration of cement produces a large amount of colloidal substances, and then lost water due to drying or "absorption effect" caused by hydration of unhydrated cement particles to make the colloidal cohesion harden. Considering the cement hydration reaction as a type of solid-phase reaction, the main difference with the above-mentioned dissolution–precipitation reaction is that the process does not need a stage of minerals dissolution in water, but solid phases directly react with water to produce hydrated products, namely topochemical reaction. Then, through diffusion of moisture, the reaction interface extends inwardly from the surface of particles to continue the hydration. Therefore, Mihalylis believed that coagulation and hardening were the aggregation process from a gel to a rigid gel.

Next, A.A. Boykov developed the above two theories, dividing the hardening of cement into three stages: dissolution, gelation and crystallization. Furthermore, P. A. Pebinder et al. proposed that the setting of cement is a cohesive-crystalline three-dimensional network structure development process. Setting is a specific process dominated by coalescence structure, and hardening process indicates the development of crystal structures with a much higher strength. Since then, many viewpoints have been put forward. For example, R. H. Bogue believed that the huge surface energy of fine particles was the main reason to adhere particles strongly. According to M. M. Szechev's opinion, the formation of slurry structure is divided into two stages. The initial structure is mainly based on the adhesion contact of electrostatic and electromagnetic properties, and the secondary structure is the crystal connection by valence bond. F. D. Tamas et al. proposed that the hydration hardening of cement was a process of continuous polymerization by forming Si–O–Si bond between $[SiO_4]^{4-}$ tetrahedrons in clinker minerals. Polymerization of silicate anions is an important factor in the formation of slurry structures.

With the development of science and technology, especially applications of many modern testing methods in the field of cement, there has been a rapid progress in understanding the structure of cement paste. Based on Figure 2.31, Taylor illustrated the development of hardened cement slurry structures in detail. He also divided the hydration into three stages: early stage, middle stage and later stage, equivalent to ordinary cement hydration at 20 °C for 3 h, 20–30 h and longer, respectively.

In the early stage of hydration, most of the cement particles as aggregates of multiple minerals (Figure 2.31(a)) form a gel-like film on the surface within a few minutes of hydration, which is rich in Al and Si, and also have considerable amounts of Ca^{2+} and SO_4^{2-}. In 1 h of hydration, outside the gel film and in the liquid phase, stubby rod-shaped AFt is formed with a length up to 0.25 μm and a width 0.1 μm (Figure 2.31(b)).

At the middle stage of hydration, about 30% of the cement has hydrated, and this stage is characterized by rapid formation of C–S–H and CH. C–S–H forms a gradually

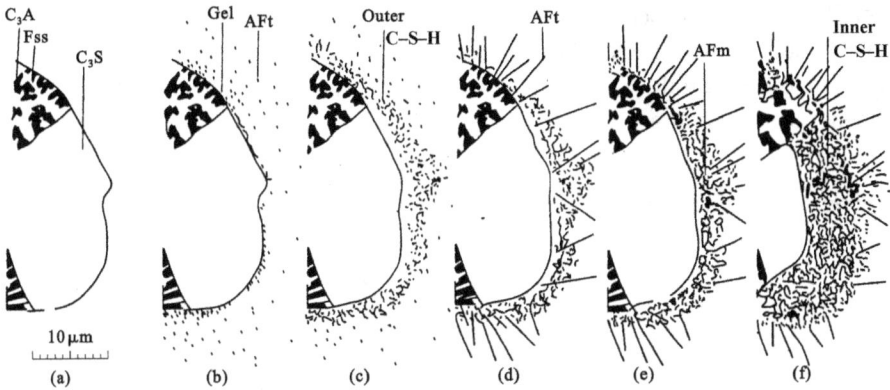

Figure 2.31: Development diagram of Portland cement paste structure during hydration. (The proportion of the mesophase in the graph is slightly enlarged.) (a) Hydration for 0 min; (b) hydration for 10 min; (c) hydration for 10 h; (d) hydration for 18 h; (e) hydration for 13 days; and (f) hydration for 14 days.

thicker coating around cement particles (Figure 2.31(c)). The amount is considerable after 3 h of hydration, and the surface of particles was completely covered after 4 h and then thickened to outside. About 12 h, the thickness is up to 0.5–1.0 μm. However, there is often a gap between the coating film and the water-free particles, which can be observed after 5 h of hydration and reaches 0.5 μm at 12 h of hydration. Until the middle stage is over, C–S–H deposits on the inside of the wrapping film to fill the gap between the wrapping film and the core (Figure 2.31(d)).

At the same time, due to the reaccelerated hydration of C_3A and the iron phase, slender AFt acicular crystals grow on the outside of the envelope, typically with 1–2 μm in length, sometimes up to 10 μm. After more products are formed, the wrapping film is not easy to penetrate, so that the liquid phase inside the wrapping film is gradually isolated from the outside and the sulfate content is reduced. Therefore, AFm is formed on the inside and AFt is on the outside. As for CH, it forms in the original water-filled space and plays an important role in filling the space.

In 20–30 h of hydration or later stage, the reaction slows down gradually. C–S–H continues to grow on both the inside and outside of the wrap. In addition, Alite partially rehydrates after the fissure is filled, and the inner C–S–H layer is continuously thickened (Figure 2.31(f)), while C_3A also reacts with the AFt on the inside of the membrane to generate AFm (Figure 2.31(e)). Similarly, AFm is formed on the outer side of the envelope as well as C–S–H due to AFt transformation. CH continues to form in space previously filled with water, up to dozens of microns in size, and can even envelop other phases.

From the above, it can be seen that with the hydration process, various hydration products continue to form, and the original water-filled space has been filled by solid phases. F. W. Locher et al. described several stages of the overall hardening process from the viewpoint of hydration formation and its development (Figure 2.32). The figure provides an overview of formation of the major hydration products and helps to provide an insight into formation of the cement paste structure.

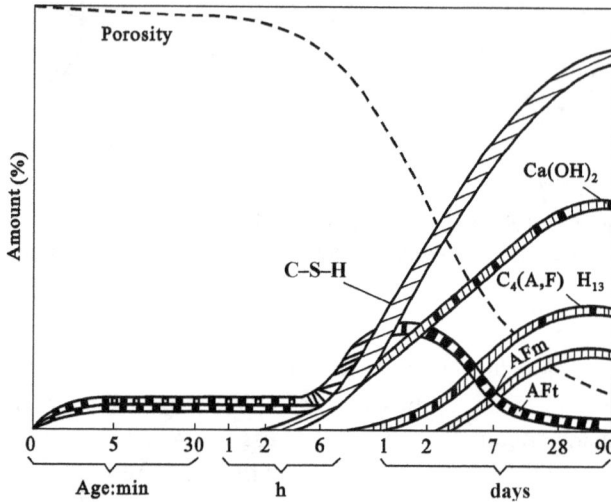

Figure 2.32: Schematic diagram of cement hydration product formation and structure development.

Although researches on the theory of hydration and hardening are far from completed, many arguments are still the subject of controversy. However, great progress has been made in the related knowledge after the long-term works, especially the application of modern testing technology. Now the more uniform opinion is the cement hydration reaction is mainly controlled by the chemical reaction; after the formation of a more complete hydration film layer around cement particles, the reaction process is affected by the diffusion rate of ions through the hydration product layer. As the hydration product layer continues to thicken, the diffusion rate of ions becomes the decisive factor to the hydration reaction kinetics. Among the generated hydration products, many of them are colloidal-sized crystals. With the process of hydration reaction, various hydration products gradually fill the space originally occupied by water, and solid particles are gradually closed to each other. Needle-shaped ettringite and rod-shaped crystals overlap each other, in particular, a large number of foil-shaped or fibrous C–S–H cross each other, which links the dispersed cement particles and hydration products together to form a three-dimensional solid combination (Figure 2.33). However, there are still various viewpoints on the nature of setting and hardening, and even the main types of bonds that form a paste structure to produce strength are unclear. In addition, the roles of Van der Waals' bonds, hydrogen bonding or atomic valence bonding also need to be further explored. In order to reveal the process and essence of cement setting, hardening and formation of paste structure more clearly, further study is necessary.

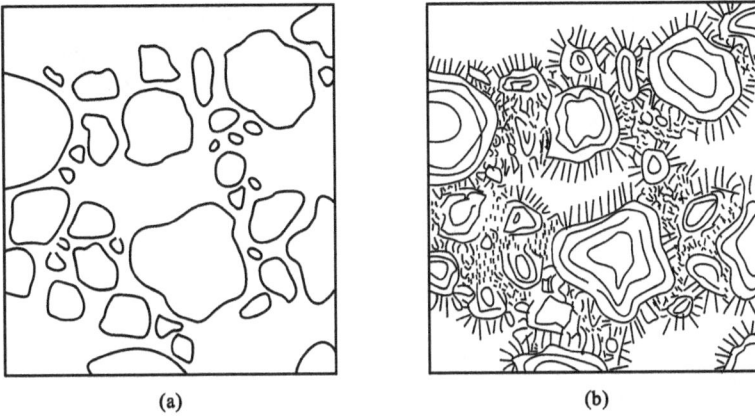

(a)   (b)

Figure 2.33: Schematic of hardened cement paste before and after its formation. (a) Before formation and (b) after formation.

2.6.5.2 C–S–H gel

(1) Composition

It is generally assumed that the molecular formula of C–S–H is $C_3S_2H_3$, that is, the mole ratio of CaO to SiO_2, $n(CaO)/n(SiO_2) = 1.5$ and the mole ratio of H_2O to SiO_2, $n(H_2O)/n(SiO_2) = 1.5$. In fact, these two ratios are not fixed values and vary with a number of factors. That is to say, the chemical composition of C–S–H is not fixed.

The $n(CaO)/n(SiO_2)$ of C–S–H increases with increasing $Ca(OH)_2$ concentration in the liquid phase. Figure 2.34 shows the balance of $n(CaO)/n(SiO_2)$ in the hydrated calcium silicate solid phase and the CaO concentration in solution.

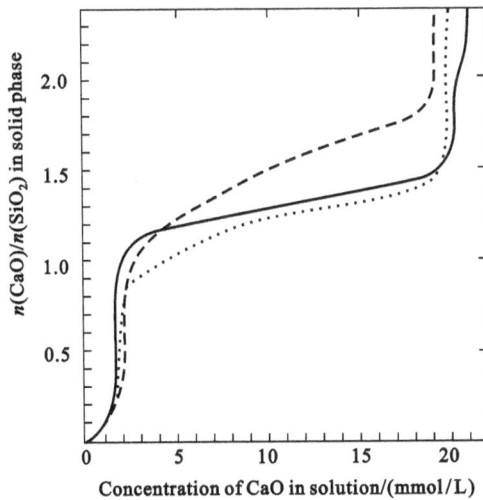

Figure 2.34: The balance between $n(CaO)/n(SiO_2)$ of C–S–H and CaO concentration in solution.

When the CaO concentration in the solution is about 2–20 mmol/L (0.11–1.12 g/L), C–S–H with $n(CaO)/n(SiO_2)$ of 0.8–1.5 is formed, named as C–S–H(I). When the concentration of CaO in the liquid phase is saturated (≥ 1.12 g/L), C–S–H with $n(CaO)/n(SiO_2)$ of 1.5–2.0 is formed, named as C–S–H (II).

Figure 2.35 shows the effects of water-to-solid ratio (w/s) on ratios of $n(CaO)/n(SiO_2)$ and $n(H_2O)/n(SiO_2)$.

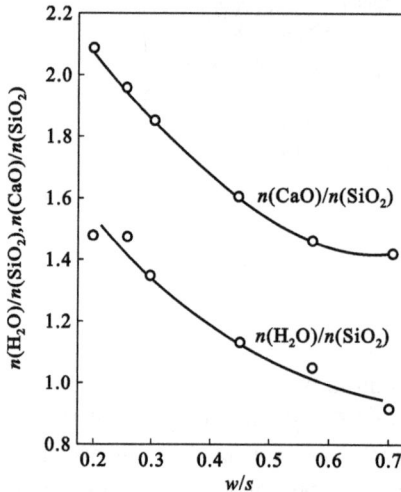

Figure 2.35: Effect of water-to-solid ratio (w/s) on $n(CaO)/n(SiO_2)$ and $n(H_2O)/n(SiO_2)$ (hydrated longer than 180 days).

At room temperature, the water-to-solid ratio (w/s) increases and the $n(CaO)/n(SiO_2)$ of C–S–H decreases (Figure 2.35). At the same time, $n(H_2O)/n(SiO_2)$ also decreases accordingly, and it can be seen from Figure 2.35 that it is about 0.5 lower than $n(CaO)/n(SiO_2)$. Thus, under normal hydration conditions, the composition of C–S–H can be roughly expressed as $C_xSH_{x-0.5}$. Most researchers also believe that the composition of C–S–H changes with hydration process and its $n(CaO)/n(SiO_2)$ decreases with age; for example, it reduces to 1.4–1.6 at 2–3 years after hydration from 1.9 at 1 day of hydration.

In addition, there are many other types of ions in C–S–H gel. Almost all C–S–H gels contain significant amounts of Al, Fe and S, as well as small amounts of Mg, K, Na and individual traces of Ti and Cl, and the data are very scattered. There is a rather clear difference in the composition of the individual particles. Table 2.27 is an example of a measurement result, and shows the average value of the number of atoms of each element contained with respect to the ratio of 10 (Ca + Mg) atoms.

(2) Structure

C–S–H is in the form of an amorphous colloidal and spherically shaped particle, whose diameter may be less than 10 nm. The degree of crystallization is very

Table 2.27: The composition of hydration products such as C–S–H from the characterization of the examples.

Hydration products	Ca	Si	Al	Fe	S	Mg	K	Na	Ti	Cl
C–S–H	10.0	5.7	0.5	0.1	0.8	<0.1	0.1	<0.1	<0.1	<0.1
AFt	9.94	0.63	3.44	0.13	3.35	0.06	0.19	–	–	–
AFm	9.9	0.55	5.30	0.15	0.75	0.05	0.1	–	–	–

poor, and even after a long time, the crystallinity only increases a little. Taylor et al. proposed that because the weak peak at 0.18 nm and the diffusive peak at 0.30–0.31 nm in the X-ray diffraction pattern of C–S–H and the surface network spacing of Ca(OH)$_2$ were basically equal, these short-range ordered C–S–H gel also had a layered structure like Ca(OH)$_2$. Alternatively, it is also believed that the structure of C–S–H is dominated by Ca(OH)$_2$, incorporating several Si–O groups on the Ca–O layer.

Hydration is the continuous polymerization of silicate anions from the viewpoint of silicate anions. C–S–H is a solid gel composed of hydrates of different degrees of polymerization. The silicate anions in C$_3$S and C$_2$S minerals are all isolated as [SiO$_4$]$^{4-}$ tetrahedra. As the hydration progresses, these monomers gradually polymerize into dimers [Si$_2$O$_7$]$^{6-}$ and higher degree of polymerization. Table 2.28 shows the distribution of polymerization degree of cement during hydration for 28 days as measured by nuclear magnetic resonance by G. Parry-Jones.

As can be seen from the data in Table 2.28, with hydration age, the amount of monopoly decreases rapidly and the amount of polymer increases. At the same time, it also shows that the higher the curing temperature, the faster the reduction of the monomer and the faster the increase of the polymer, and the faster the polymerization rate is.

E. E. Lochowski et al. used the data from the trimethylsilylization method to show that in cement paste of long-term hydration (1.8–6.3 years), monomer accounted for 9–11%, dimers 22–30% and polymers up to 44–51%. As for the type of polymers, all aspects of the measurement results have a certain discrepancy, more generally it was considered that most is linear pentamers [Si$_5$O$_{16}$]$^{12-}$. Thus, in the early stage of hydration, silicate anions in C–S–H gels exist mainly as dimers. However, the proportion of polymer with high degree of polymerization increases accordingly. In a fully hydrated cement paste, about 50% of silicon is in the form of polymers.

Therefore, the structure of many C–S–H crystals can be assumed to be composed of [CaO$_6$]$^{10-}$ octahedrons and [Si$_2$O$_7$]$^{6-}$, which are then linked into a sheet to form a layered structure. Many researchers have proposed a structural model of C–S–H gel in an attempt to demonstrate some of the properties of C–S–H gel. It is believed that

Table 2.28: Results of the polymerization degree distribution during the hydration of Portland cement. (%, $w/c = 0.4$).

Hydration time(days)	Curing temperature (°C)											
	21			35			45			55		
	Monomer	Polymer	Dimer	Monomer	Polymer	Dimer	Monomer	Polymer	Dimer	Monomer	Polymer	dimer
3	65	29	6	67	24	9	59	25	16	56	26	18
7	61	30	9	55	33	12	51	31	18	51	27	22
14	53	36	11	50	33	17	44	31	25	41	31	28
28	45	39	16	41	34	25	34	31	35	34	31	35

the C–S–H gel has a degenerated clay structure consisting of C–S–H flakes as the main part of the layered structure.

A schematic diagram of the structure of C–S–H gel is shown in Figure 2.36.

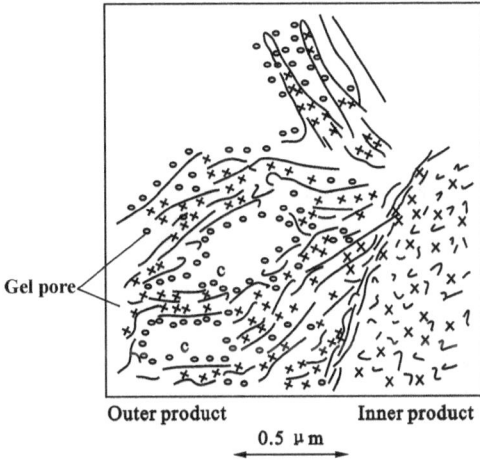

Outer product   Inner product
0.5 μm

Figure 2.36: Schematic diagram of the structure of C–S–H gel.

However, it is quite different from the well-crystallized clay minerals. The C–S–H flakes are neither flat nor regularly stacked up and down, with very poor crystallinity. In this way, the space formed between the sheets is very irregular to form pores in various sizes.

It is mentioned in many papers that C–S–H formed in cement paste at room temperature is called tobermorite gel. This is because in its X-ray diffraction pattern, *d* values of the three strong peaks roughly equal the natural mineral tobermorite. There is some additional evidence that certain C–S–H has a degenerated structure of hydroxyl jennite. However, in fact, natural minerals of tobermorite and jennite have fixed compositions and high degree of crystallization, and have some differences with the general C–S–H in cement paste. Taylor believed that C–S–H formed in early hydration was mostly composed of hydroxyl jennite with an imperfect structure, while the rest was tobermorite. So C–S–H is a mixture of two-layered structures of hydroxyl jennite and tobermorite, and the contained silicate anion is a dimer. In the later stage of hydration, hydroxyl jennite structure containing larger anion (especially pentamer) is even more dominant. Therefore, because the composition is not stable and the degree of crystallization is very poor, it is generally referred to as C–S–H gel more appropriately.

(3) Morphology

C–S–H gel in cement paste shows a variety of morphologies, fibrous, granular, network, flake, radial and so on. Figures 2.37–2.40 show the C–S–H phases in various

Figure 2.37: Fibrous C–S–H phase (scanning electron microscopy, SEM).

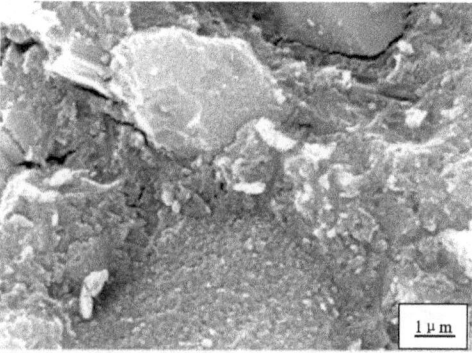

Figure 2.38: Granular phase of C–S–H (SEM).

forms. The sample used in Figure 2.37 was prepared by using powders of diatomite and calcium oxide with an initial molar ratio of calcium to silicon of 1.7, and a water-to-solid ratio of 10, at 100 °C for 15 days. The other specimens are hydrated cement samples.

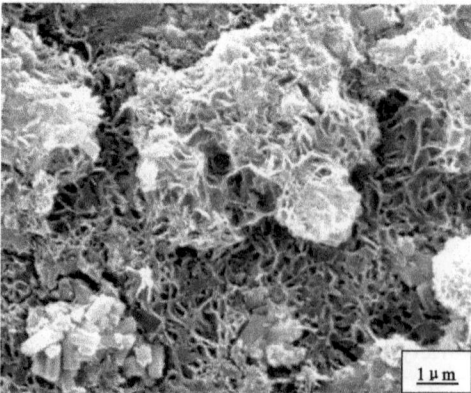

Figure 2.39: Network-like C–S–H phase (SEM).

In general, the morphology of hydration products has a close relationship with the growth space it may obtain. Except the above basic forms, C–S–H may also be observed on different occasions to be flaky, wheat-tube shaped, coral and flower-like shapes. What needs to be asked is that observation of wet cement paste by high-pressure transmission electron microscopy reveals that the one or the other topography of the C–S–H gel does not have to be clearly distinct and they all evolve from lamellae, rolled along a certain orientation, wrinkled and fragmented. To a large extent, it depends on the formation of occupied space and the rate of formation. In the process of drying or fracture after formation, there will be further changes. Therefore, it is often the primary reason for the formation of a variety of "illusions" that samples are prepared and dried to lose water for observation by using normal methods.

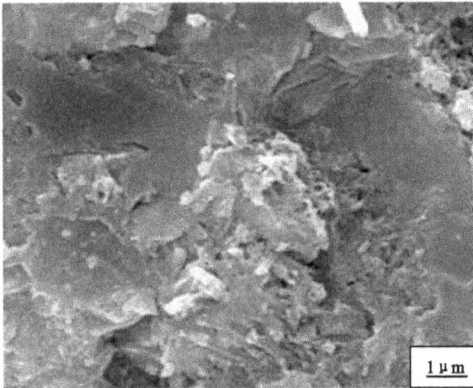

Figure 2.40: Inner hydrate phase of C–S–H (SEM).

2.6.5.3 Calcium hydroxide

Unlike C–S–H, $Ca(OH)_2$ has a fixed chemical composition, high purity and may only contain very small amounts of Si, Fe and S. The crystal is well crystallized and belongs to the tripartite crystal system with a layered structure composed of $[Ca(OH)_6]^{4-}$ octahedrons bonded to each other. The $Ca(OH)_2$ layers are ionic bonds with strong bonding, while the interlayer is a molecular bond with a weak interlayer connection, which may be a source of cracks in the hardened cement paste under stress.

When the hydration process reaches the accelerated stage, more $Ca(OH)_2$ crystals nucleate and crystallize in the water-filled space. The characteristic is that they are growing only within the existing space. If barriers are encountered, they will turn to the other direction to grow up, and even bypass the hydrating cement particles to wrap them up completely leading to increase in the occupied volume. In the initial stage of hydration, $Ca(OH)_2$ often has a thin hexagonal plate shape, with a width of dozens of microns, and can be clearly observed by ordinary optical microscope. Ca $(OH)_2$ crystals grow in pores in cement paste, sometimes grow big to be visible to

Figure 2.41: Hexagonal plate-shaped Ca (OH)$_2$ crystals (SEM).

naked eyes. Subsequently, Ca(OH)$_2$ grows thicker into a stack. Figure 2.41 shows the morphology of Ca(OH)$_2$ crystals.

In addition, part of Ca(OH)$_2$ in cement paste is amorphous or cryptocrystalline. It has been reported that the degree of crystallization of Ca(OH)$_2$ decreases correspondingly to a low water–cement ratio.

2.6.5.4 Ettringite

Ettringite is well crystallized. It belongs to a tripartite crystal system and has a columnar structure. According to the structural model proposed by Taylor et al., the basic unit of columnar structure is $[Ca_3Al(OH)_6 \cdot 12H_2O]^{3+}$, which is composed of $[Al(OH)_6]^{3-}$ octahedron combined with three calcium polyhedrons, as shown in Figure 2.42. Each Ca polyhedron associates with four OH$^-$ and four water molecules. There are three SO_4^{2-} in the trenches between the columns that act as a counterbalance to the charge to connect the adjacent columns to one another. In addition, there is a water molecule. Therefore, the structural formula of ettringite can be written as $[Ca_3Al(OH)_6 \cdot 12H_2O](SO_4)_{1.5} \cdot H_2O$, in which the structural water occupies 81.2% of the total volume of ettringite, and is as high as 45.9% in mass.

In fact, under appropriate conditions, a wide range of anions can be combined with CaO, Al$_2$O$_3$ and H$_2$O to form tri- or higher-salt quaternary compounds of the general formula:

$$C_3A \cdot 3CaX \cdot mH_2O$$

where X represents dianions such as SO_4^{2-} and CO_3^{2-}, and monovalent anions such as Cl_2^{2-} and $(OH)_2^{2-}$; and m represents usually 30–32 in a fully hydrated state. Thus, when X is SO_4^{2-}, it is the most common trisalt calcium sulfoaluminate, that is, ettringite ($C_3A \cdot 3CaSO_4 \cdot 32H_2O$).

When X is CO_3^{2-}, it is the trisalt type of calcium aluminate $C_3A \cdot 3CaCO_3 \cdot 30H_2O$. Its refractive index and X-ray diffraction pattern are very similar to ettringite, and the

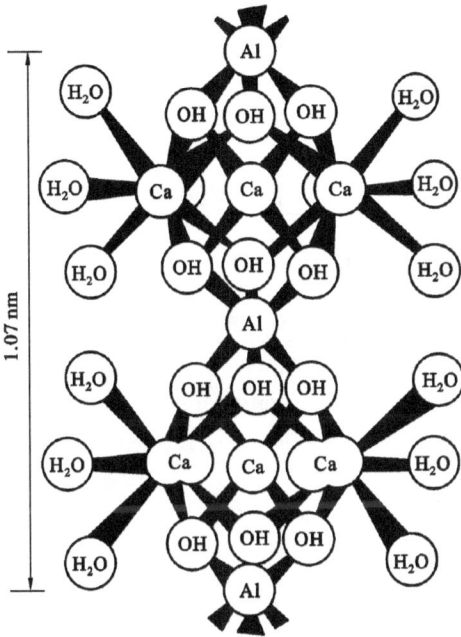

Figure 2.42: Structural unit of ettringite.

two are difficult to distinguish. If X is Cl_2^{2-}, it forms the tritype calcium chloroalumi-nate $C_3A \cdot 3CaCl_2 \cdot 30\ H_2O$. Testing results show that the structure is also roughly the same as ettringite.

In addition, among the above ternary quaternary compounds, Al_2O_3 can be replaced by Fe_2O_3 to obtain the corresponding double salts of ferrite, that is, calcium sulfoferrite, calcium carbonate ferrite, calcium chloride ferrate and so on.

Because these hydrates have very similar structures, when two or more anions are present, it is possible to generate a complex series of solid solutions and it is also possible to replace some of the Al_2O_3 with Fe_2O_3 to form solid solutions of trisalt aluminates and ferrite. For example, the ettringite phase observed in Portland cement pastes has been measured as

$$3CaO \cdot (Al, Fe)O_3 \cdot 3[CaSO_4, Ca(OH)_2] \cdot (30 - 32)H_2O.$$

The solid solution of the two series of quaternary double salt can be synthesized as shown in Figure 2.43.

In the Portland cement paste, ettringite is the most frequently produced hydra-tion product in the above series of solid solutions. The $n(SO_3)/n(CaO)$ ratio is often lower than the stoichiometry and it contains a significant amount of Si. Ettringite generally is a hexagonal prism-shaped crystal, and the shape depends on the actual growth of space and the supply of ions. In hydration within a few hours, it often precipitates in gel-like form, and then grows into a needle bar shape with clear facets.

$$C_3A \cdot 3CaSO_4 \cdot 32H_2O \longrightarrow C_3F \cdot 3CaSO_4 \cdot 32H_2O$$
$$|\qquad\qquad\qquad\qquad\qquad |$$
$$C_3A \cdot 3CaCO_3 \cdot 32H_2O \longrightarrow C_3F \cdot 3CaCO_3 \cdot 32H_2O$$
$$|\qquad\qquad\qquad\qquad\qquad |$$
$$C_3A \cdot 3CaCl_2 \cdot 32H_2O \longrightarrow C_3F \cdot 3CaCl_2 \cdot 32H_2O$$

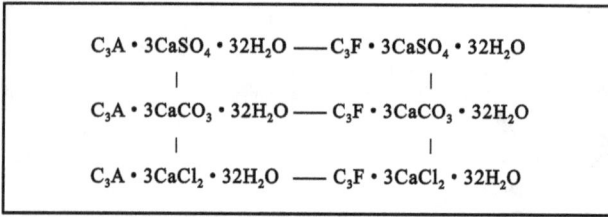

Figure 2.43: Trisalt aluminate and ferrite solid solution series.

Although size and diameter changes, both ends are straight, one side does not become thinner and there is no bifurcation phenomenon (Figures 2.44 and 2.45). According to TEM (Transmission Electron Microscope) observations, some ettringite appears as a hollow tube, which may differ in composition.

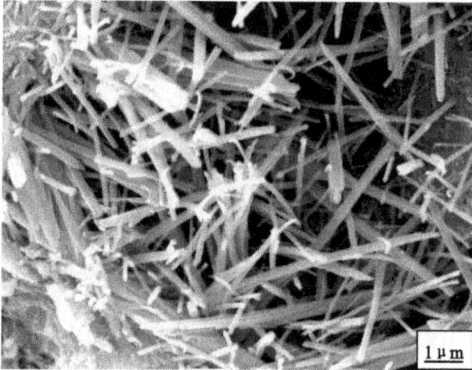

Figure 2.44: Ettringite crystals (SEM).

Figure 2.45: Rod-like ettringite (SEM).

Ettringite dehydration temperature is low. Figure 2.46 shows ettringite dehydration curve between 50 °C and 144 °C. It can be seen that a small amount of crystal water

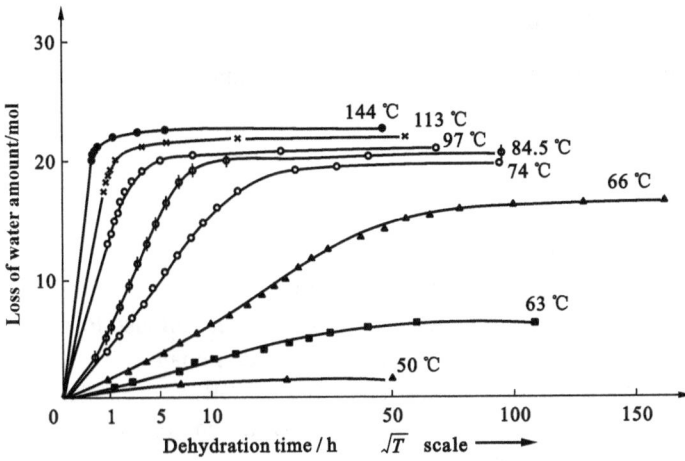

Figure 2.46: Ettringite dehydration curve at atmospheric pressure and normal humidity conditions.

has been eluted at 50 °C; the dehydration at 74 °C is quite strong; 20 mol of combined water will be lost after about 5 h at 97 °C. When the temperature reaches 113–144 °C, ettringite with eight water molecules will be formed. Further experiments also show that when the temperature rises to 160–180 °C, the combined water continues to lose and the temperature for complete dehydration is 900 °C. According to the X-ray diffraction analysis, the crystal structure of ettringite has been destroyed at 74 °C. However, some data also show that ettringite is stable below 100–110 °C and decomposes into monosulfate hydrated sulfoaluminate and hemihydrate gypsum at higher temperatures when the temperature is higher:

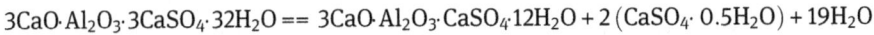

$$3CaO \cdot Al_2O_3 \cdot 3CaSO_4 \cdot 32H_2O == 3CaO \cdot Al_2O_3 \cdot CaSO_4 \cdot 12H_2O + 2\,(CaSO_4 \cdot 0.5H_2O) + 19H_2O$$

which is

$$C_3A \cdot 3C\overline{S} \cdot H_{32} == C_3A \cdot C\overline{S} \cdot H_{12} + 2C\overline{S}\,H_{0.5} + 19H$$

In addition, the temperature of dehydration of ettringite is related to the environment in which it is located. Therefore, due to the difference in experimental conditions, the results obtained in all aspects are also not very consistent. It is noteworthy that ettringite in cement paste coexists with other hydration products. Especially, due to the enclosure of C–S–H gel, its thermal stability is improved with different degrees as compared to pure ettringite.

2.6.5.5 Monosulfate hydrated calcium sulfoaluminate and solid solutions

Monosulfate is also a tripartite crystal system, but with a layered structure. The basic unit is $[Ca_2Al(OH)_6]^+$, and the interlayer is half SO_4^{2-} and three water molecules. So its structural formula should be

$$[Ca_2Al(OH)_6](SO_4)_{0.5} \cdot 3H_2O$$

Similar to ettringite, monosulfate also has a wide variety of anions that occupy the interlayer positions, making up the so-called mono- or low-salt quaternary hydrates represented by the following general formula:

$$C_3A \cdot CaY \cdot nH_2O$$

where Y is SO_4^{2-}, CO_3^{2-}, Cl_2^{2-}, $(OH)_2^{2-}$, $Al(OH)_4^-$ and so on, and n is the number of water molecules, usually 10–12 in case of complete hydration.

Therefore, in addition to monosulfate $C_3A \cdot CaSO_4 \cdot 12H_2O$, it will also form calcium carbonate aluminate $C_3A \cdot CaCO_3 \cdot 11H_2O$, calcium chloroaluminate $C_3A \cdot CaCl_2 \cdot 10H_2O$ and other single-salt quaternary double salts. If Y is $(OH)_2^{2-}$, it is hydrated calcium aluminate C_4AH_{13}; while Y is $Al(OH)_4^-$, it is C_2AH_8, which are all important members in the hydration products family.

Similarly, Fe_2O_3 will replace some or all of Al_2O_3 to form double salts of single-salt ferrites such as calcium sulfoferrite, calcium carbonate ferrite and calcium ferrochloride. These two series of single-salt hydrates, due to structural similarities, easily form complex solid solutions with each other, in the nearly same way with trisalt hydrates. Taking into account the replacement of iron in the structural unit layer of aluminum, the interlayer position is not necessarily occupied by an anion. The amount of structural water is variable, and the compositions of particles are very different, so the single-salt hydration product formed in cement paste is actually both a solid solution and a mixture. The ratio of $n(Al_2O_3)/n(CaO)$ is lower than 0.5 and the ratio of $n(SO_3)/n(CaO)$ is generally 0.10–0.15, while the ratio of $n(SiO_2)/n(CaO)$ is about 0.05. The composition of AFm, which is closely intermixed with C–S–H in the external products, is similar to that described above, and it is possible that $n(SO_3)/n(CaO)$ is relatively high, up to a theoretical value of 0.25 for monosulfates. In rare occasions, AFm will contain limited amount of S, or even without detected S. It was also reported that the sulfate content decreases with the progress of hydration.

Compared with ettringite, monosulfate has less structural water, accounting for 34.7% of the total, but a higher density of 1.95. So when various sources of SO_4^{2-} are converted to ettringite, the structural water increases and the density decreases, resulting in considerable volume expansion, which will be a major contributor to the volume change of the hardened cement paste.

The monosulfate in the cement paste initially is irregularly plate like, clustered or flower shaped and then gradually turns into a well-developed hexagonal plate. The width of the plate is a few microns, but thickness does not exceed 0.1 μm, creating special edge-to-surface contact with each other. The morphology of flower-shaped monosulfate is shown in Figure 2.47.

In addition, C_4AH_{13}, C_2AH_8 and monosulfate are similar in crystal structure, so their morphologies are difficult to distinguish and it is sometimes difficult to distinguish monosulfate with $Ca(OH)_2$.

2 μm

Figure 2.47: Petal-shaped monosulfate (SEM).

2.6.5.6 Pores and their structural features

Pores of various sizes are also an important component of the hardened cement paste. The porosity, pore size and pore distribution, pore shape and the huge internal surface area formed by the pore walls are all important structural features of the hardened cement paste.

(1) Inner surface area

Due to the highly dispersible nature of hydration products, especially C–S–H gels, which contain so many fine pores, the hardened cement paste has an extremely large internal surface area and thus constitutes another structural factor that has a significant influence on the physical and mechanical properties. The internal specific surface area is usually measured by the water vapor adsorption method. The method of measurement is as follows:

(i) The dried sample is subjected to different vapor pressures to determine the volume of absorbed gas V at the equilibrium pressure P.

(ii) According to the BET formula (BET formula is a multi-molecular layer adsorption formula, and BET is the acronym for the names of three scientists, Brunauer, Emmett and Teller.) that describes multimolecular-layer adsorption theory, when the single molecule adsorbed layer is formed on the solid surface and saturated, the volume of adsorbed gas V_m is calculated by

$$V_m = V \frac{(p_0 - p)}{Cp} \left[1 + (C - 1) p/p_0\right] \tag{2.80}$$

where V_m is the volume of adsorbed gas when the single molecule adsorbed layer is formed on the solid surface and saturated, m^3; V is the volume of absorbed gas at equilibrium pressure p, m^3; p_0 is the saturated vapor pressure of adsorbed gas at the same adsorption temperature, Pa; and C is the constant related to vaporization heat of adsorbed gas.

(iii) The specific surface area of the hardened cement paste is calculated according to the following formula:

$$S = aN\frac{\rho V_{m}}{M_{r}} \qquad (2.81)$$

where S is the specific surface area, m^2/g; a is the covered area per molecule of adsorbed gas, nm^2, $a = 0.114\,nm^2$ (25 °C) for water vapor, $a = 0.162\,nm^2$ (−195.8 °C) for nitrogen; N is the Avogadro number, $N = 6.02 \times 10^{23}$; ρ is the density of the absorbed gas, g/m^3; V_m is the volume of adsorbed gas when the single molecule adsorbed layer is formed on the solid surface and saturated, m^3; and M_r is the relative molecular mass of the adsorbed gas.

The hardened cement paste has a specific surface area of about $210\,m^2/g$, measured by the above method, and three orders of magnitude is increased compared to unhydrated cement. The surface effect of such a large specific surface area must be an important factor that determines the performance of cement paste. For cement with different minerals, the hydration hardening specific surface area is slightly different, as shown in Table 2.29.

Table 2.29: Specific surface area of cement paste measured by water vapor method.

No.	Composition (mass, %)				Specific surface area (m^2/g)	
	C_3S	C_2S	C_3A	C_4AF	S_c	S_g
A	45.1	27.7	13.4	6.7	219	267
B	48.5	27.9	4.6	12.9	200	253
C	28.3	57.5	2.2	6.0	227	265
D	60.6	11.6	10.3	7.8	193	249
Average					210	258
E	100	0	0	0	210	293
F	0	100	0	0	279	299

In the table, S_c shows the specific surface area measured on the whole hardened cement paste, including a certain number of crystal phases such as $Ca(OH)_2$, AFt or AFm and C_4AH_{13}, but their sizes are larger than that of C–S–H gel. So the proportion of S_c is very small, while S_g is the specific surface area of C–S–H gel, so it is larger than S_c. S_c of pure C_3S after hydration is less than that of β-C_2S, which is also due to higher $Ca(OH)_2$ content in the product. If only the specific surface area of C–S–H gels is calculated, they are close to $300\,cm^2/g$, and the two are basically the same.

Also it should be noted that the specific surface areas measured by different methods can vary widely.

For example, when nitrogen is used as the adsorbent gas, the result is only one-fifth to one-third of that from the water vapor adsorption method. Some think that is because the cross-sectional area of nitrogen molecule is too large, when the aperture is too small or too narrow access, nitrogen cannot enter. Another example is the use of small-angle X-ray method. It is found that the specific surface area of the hardened cement paste is also related to humidity. It has been measured that under water-saturated conditions, the specific surface area of C–S–H gel can be as high as $900 \, \text{m}^2/$g, which is much larger than that after drying of C–S–H gel.

(2) Pore size distribution and total porosity

During hydration, the volume of hydration product is greater than the volume of clinker minerals. About $2.2 \, \text{cm}^3$ is needed for hydration of $1 \, \text{cm}^3$ of cement, that is, about 45% of the hydration product is within the original perimeter of the cement particles and becomes an inner hydration product; another 55% of the hydration product is outer hydration product, occupying the original water-filled space. In this way, as the hydration proceeds, the space originally filled with water is reduced and the space unfilled with hydration products is gradually divided into extremely irregularly shaped pores.

In addition, there are pores in the space occupied by the C–S–H gel, which is extremely small in size and difficult to distinguish by scanning electron microscopy. Various structural models have descripted the pores in C–S–H. For example, Bowie et al. assumed that the diameter of gel particles was about 10 nm, with 28% of the gel pores and 1.5–3 nm of pore size. Forderman et al. emphasized the existence of interlayer pores and determined the hydraulic radius between 0.095 and 0.278 nm. Kondo et al. divided pores of C–S–H gel into pores between microcrystals and pores of inner microcrystals.

The model of pore structure of C–S–H gel is shown in Figure 2.48.

The pore classification methods is listed in Table 2.30, where the gel pores are divided into three types: micellar pores, micropores and interlayer pores.

It can be seen from Table 2.30 that the pore size varies within a very large range. Even if the coarse pores are not encountered, the pore sizes of the capillary pores and the gel pores should be as small as 15 μm to smaller than 0.5 nm. The difference in size can reach about five orders of magnitude. There are many classification methods for pores, and the views are not exactly the same. The pores are divided into three categories: coarse pores, capillary pores and gel holes, which is also arbitrary. In fact, the distribution of pores is continuous. There is no clear distinction between boundaries, and from a capillary effect point of view, the intergranular pore belonging to gel pore is actually a small capillary pore.

After 24 h of hydration, most of the pores in the hardened paste (70–80%) have been below 100 nm. With the hydration process, the pore size is less than 10 nm, the number of gel pores increases due to the increase in hydration products, the pores are

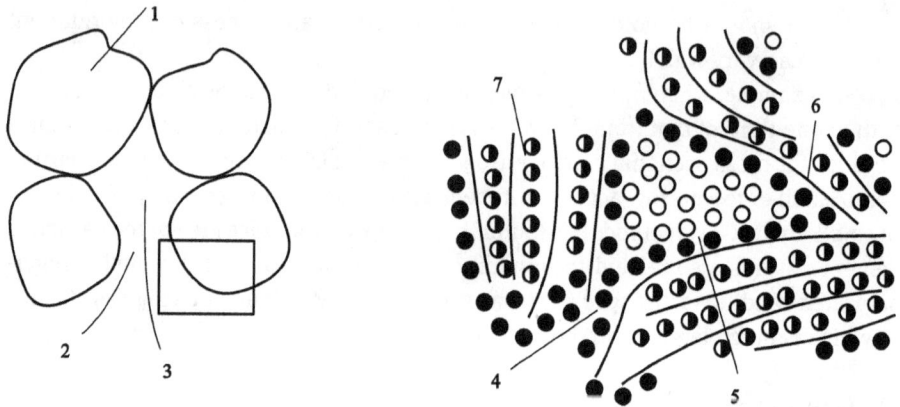

Figure 2.48: Model of C–S–H gel pore structure.
1, Gel particles; 2, 4, narrow channels; 3, intergranular pores; 5, intermeshing of microcrystals; 6, monolayer of water; 7, crystallite internal pores.

Table 2.30: Examples of pore classification methods.

Type	Name	Diameter	The role of water	Impact on properties of cement paste
Large pore	Large spherical hole	1,000–15 μm	Same as normal water	Strength, permeability
Capillary pore	Large capillary pores	10–0.05 μm	Same as normal water	Strength, permeability
	Small capillary pores	50–10 nm	Produces moderate surface tension	Strength, permeability, shrinkage under high humidity
Gel pore	Intergranular pore	10–2.5 nm	Produces strong surface tension	Shrinkage when relative humidity is below 50%
	Micropore	2.5–0.5 nm	Strong adsorption of water, cannot form a crescent-shaped liquid	Shrink, creep
	Interlayer pore	<0.5 nm	Structure of water	Shrink, creep

gradually filled and the total porosity decreases accordingly. The volume change of cement paste with age is shown in Figure 2.49.

The pore size distribution of the hardened paste, that is, the volume occupied by pores of various sizes can be calculated from the adsorption data of water vapor or

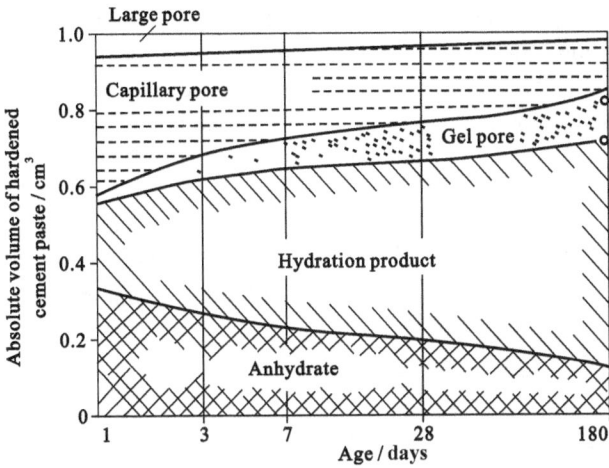

Figure 2.49: Diagram showing volume change of cement paste with age.

nitrogen. However, it is generally limited to pores with a diameter of less than 30 nm. Therefore, it is only suitable for gel pores such as intergranular pores and micropores. For the capillary system, mercury intrusion porosimetry is commonly used. When mercury is pressed into the dried cement paste with pressure, since the surface of the solid phase is not wetted with mercury, and the pore diameter to which mercury can enter is only inversely proportional to the applied pressure. By measuring the volume of mercury into the cement paste under various pressures, the pore size distribution curve can be drawn. This method can measure pores up to 1 μm in diameter, but there is a limit to the minimum pore size due to equipment conditions. It can be used together with the adsorption method if necessary. Figure 2.50 shows an example of pore size distribution curves for cement paste. It can be seen that as the age increases, the pore volume decreases and the pore size becomes smaller.

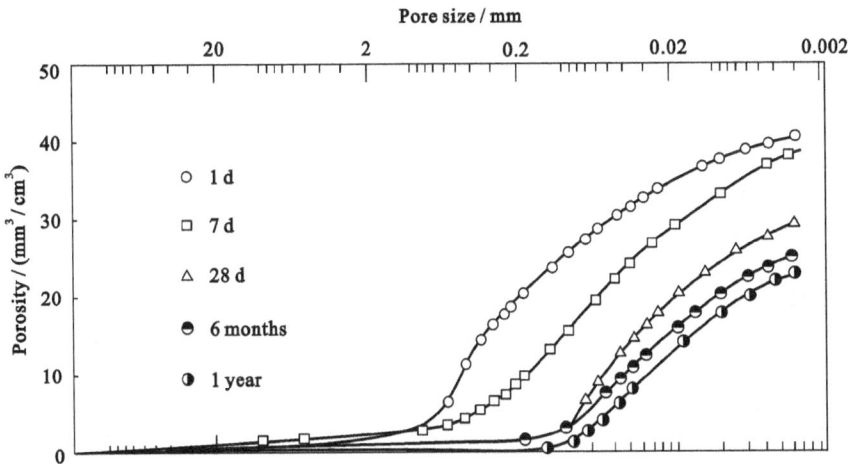

Figure 2.50: Cement paste pore size distribution.

With the determination of pore size distribution by mercury intrusion porosimetry and adsorption method, the total porosity of the cement paste can be obtained. But the total porosity can also be determined by the method of determining the density, that is, when the volume of the cement paste is known, after the density is measured, the porosity can be calculated. However, when different media were used, the measured porosity had a greater discrepancy. For example, for a hardened cement paste with a w/c ratio of 0.4, the measured porosities using water, helium and methanol were 37.8%, 23.3% and 19.8%, respectively. Therefore, the data measured under nonspecified conditions is difficult to compare with each other.

2.6.5.7 Water and its existence

There are various forms of water in cement-hardened slurries. According to the different action of water and solid phase, it can be divided into three basic types: crystal water, adsorbed water and free water.

(1) Crystal water

Crystal water, also known as the chemically bound water, is divided into two kinds: strong water and weak crystal water, according to the strength of its binding force.

Strong crystal water can also be called crystal coordination water, which exists in the form of OH^- and occupies a fixed position on the lattice. It is strongly bonded and there is a definite content ratio with other elements. Only when the lattice is destroyed at high temperatures can it be removed. For example, $Ca(OH)_2$ is a strong crystal water in the form of OH^-.

Weak crystal water occupies a fixed position in crystal lattice in the form of neutral water molecules, and it is mainly bonded by hydrogen bonding. It is not as strong as crystal water. Its dehydration temperature is not high. It can be removed above 100–200 °C, and it will not lead to damage of the lattice. When the crystal is a layered structure, such water molecules often exist between the layered structures. This water is also known as interbedded water. The level of interlayer water in the mineral varies with the outside temperature and humidity. When the temperature rises and the humidity decreases, part of the interlamellar water will be released and the distance between adjacent layers will decrease, causing corresponding changes in certain physical properties.

(2) Adsorbed water

The adsorbed water is present in the form of neutral water molecules, but does not participate in the crystal structure that forms the hydrates and is mechanically adsorbed on the surface or pores of the solid particles under the effect of adsorption or capillary force. Therefore, according to the location, adsorbed water can be divided into two kinds: gel water and capillary water.

Gel water also contains the moisture contained in the pores of the gel and the moisture adsorbed on the surface of the micelles. Due to the strong orientation of the gel surface, the binding strength can be quite different. The dehydration temperature also has a large range. The amount of gel water is substantially proportional to the amount of gel. Bowes believes that the gel water accounts for 28% of the gel volume, which is essentially constant.

Capillary water is only subjected to the capillary force. Its binding force is weak, its dehydration temperature is low and the amount of capillary water depends on the number of capillary holes depending on the number of pores.

(3) Free water

Free water exists in coarse pores, with the same nature of ordinary water.

It is difficult to quantify the different forms of water in the cement-hardened slurry described earlier. Therefore, from a practical point of view, the water in the hardened slurry is often divided into evaporating water and nonevaporating water. Evaporable water (whose physical quantity is represented by w_e) refers to the water that can be removed under specified reference conditions, while the remaining is non-evaporable water (its physical quantity is expressed by w_n). All of the adsorbed water and free water can be removed under the test conditions, and only crystal water remains.

The effect of the various drying methods on the amount of residual water in the Portland cement slurry is shown in Table 2.31. The method of drying using dry ice (-79 °C) is usually called *D*-drying method, and the method of drying using magnesium perchlorate is called *P*-drying method.

Table 2.31: Effect of drying method on the amount of water remaining in the Portland cement paste.

Drying method	Vapor pressure (mmHg, 25 °C)	The relative amount of water left (%)
$Mg(ClO_4)_2 \cdot (2-4)H_2O$	0.008	1
P_2O_3	0.00002	0.8
Tick H_2SO_4	<0.003	1
Dry ice (−79 °C)	0.0005	0.9
Heating at 50 °C	—	1.2
Heating at 105 °C	—	0.9

It is apparent from Table 2.31 that the amount of evaporated water and non-evaporable water is affected to a considerable extent by the drying method. Under more intense dry conditions, the amount of evaporated water will increase, while the corresponding nonevaporative water is reduced. However, when the capillary water and gel water are moved out, calcium sulfoaluminate, hexagonal crystal hydrated calcium aluminate and C–S–H will also lose part of the crystal water because of the weak binding force. Therefore, the measured non-evaporable water is not exactly the true crystal water, just an approximation.

The amount of non-evaporable water is related to the cement clinker composition, water–cement ratio and hydration time. According to the statistical analysis of the experimental results, the relationship between cement clinker mineral composition and the amount of non-evaporable water (w_n) can be expressed by the following formula:

$$\frac{W_n}{c} = a_1 w(C_3 S) + a_2 w(C_2 S) + a_3 w(C_3 A) + a_4 w(C_4 AF) \tag{2.82}$$

where, w_n is the amount of nonevaporated water, in kg; c is the cement consumption, in kg; $w(C_3 S)$, $w(C_2 S)$, $w(C_3 A)$, $w(C_4 AF)$ are the calculated mineral compositions (mass fraction, %) of $C_3 S$, $C_2 S$, $C_3 A$ and $C_4 AF$ in clinker; and a_1, a_2, a_3, a_4 are the experimental coefficients of the amount of water bound to each clinker mineral, as shown in Table 2.32.

Table 2.32: Experimental coefficients for the combined water content of each clinker mineral.

Coefficient	Water–cement ratio (w/c)				
	0.4	0.4	0.6	0.8	0.4
	1 year		6.5 years		13 years
a_1	0.228	0.234	0.238	0.234	0.28
a_2	0.168	0.178	0.198	0.197	0.196
a_3	0.429	0.504	0.477	0.509	0.522
a_4	0.132	0.158	0.142	0.184	0.109

Therefore, the above formula can be used to calculate the amount of nonevaporated water. For example, if the mineral composition of cement clinker is given as follows: $C_3 S$ 47%, $C_2 S$ 25%, $C_3 A$ 12% and $C_4 AF$ 8%, 13 years later, the content (mass fraction, %) of nonevaporative water is calculated to be 22.8%.

Experiments show that the volume of evaporated water can be roughly used as a measure of the pore volume in the slurry. If the amount is larger, the more capillary pores will appear under certain drying conditions. And the amount of non-evaporable water is related to the number of hydrated products. Since the amount of non-evaporable water is fixed when a certain amount of cement completely hydrates the amount of nonevaporated water measured at different ages can be used as a representation of the degree of hydration.

In summary, the hardened cement slurry has both solid-phase cement hydration products and unhydrated residual clinker, and water or air is filled in various types of pores. Hence, it is a three-phase heterogeneous system. Among them, as the most important part, hydration products not only have different chemical composition but also have different morphologies. It may be fibrous, prismatic, needle rod, tubular, granular, plate, flaky or scaly, and amorphous or other basic forms. It is more noteworthy that from the distribution of hydration products, hardened cement slurry is also micro-uneven. Therefore, the structure of hardened cement slurry is quite complicated. At present, its structure cannot be fully clarified in real, but it must be understood from various aspects, composition, morphology, structure and so on. In particular, the mixing of mixed materials, admixtures and other components, coupled with the formation of special methods such as molding, curing and to study and predict the performance of hardened cement slurry, it should have a more comprehensive understanding concerning the slurry structure.

2.6.5.8 Rheology of grout

The processing characteristics of cement slurry and cement concrete, especially what is commonly referred to as workability, are of great concern to the technical technicians engaged in concrete construction and cement product processing. The workability of cement paste or cement concrete is a comprehensive index for assessing its process properties. It is a unified requirement of various contradictory process requirements such as mobility, plasticity, stability and bulk density. So far, there has been no accurate and unified description, but with the application of rheology in cement slurry and concrete, it is possible to make a scientific description of it.

Rheology is the science that studies the flow and deformation of particles in objects due to relative motion. Because rheology can describe the relationship between the internal structure of the material and the macroscopic mechanical properties, it gradually becomes an important part of the basic theory of materials science and involves various types of materials.

Rheology links each of the three ideal objects with their parameters of stress and strain, and is expressed as a rheological equation. The constant in the formula expresses the rheological properties of a class of matter. Now the three ideal objects and their rheological equations are introduced separately:

(1) Hooker elastic solid

It refers to a certain flexible solid. When it undertakes external force, the deformation occurs immediately. The deformation of the size and the force is proportional. When the external force is removed, the object can restore the original shape. If it is

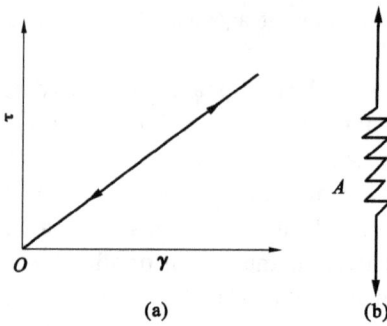

Figure 2.51: The rheological curve and model of the ideal elastomer.

(a)   (b)

assumed that the tiger gages produce a displacement gradient under the action of shear stress τ, then they have a linear relationship as shown in Figure 2.51. The rheological equation is

$$\tau = E\gamma \tag{2.83}$$

where E refers to elastic modulus.

In order to visually represent the above rheological equation, a fully elastic spring (A) can be used as the original model of the ideal elastomer, as shown in Figure 2.51(b). But when the external force exceeds a certain value, the shear stress and strain rate are no longer subject to the above rheological equation (or Hooker equation), and the deformation is no longer restored. This limit is called the elastic limit, and the corresponding shear stress is called the ultimate shear stress (or yield stress) τ_0. When the force exceeds the ultimate shear stress τ_0, the solid will lose its elasticity and cannot present plastic deformation.

(2) St Venan plastic solid

It refers to a hypothetical ideal plastic body. When the deformation of the solid force exceeds the yield stress τ_0, the plastic flow occurs when the shear stress is constant; when the additional stress is equal to the yield stress τ_0, the object flows at a constant speed. As shown in Figure 2.52, the rheological equation is

$$\tau = \tau_0 \tag{2.84}$$

The ideal plastic body model can be represented by a weight (B) placed on the table, as shown in Figure 2.52(b). There is friction between the weight and the desktop. When the force P reaches and exceeds the static friction force, the weight begins to move; when the P is reduced to the same as the dynamic friction, the weight moves at a constant speed. This is the St Venen plastic model.

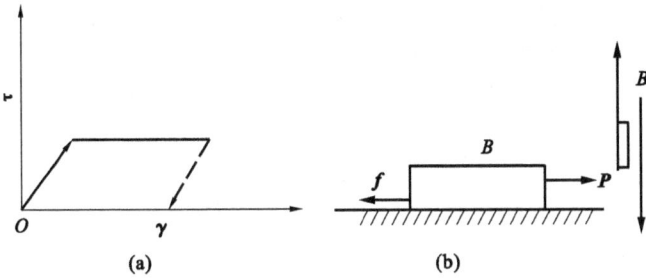

Figure 2.52: Rheological curve and model of ideal plastic body.

(3) Newton viscous liquid

When the liquid flows, it can be divided into several layers with different flow speeds along the direction of liquid flow. There is a viscous resistance opposite to the direction of flow between the adjacent two layers (the friction generated between the molecules inside the liquid). Viscosity (or viscosity coefficient) is usually used to characterize the viscous resistance of a liquid during flow. The viscosity (η) is the ratio of the shear stress (τ) to the strain rate (γ) of the liquid. At a certain temperature, the greater the shear stress experienced by the liquid during the flow, the greater its strain rate

$$\tau = \eta \gamma \tag{2.85}$$

The Newtonian fluid model is shown in Figure 2.53, which is moved with a perforated piston in a cylindrical cuvette filled with viscous liquid. If this liquid obeys the Newtonian liquid equation, this element (C) is the Newtonian liquid model.

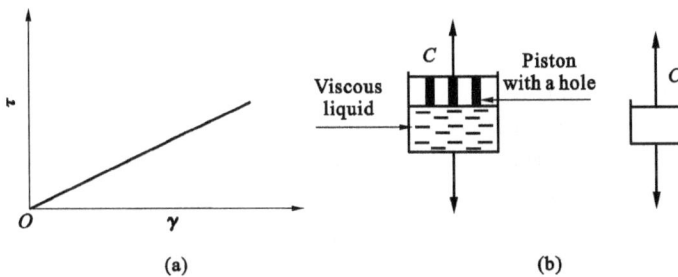

Figure 2.53: Newtonian fluid rheological curves and models.

The above three basic ideal objects are strictly nonexistent. In fact, objects are nonhomogeneous body between the bomb, plastic and sticky, but with these three basic rheological equations and model components, it is possible to extend the development and study the rheological properties of various more complex objects.

In the study of the deformation process of diatomaceous earth, porcelain clay and paint, E. C. Bingham found that when the applied external force was small, the shear stress produced was less than the ultimate shear stress or the yield stress τ_0. The object will remain in the original state and flow does not occur. When the shear stress exceeds τ_0, the object will produce flow deformation. Such objects are called Bingham body.

The rheological equation is

$$\tau = \tau_0 + \eta_0 \gamma \tag{2.86}$$

where τ is the shear stress applied on an object, Pa; τ_0 is the ultimate shear stress on an object, Pa; y is the plastic viscosity of an object, Pa·s.

The rheological curve and rheological model of the Bingham body are shown in Figure 2.54. It can be seen from the figure that the rheological model consists of the above three basic elements.

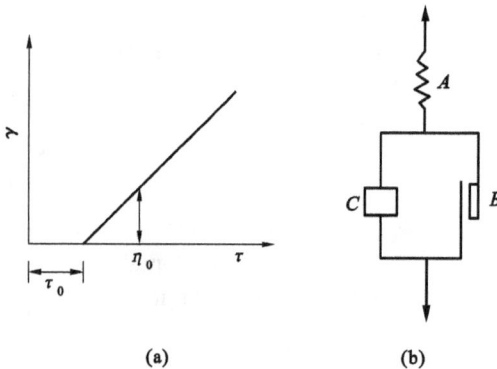

(a) (b)

Figure 2.54: Curve and model of Bingham body.

Some tests have shown that asphalt, soil, cement paste and cement concrete belong to Bingham model. Therefore, the rheological properties of the cement slurry are mainly determined by the relationship between the strain rate (y) and the shear stress (τ), the values of τ_0 and η_0 of the cement paste can be then determined.

The rheological properties of the slurry are usually measured by using a rotary cylinder viscometer method. The working part of the viscometer includes two cylinders inside and outside, and the sample to be measured is placed between the inner and outer cylinders, as shown in Figure 2.55.

The radius of the outer cylinder of the rotary cylinder viscometer is R_c and the radius of the inner cylinder is R_b. The effective height of the specimen is h. When the outer cylinder is rotated at different angular velocity ω, the internal friction force of the cement slurry can rotate the cylinder and the torque T can be obtained from the rotation angle of the inner cylinder. Therefore, the rheological properties of cement

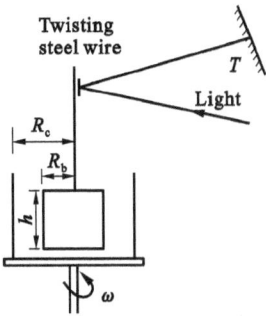

Figure 2.55: A schematic diagram of a rotary cylinder viscometer.

slurry can be measured directly by the two parameters of w and T. According to these two parameters, the equation derived by Reiner–Riwlin is used for the calculation. The mathematical expression is as follows:

$$w = \frac{T}{4\pi\eta_0 h}\left(\frac{1}{R_b^{2}} - \frac{1}{R_c^{2}}\right) - \frac{\tau_0}{\eta_0}\ln\frac{R_c}{R_b} \tag{2.87}$$

With mapping from w to T, the values of τ_0 and η_0 can be calculated from the slope and tangent points of the obtained curve.

In the study of rheological properties of cement paste with hydration times of 15 min, 45 min and 3 h, M. Ish-Shalom et al. found that there were three different rheological curves, as shown in Figure 2.56.

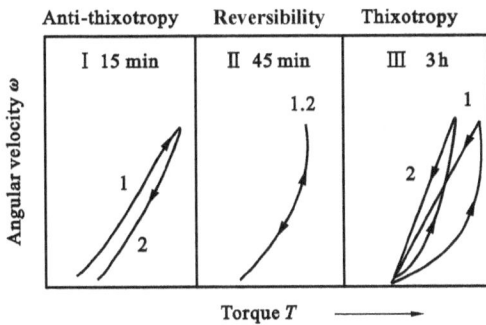

Figure 2.56: Three new flow characteristics of freshwater slurry.

In Figure 2.56, the ordinate is the angular velocity w of the cylinder rotation, and the abscissa is the torque T. From the curve slope, tangent point and instrument constant can calculate the parameters of τ_0, η_0 and A. Among them, A is the area surrounded by the ascending and descending rheological curves, called the hysteresis loop, showing the thixotropy of the cement slurry. The so-called thixotropy refers to some colloidal systems in the external force under the temporary increase in liquidity, external force removed with a slow reversible recovery performance.

In Figure 2.56, I is hydrated for 15 min, and the descending curve (2) is on the right side of the ascending curve (1). The second measurement is also on the right side of the first measurement, indicating that at the same speed, the torque increases, namely the viscosity increases. This phenomenon is a reverse thixotropic phenomenon.

II is hydrated for 45 min, and it exhibits a reversible curve.

III is hydrated for 3 h, and the descending curve is on the left side of the ascending curve and it is a straight line. The second cycle is also moved to the left of the first time. It shows that at the same speed, the moment decreases. This phenomenon is thixotropic phenomenon.

The antithixotropic phenomenon is the characteristic of some coarse particle suspensions, and the thixotropic phenomenon is the characteristic of some colloidal systems. The cement paste gradually forms a hydrate gel during the hydration process, so the rheological properties of the cement slurry also transition from the antithixotropic phenomenon to the thixotropic phenomenon, which indicates the cement particles from the initial dispersion to the colloidal size particles of the cement slurry suspension and the structure of the transition process.

There are many factors that could influence the rheological properties of cement slurry, including hydration time, water–cement ratio (w/c), hydration temperature, mineral composition of cement, stirring system and so on. In order to investigate the influence of these factors, M. Ish-Shalom et al. compared the Portland cement with a composition of C_3S 44.8%, C_2S 26.9%, C_3A 13.6% and C_4AF 6.7% under different conditions. The experimental results are shown in Table 2.33.

Table 2.33: Rheological indicators of cement slurry.

Hydration time	τ_0 (Pa)	η_0 (Pa·s)
15 min	44.5	2.4
45 min	56	2.5
2 h	99	5.4
3 h	134	6.3

(1) The impact of hydration age
The test results in Table 2.33 show that the change of $\tau 0$ and $\eta 0$ is not significant before 45 min, and the growth of $\tau 0$ and $\eta 0$ is faster after 2 h.

(2) The effect of water–cement ratio
The water–cement ratio has a greater effect on the rheological properties of the cement slurry. Figure 2.57 shows the cement paste with different water–cement ratios

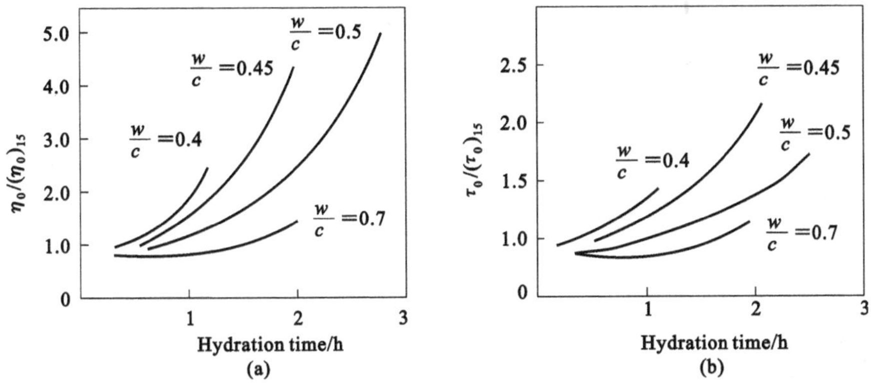

Figure 2.57: Effect of water–cement ratio on rheological properties of cement slurry.

(0.4, 0.45, 0.5 and 0.7, respectively), the plastic viscosity and ultimate shear stress, and the ratio after 15 min of hydration $[\eta_0/(\eta_0)_{15}, \tau_0/(\tau_0)_{15}]$ varies with time. It shows that the plastic viscosity and ultimate shear stress increase with the decrease in water–cement ratio.

(3) The effect of hydration temperature

The effect of hydration temperature on the rheological properties of the cement slurry is shown in Table 2.34. It can be seen from the table that as the hydration temperature increases, the rheological properties of the slurry increases. However, the increase was not significant within 45 min, and it increased significantly after more than 45 min.

Table 2.34: Effect of hydration temperature on rheological properties of cement slurry.

Indicators	Temperature (°C)	Hydration time			
		15 min	45 min	2 h	3 h
τ_0 (Pa)	25	44.5	56	99	134
	30	50	–	124	Too big to be measured
	35	48	68	134	Too big to be measured
η_0 (Pa·s)	25	2.4	2.5	5.4	9.3
	30	2.5	3.2	15	Too big to be measured
	35	3.3	3.5	15.7	Too big to be measured

(4) The effect of cement clinker mineral composition

Experiments show that the compositions of cement clinker mineral also have an impact on its rheological properties, and the most important factor is the content of C_3A. When the water–cement ratio is 0.5 and the temperature is 25 °C, the relationship between the rheological properties of the slurry and the content of C_3A (mass fraction, %) is shown in Figure 2.58.

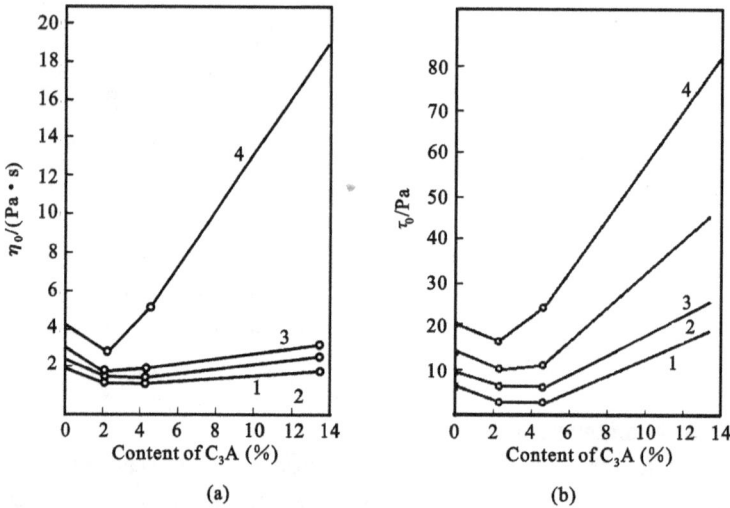

Figure 2.58: Effect of C_3A content on the rheological properties of the slurry.
1, hydration time, 15 min; 2, hydration time, 45 min; 3, hydration time, 2 h; 4, hydration time, 3 h.

It shows that as the content of C_3A increases, the plastic slurry viscosity and ultimate stress also increase. This effect was particularly evident after 2 h of hydration.

The process characteristics of the cement slurry from the rheological point of view are discussed here.

The compatibility of the cement slurry and the fresh concrete requires both good flow ability and plasticity, and also requires good stability and compactness.

The fluidity of the cement slurry refers to the performance of the cement paste under the effect of external force to overcome the interaction between internal particles and produce deformation. The smaller the internal force between particles, the higher the fluidity. From the rheological point of view, the fluidity of the slurry can be characterized by the ultimate shear stress (or yield stress). The smaller the ultimate shear stress, the greater the fluidity. From the point of the cement slurry structure, the ultimate shear stress corresponds to the strength of the cohesive structure. Increasing the water–cement ratio, expanding the distance between the particles in the slurry and weakening the interaction between the particles can increase the mobility.

The plasticity of the cement slurry refers to the performance of the slurry to overcome the ultimate shear stress (yield stress) and produce a plastic deformation (flow) without "breaking." From the rheological point of view, the smaller the plastic viscosity, the better the plasticity.

The stability of the cement slurry refers to the ability of the cement slurry to maintain the stability of the solid–liquid system after plastic deformation occurs. The slurry with good stability refers to the phenomenon that the solid particles do not come into contact with or separate from each other when the plastic deformation occurs, that is, the dispersion system of the entire flow remains stable. The greater the cohesion of the dispersion, the better the stability. From the rheological point of view, the stability requires that the slurry has a large strain value under a certain shear stress and can maintain continuous stability, that is, it has a higher viscosity. The ease of cement slurry refers to the consumption of the smallest work, and finally makes the slurry to achieve a dense ability.

However, the above requirements are contradictory; good mobility and plasticity require the slurry to have a small limit stress and plastic viscosity, and good stability and ease of density also require a higher ultimate shear stress and plastic viscosity. Therefore, the task is based on specific requirements to take a variety of means to adjust and control those performances in order to determine reasonable rheological parameters, to meet the requirements of process performance and also to access high-performance cement or concrete.

2.7 Properties of ordinary Portland cement

With the development of economy, cement is widely used in modern construction projects. Therefore, studying and improving its performances is of great significance for the development of cement varieties, building efficiency and project quality. The properties of Portland cement include physical properties such as density, bulk density and fineness, and building properties such as setting time, bleeding, strength, volume change, hydration heat and durability.

2.7.1 Density

Density is the property of a material and does not change with the mass and volume. Density only changes with the state, temperature and pressure. The mass (m) to volume (V) ratio (the mass of a certain material per unit volume, $\rho = m/v$) of a material that contains solid volume, open pore and closed pore is called as the volume mass or mass density of the material, or density. The density is usually expressed in symbol ρ, and the unit is kg/m^3.

Under different structural conditions, the density of building materials can be divided into three types: true density, apparent density and bulk density.

(1) True density

The true density refers to the ratio of the mass to the volume of the solid material in an absolutely tight state, $\rho = m/v$, that is, the density after excluding the inner pores or the void between the particles, expressed in symbol ρ.

(2) Apparent density

The apparent density refers to the ratio of the mass (m) to the apparent volume (V_0) in the natural state (the dry state stored in the air for a long time), and $\rho_0 = m/v_0$, represented by the symbol ρ_0. The apparent volume is the volume of the water outflowed by the material (apparent volume = solid volume + closed pore volume).

(3) Bulk density

The bulk density refers to the free filling of granular materials or powdery materials in a certain container, and the ratio of the mass (m) to the bulk volume (v_0) after the completion of the filling has been measured, and $\rho_0 = m/v_0$ is expressed as a symbol of ρ_0.

The determination of bulk density of cement is mainly used for mortar or concrete mixing by volumetric method in engineering, and designing the capacity of cement silos or estimating the storage in cement silos.

For Portland cement, if it is in different tectonic state, it has different density. For example, the bulk density of Portland cement is 900–1,300 kg/m³ at a loose state, and 1,400–1,700 kg/m³ at a compact state.

The density of cement generally refers to the mass (i.e., true density) of the volume per unit volume of cement in an absolutely tight (no void) state.

For cement in an absolutely tight (no gap) state, the mas of unit volume is named as density, whose unit is kg/m³ or g/cm³.

Different cement has different density, whose general ranges are as follows:

For Portland cement and ordinary Portland cement, $3.1\,g/cm^3 \le \rho \le 3.2\,g/cm^3$

For slag Portland cement, $3.0\,g/cm^3 \le \rho \le 3.1\,g/cm^3$

For pozzolanic Portland cement and fly ash Portland cement, $2.7\,g/cm^3 \le \rho \le 3.1\,g/cm^3$

It can be seen that the density of cement (pozzolanic cement and slag cement) mixed with other materials is only about 3.0 g/cm³, so according to the cement density, we can indirectly identify Portland cement or pozzolan, fly ash or slag cement.

The density of cement is one of the important building parameters for some special construction projects such as protection of atomic energy radiation and oil well blockage engineering. Because these projects need dense cement materials and requires larger density of cement. For example, density of spar-added high-density well cement can reach 3.27 g/cm^3. In the determination of specific surface area of cement, the density of cement is an essential factor to numerical calculation.

The main factors affecting the density of cement include clinker mineral composition, calcination degree of clinker, cement storage time and conditions, as well as the dosage and type of mixed materials. The C$_4$AF content in clinker increases, and the density of the cement increases. After long-term storage, cement density will decline. The density of underburnt clinker is too low, and the density of overburnt clinker is too high. Factors affecting the density of cement are also the factors that affect the bulk density of cement. In addition, the cement bulk density has a great relationship with grinding fineness; the higher the fineness, the smaller the bulk density.

2.7.2 Fineness

Cement grinding fineness is closely related to setting time, strength, dry shrinkage and hydration heat rate and a series of properties, and must be controlled within the appropriate range. Cement fineness can be used as different indicators, such as the residual amount of sieve, the specific surface area, the average particle diameter or particle size and so on. China's national standard regulations that cement fineness is expressed by the residual amount of sieve (%) or the specific surface area measured by the air permeability. For example, Portland cement and ordinary Portland cement, expressed in terms of specific surface area, require not less than 300 m^2/kg; slag Portland cement, pozzolanic Portland cement, fly ash Portland cement and composite Portland Cement, expressed by the residual amount of sieve R (mass,%), require not more than 10% for 80 μm square hole sieve, or not more than 30% for 45 μm square hole sieve.

The specific surface area of Portland cement and ordinary Portland cement is usually controlled to 320–350 m^2/kg. The residual amounts of 80 μm square hole sieve for other cement R are usually controlled from 1% to 4%.

The finer grinding the cement, the faster its hydration, and the more complete it hydrates. In the hydration process, the reaction rate is gradually controlled by the diffusion due to cement particles being encapsulated by the C–S–H gel. According to the study, when the coating thickness reaches 25 μm, the diffusion is very slow and the hydration actually stops. Therefore, cement particles with a particle size above 50 μm may leave the unhydrated core. It has been reported that after 23 years, on concrete cross section, the presence of C$_3$S and C$_2$S can also be clearly seen in the unhydrated coarse clinker particles. In addition, there are even reports of unhydrated clinker found in 136-year-old concrete. Therefore, the cement must be finely ground to fully exert its activity.

When other conditions are the same, the strength of cement increases with the increase in specific surface area, and the influence on the early strength is the most significant. Subsequently, the diffusion gradually controls the hydration process and the specific surface area takes the second place. Therefore, the fineness has almost no effect on the strength after 90 days, especially up to 1 year. At the same time, the effect of increasing fineness is more obvious to the original coarse cement. When the fineness is increased to more than 500 m²/kg, the strength increment at other ages is less except for the strength within 1 day.

The finer the cement, the higher the water demand to reach the standard consistency. This is mainly due to the larger surface area, which needs more water to cover.

The effect of specific surface area on the dry shrinkage of cement paste is shown in Figure 2.59.

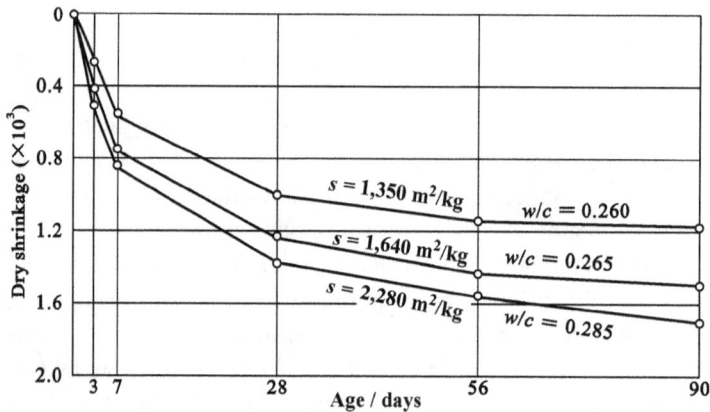

Figure 2.59: Effect of grinding fineness on dry shrinkage of the paste.

It can be seen that the dry shrinkage of cement increases with increasing specific surface area, but it may also include the effect of water-to-cement ratio on the dry shrinkage. However, by considering finer cement leads to a faster hydration of cement, and the increase in the gel content of the cement paste should be a major cause of increasing shrinkage.

In addition, the amount of gypsum should correspond to fineness of grinding. The larger the specific surface area of cement, the higher the amount of C_3A would interact with water at an early age. Therefore, when the fineness increases, the amount of gypsum needs to be increased accordingly.

It should also be noted that cements with the same specific surface area may have different particle size distributions. And due to the difference in the grinding ability of some clinker minerals, the particles with different sizes are different in compositions. It has been measured that the C_3S content is high in the fine particles, while the

C_2S content is high in the coarse particles. The distribution of C_3A and C_4AF in the various sizes of particles is roughly the same. Table 2.35 lists the analysis data for cements with different particle sizes.

Table 2.35: Analysis on particle size of various cement clinker (mass fraction, %).

Size (μm)	LOI	C_3S	C_2S	C_4AF	C_3A
All	24	56	19	11	12
0–7	64	59	14	11	13
7–22	25	62	11	11	13
22–35	1.5	52	22	11	13
35–55	1.1	49	24	11	13
> 55	0.9	47	25	11	14

Research shows that with the same specific surface area, when the range of particle size is narrowly distributed, that is, both large and small particles are with limited fractions, which is also called the "narrow grading," the cement strength will increase.

The size distribution of cement particles follows the RRB (Rosin–Rammler–Benner) equation, that is

$$R(x) = \exp\left[-\left(\frac{x}{\bar{x}}\right)^n\right] \qquad (2.88)$$

where R is the residual amount of sieve of cement particles (mass percentage, expressed in decimal fraction); x is the particle size of cement particles (random variable), μm; $\bar{x}$ is the characteristic particle size of cement particle distribution (equivalent to the particle size of sieve residue at 36.79%), μm; n is the uniformity coefficient (or distribution index) of the cement particle distribution that indicates the width of the particle size distribution.

$\bar{x}$ and n are the parameters of the particle distribution in the cement system. If the n value increases, the distribution of cement system will be more concentrated. The $\bar{x}$ value reflects the size of most cement particles. $\bar{x}$ and n are the only parameters that determine the distribution of particle size of cement system.

Figure 2.60 presents an example of the relationship of the uniformity coefficient of cement particle distribution (n) with the specific surface area (s) and the compressive strength (R_c) of cement. It can be seen that as the value of n increases, that is, on the RRB graph describing the particle size distribution of the material, the greater the slope of the line n, the more uniform the particle size, the higher the strength. This is mainly with the same specific surface area, the narrow-graded cement hydrates faster and the formed hydration products are increased.

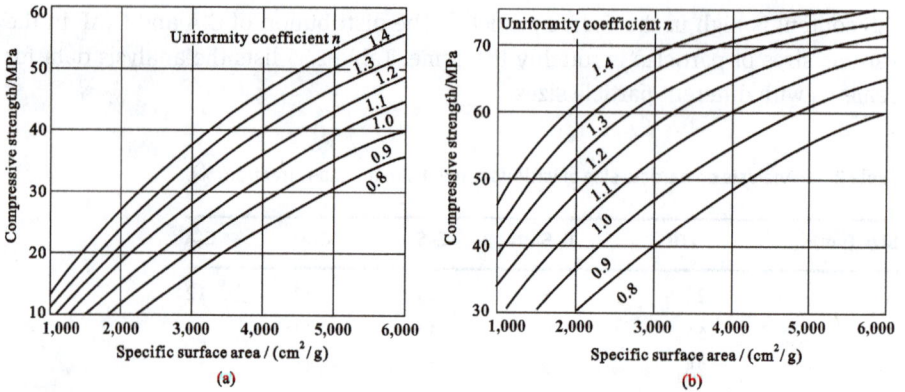

Figure 2.60: Relationship of the uniformity coefficient of cement particle distribution (n) with the specific surface area (s) and the compressive strength (R_c) of cement: (a) 2 days strength and (b) 28 days strength.

Figure 2.61 shows the relationship between hydration volume change rate $\Delta V/V$ (mass, %) and the uniformity coefficient n at different ages.

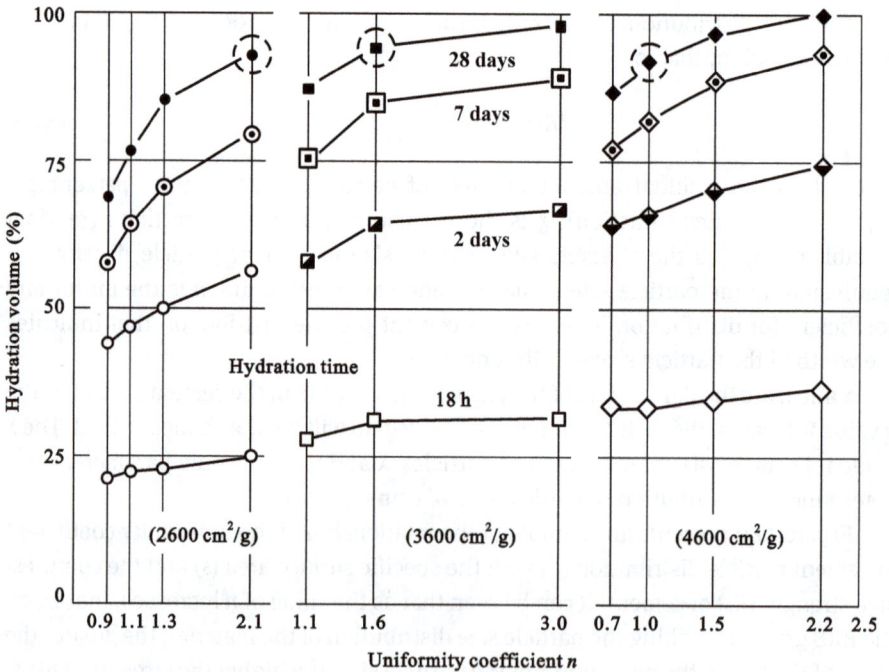

Figure 2.61: Relationship between the uniformity coefficient of cement n and the hydration volume change rate $\Delta V/V$ of cement.

In Figure 2.61, when the uniformity coefficient n is constant, the cement with a coarse fineness has a small hydration volume change rate $\Delta V/V$. In addition, it can be seen from the three points specially marked in the figure that for the cement with a coarse fineness, the hydration volume change rate $\Delta V/V$ is same as cement with a fine fineness and a wide gradation if a large uniformity coefficient is applied. Therefore, by using narrow-graded cement, the same strength may be obtained if grinding is not sufficient. However, when the cement particle size distribution is narrowed, the standard consistency of water demand will be increased. Moreover, some tests have also shown that hydration process is rather unfavorable with less particles smaller than 3 μm, and it must be considered together.

Finally, it should be pointed out that, while improving the fineness of grinding, the output of the mill will be reduced; the consumption of electrical power, the ball forging and the lining will inevitably increase accordingly. Moreover, as the specific surface area of cement increases, the rate of dry shrinkage and hydration heat become higher, and the cement can be more easily damped during storage. However, for slag cement, improving cement grinding fineness can significantly increase the strength of cement or increase the amount of slag in cement, and will help to reduce the cost of cement production. Therefore, the appropriate fineness of grinding should be carefully selected to make the cement quality to meet the specified requirements by considering all concerned technical and economic indicators, such as mill output and cost. If possible, a better particle gradation should also be used to meet the different performance requirements.

2.7.3 Water demand (consistency and fluidity)

When cement paste, mortar and concrete are produced from cement, it is needed to add the necessary amount of water. On the one hand, the water can participate into the cement hydration to make it hardened. On the other hand, paste, mortar and concrete require a certain fluidity, in order to fill the mold during the construction. Therefore, water demand is also one of the important properties of cement. When other conditions are fixed, the lower the water demanded by cement, the higher the quality of cement. Consistency and fluidity are the physical quantities that characterize the water requirement for cement, the former is for cement paste and the latter is for mortar and concrete.

In order to achieve the accurate comparability of the setting time of cement, and the determination of volume stability, the cement paste in a particular plastic state is stipulated, known as the standard consistency. The standard consistency of water demand refers to the amount of water per unit weight of cement that makes cement paste to reach the standard consistency, expressed by percentage.

Similarly, the fluidity of mortar is a measurement of the fluidity of mortar. For a certain amount of water, fluidity depends on the amount of water required by the cement. The fluidity is usually expressed as the mean value of the expanded diameter of the paste/mortar on the flow table (mm).

Cement paste has a certain resistance to the sinking of the standard test rod (or cone), the national standard is to determine the standard consistency of water consumption through testing cement paste permeability with different water contents. There are two methods: fixed amount of water and adjusted amount of water.

In general, the ranges of water consumption for the cement standard consistency are:

Portland cement, 21–28%
Ordinary cement, 23–28%
Slag cement, 24–30%
Pozzolanic cement, fly ash cement, 26–32%.

There are many factors that affect the standard consistency of cement, of which the most important ones are the fineness of grinding, mineral composition, the form of gypsum (such as hemihydrate gypsum) and the types and amounts of mixed materials.

2.7.4 Setting time

The setting time of cement paste is of great significance to the construction projects. Cement paste condensation can be divided into initial setting and final setting. Initial setting indicates that the cement paste loses its fluidity and plasticity and begins to set. The final setting indicates that cement paste gradually hardens, completely losing plasticity, and has a certain mechanical strength, which can resist certain external forces. The time from mixing with water and cement to initial setting is called the initial setting time, and the time until the final setting is called the final setting time.

In the construction process, if the initial setting time is too short, usually there is no enough time to carry out construction, and the cement paste has set. Conversely, if the final setting time is too long, strength is not enough, which will delay the construction progress. Therefore, there should be enough time to ensure the smooth operation of concrete mortar mixing, conveying, casting, molding and other operations. At the same time, it should also speed up the demolding and construction progress, to ensure the progress of the project requirements. To this end, the cement standards in all countries have set the cement setting times. According to China's national standard *Ordinary Portland cement* (GB 175-2007), initial setting time of Portland cement is not less than 45 min and the final setting time is no longer than 390 min.

2.7.4.1 Setting rate
The setting time of cement paste depends on its setting rate. It can be seen from the process of cement hydration and hardening that various hydration products are

produced after the cement is mixed with water. When the hydration products gradually grow up and initially form a network, they lose their fluidity and start to coagulate. Therefore, all factors that affect the rate of hydration, basically also affect the rate of condensation of cement, such as mineral composition, fineness, water–cement ratio, temperature and additives. However, there are some differences between hydration and coagulation. For example, the higher the water–cement ratio is, the faster the hydration and the slower the setting. This is because too much water leads to the increase of spacing, and it is not easy to form dense cement paste structure and the network structure.

The setting rate of cement paste is not only related to the hydration of clinker minerals, but also related to the content of each mineral. The main minerals that determine the setting speed are C_3A and C_3S. Bougue and others believed that C_3A content is the key factor in determining the initial setting time. When the C_3A content is high, or gypsum and other retarder dosages are too small, after Portland cement is mixed with water, C_3A rapidly hydrates and produces a large number of flake hydrated calcium aluminate, which are connected to form a loose network structure. The irreversible curing phenomenon appears, known as quick setting or flash setting. Due to this abnormal fast setting, a large amount of heat is rapidly released, and the temperature rises sharply. However, if C_3A content is less than 2% or with retarders such as gypsum, the rate of setting of the cement is mainly determined by C_3S. Therefore, fast setting is caused by C_3A, while normal setting is controlled by C_3S.

In fact, the setting rate of cement paste is also related to the morphological structure of clinker minerals and hydration products. Experiments show that even if the chemical composition and specific surface area of cement are the same, but due to differences in calcination system, clinker structure is still different, the setting time will change accordingly such as the setting of quenched clinkers is normal and the setting of slow-cooled clinkers is often rapid. This is because slowly cooled C_3A can fully crystallize, C_3A crystals are relatively abundant and hydration is accelerated. With rapid cooling, C_3A dissolves in the glass system, its structure is compact, so the hydration rate is slower. Likewise, if the hydration product is gelatinous, a thin film is formed that wraps around the unhydrated cement, impedes further hydration and thereby retards the setting of the cement.

Temperature change will also affect the setting rate of cement paste. As temperature increases, hydration accelerates, setting time is shortened, whereas setting time is longer, as shown in Figure 2.62. Therefore, the construction in hot seasons or at high-temperature condition, attention should be paid to the change of the initial setting time. In winter or low-temperature condition, attention should be drawn to adopt appropriate thermal insulation method to ensure the normal setting time of cement paste.

Figure 2.62 shows an example of the effect of temperature on the setting time.

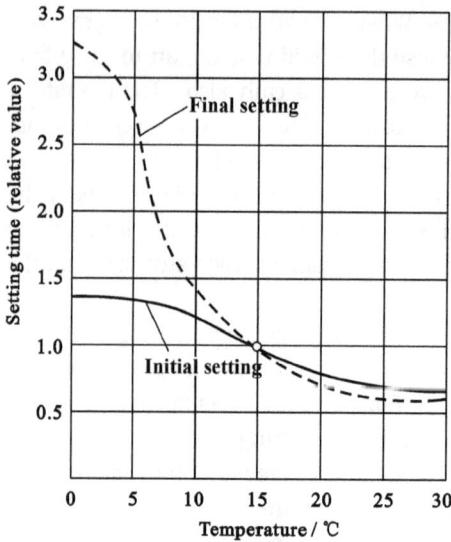

Figure 2.62: An example of the effect of temperature on the setting time.

2.7.4.2 The effect of gypsum and determination of the appropriate dosage

In general, the content of C_3A in cement clinker is high. Without setting retarders, when the cement is used, it will set quickly after it is mixed with water and construction work cannot be conducted. Addition of appropriate amount of gypsum can control the hydration rate and adjust the setting time; also the incorporation of gypsum can improve early strength, reduce shrinkage deformation and improve the durability and a series of other properties.

Gypsum is a retarder commonly used in cement. There are different viewpoints regarding its retarding mechanism. It is generally believed that C_3A produces ettringite with extremely low solubility in a saturated solution of gypsum and lime, and these prismatic crystals grow on the surface of the particles to form a film that covers and seals the surface of the cement particles, thereby blocking the water molecules and the diffusion of ions, hindering further hydration of cement particles, especially C_3A, and therefore preventing quick setting. As diffusion continues, when the surface coating by ettringite increases sufficiently thick, SO_4^{2-} penetrates into the interior gradually decreasing to insufficient to form Ettringite, monosulfate, C_4AH_{13} and its solid solution, accompanied by an increase in volume. When the pressure of the crystallization generated by the volume increase of the solid phase reaches a certain value, the ettringite membrane will crack locally, the water and ion diffusion will not hinder and the hydration will continue.

Young believed that due to the presence of a retarder on the surface of $Ca(OH)_2$ nuclei during the hydration process of cement, the further growth and growth of the Ca $(OH)_2$ nucleus prevent the $Ca(OH)_2$ crystals from precipitating in time, hydration rate of calcium silicate, which lead to retarding. This is the so-called crystal damage theory.

Locher proposed that the setting of cement is due to the formation of a network of structures inside the cement paste. Gypsum does not change the hydration rate of C_3A.

Figure 2.63 shows the diagram of the relationship between the structure formation during Portland cement setting and the contents of C_3A and gypsum.

Reactivity	Sulfate efficieny in solution	Hydration time		
		10 min	1 h	3 h
		Recrystallization of ettringite ⟶		
Low (I)	Low	Covered by ettringite / Workabler	Workabler	Setting
High (II)	High	Covered by ettringite / Workabler	Setting	Setting
High (III)	Low	Covered by ettringite, C$_4$AH$_{12}$ and Afm in pores / Setting	Setting	Setting
Low (IV)	High	Covered by ettringite, and gypsum in pores / Setting	Setting	Setting

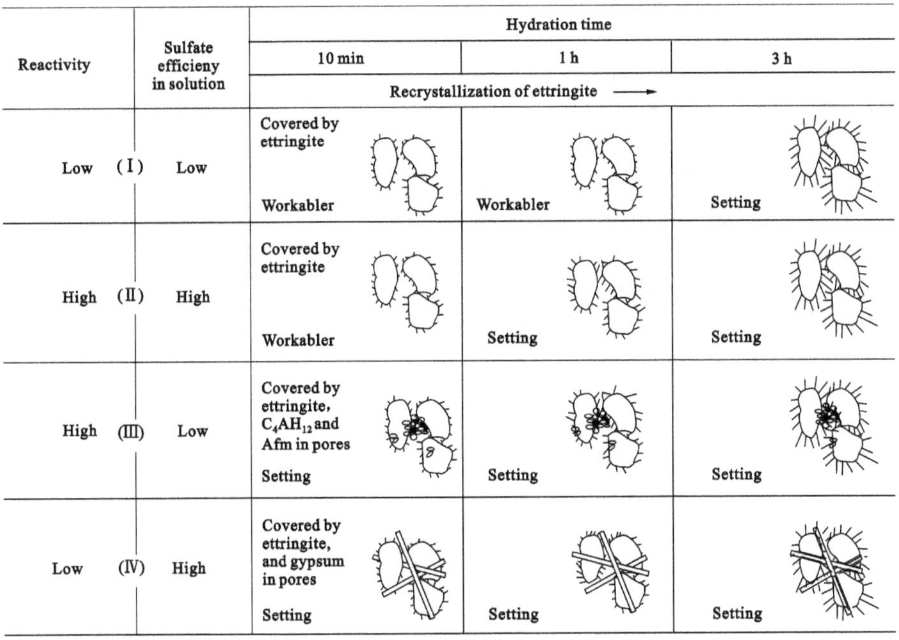

Figure 2.63: Diagram of the relationship between structure of Portland cement and the contents of C_3A and gypsum.

When C_3A content in the clinker is low (i.e., low reactivity), sulfate content is low, the hydration begins to form a fine ettringite film. It does not prevent the cement particles from moving toward each other and the cement paste is still plastic. A few hours later, ettringite increases to a sufficient number and the crystal grows into slender acicular cross-linked cement particles, forming a network structure to achieve a normal setting, as shown in Figure 2.63(I).

If the C_3A content is high, the amount of sulfate also increases correspondingly, the amount of ettringite formed in the initial stage of hydration increases correspondingly and the setting is slightly faster, but still normal, as shown in Figure 2.63(II).

If the C_3A content is high and the sulfate amount in the solution is small, in addition to the formation of ettringite, the remaining C_3A soon generated sheet-shaped C_4AH_{13} and monosulfate in the gap of particles and precipitate crystal, the cement particles connect with each other to form a network structure resulting in rapid setting, as shown in Figure 2.63(III).

If the C_3A content is low and the sulfate concentration is relatively high, the remaining sulfate from the reaction will crystallize immediately to form slab-shaped secondary gypsum, which will also result in rapid setting, as shown in Figure 2.63(IV).

Therefore, the appropriate dosage of gypsum is the key to determine the setting time of cement.

Therefore, too much or too little gypsum dosage (represented by SO_3 amount) will lead to abnormal setting. Generally, the content of gypsum is not high enough to cause a rapid setting, and when its content is increased to a certain level, the impact on the setting time will become very small. The effect of gypsum on the setting time of cement is shown in Figure 2.64.

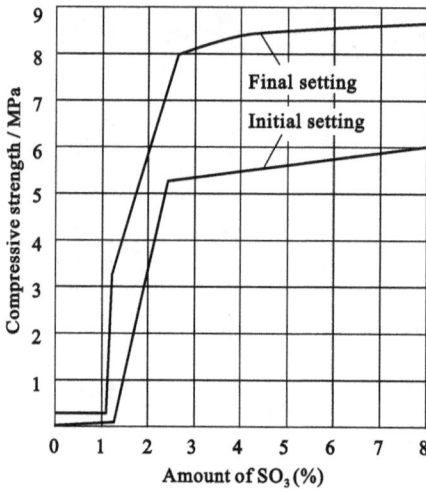

Figure 2.64: Effect of gypsum on the setting time of cement.

When the content of SO_3, $w(SO_3) < 1.3\%$, the gypsum content is too low, the fast setting of the cement will happen. Further increasing the content of SO_3, gypsum shows significant retarding effect, but when $w(SO_3) > 2.5\%$, there is little increase in setting time. There are also many researchers pointing out that the gypsum with an appropriate dosage can be depleted after about 24 h of mixing with water.

It should be pointed out that the determination of the optimum dosage of gypsum not only need to consider the setting time, but also need to pay attention to strength and stability at different ages. According to the relevant statistical analysis, the suitable ratio of SO_3 to Al_2O_3 in modern Portland cement $w(SO_3)/w(Al_2O_3)$ is 0.5–0.9, with an average of about 0.6. In general, the amount of gypsum is hard to be calculated empirically. A reliable method of determining the optimum amount of gypsum is the test of strength and related properties, as shown in Figure 2.65.

In fact, many factors affect the dosage of gypsum, mainly as listed in the following.

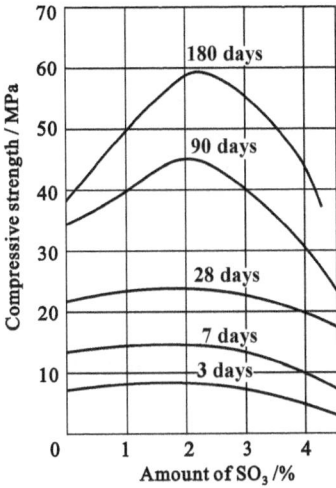

Figure 2.65: Relation between cement strength and SO_3 content.

(1) Type of gypsum

In addition to the commonly used dihydrate gypsum, there are anhydrite gypsum and industrial by-product gypsum. For example, the solubility of anhydrite gypsum at normal temperature is higher than that of dihydrate gypsum, but the dissolution rate of anhydrite gypsum is very slow, as shown in Table 2.36. Therefore, the dosage (represented by SO_3 amount) should be increased appropriately compared to the dihydrate gypsum.

Table 2.36: Solubility, dissolution rate and retarding effect of sulfate.

Type	Formula	Dissolution (g/L)	Relative dissolution rate	Relative retarding effect
Semihydrated gypsum	$CaSO_4 \cdot 0.5H_2O$	6	Fast	Very strong
Dihydrate gypsum	$CaSO_4 \cdot 2H_2O$	2.4	Slow	Strong
Soluble anhydrous gypsum	$CaSO_4 \cdot (0.001-0.5)H_2O$	6	Fast	Very strong
Natural anhydrous gypsum	$CaSO_4$	2.1	Slowest	Weak

The amount of gypsum added in general Portland cement and ordinary cement (SO_3) is between 1.5% and 2.5%.

(2) C₃A content in clinker

The content of C$_3$A is the most important factor affecting the amount of gypsum. When the content of C$_3$A is high, the amount of gypsum should be increased correspondingly and vice versa. According to the general rules, the C_3A content is less than 11% of the ordinary cement, and the best content of SO_3 is 2.3%.

(3) SO₃ content in clinker

Due to the original fuel use, and because part of shaft kilns use gypsum, barite as mineralizers, clinker often contains a small amount of SO_3, when the content of SO_3 in clinker is high, the dosage of gypsum needs to be reduced.

(4) Cement fineness

Cement composed of the same mineral, if the fineness increases, the specific surface area increases, and hydration is accelerated, the amount of gypsum should be increased properly.

(5) Variety and content of blending materials

When the cement is mixed with different kinds and quantities of blending materials, the amount of gypsum is different. Generally speaking, when slag is used as a mixed material, sulfate can stimulate slag activity and increase its reaction speed, so the amount of gypsum for slag is higher than that for other mixed materials.

In addition, the alkali content of cement is higher, the setting speeds up, the amount of gypsum should also be appropriate increased.

2.7.4.3 Pseudo-setting phenomenon

(1) Pseudo-setting and its characteristics

Pseudo-setting is referred as follows: when the cement is mixed with water, it is quickly solidified and hardened in a few minutes, but after vigorous stirring, it restores the plastic state. This is an abnormal phenomenon of early rapid solidification but is essentially different from fast solidification.

In the measurement of consistency and setting time of cement by Vicat's Apparatus, the characteristic curves of the abnormal condensation of cement displayed by the depth h penetrated by the Vicat needle and the time t are shown in Figure 2.66.

The heat release is very small, and after stirring vigorously, the cement paste restores plasticity and achieves normal setting. It has no adverse effect on strength, but increases the difficulty of construction.

Rapid setting refers to the rapid formation of irreversible solidification of the cement paste. The cement paste has produced a certain strength, and the retiring cannot restore its plasticity.

(2) Cause of pseudo-setting and preventive methods

Pseudo-setting phenomenon is related to many factors. It is generally considered that the main reason for pseudo-setting is that the cement is over-heated (≥130 °C) during

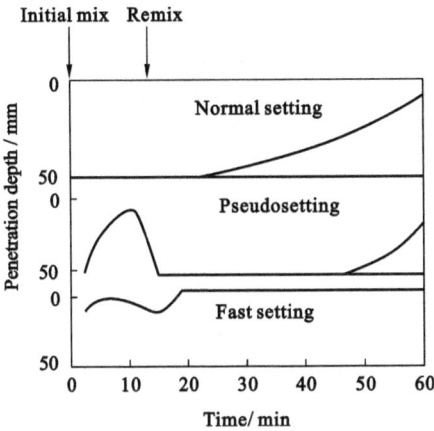

Figure 2.66: Typical curves of abnormal setting.

the grinding process, in which the dihydrate gypsum is dehydrated to form hemi-hydrate gypsum or even anhydrous gypsum. As mentioned earlier, hemihydrate gypsum and soluble gypsum are more soluble in water than the dihydrate gypsum. When the cement is added to water, the hemihydrate or anhydrous gypsum dissolves and hydrates to precipitate crystals of dihydrate gypsum, forming acicular crystals and reticular structure, causing rapid solidification of the cement paste. After re-mixing the cement paste, the structural network of the gypsum can be destroyed to restore the plastic state.

When the alkali content of cement is high, K_2SO_4 will form slab-shaped or bar-shaped gypsum $K_2C\bar{S}_2H$ crystals and grow rapidly, resulting in pseudo-setting. The reaction formula is as follows:

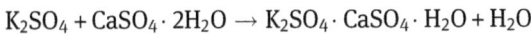

$$K_2SO_4 + CaSO_4 \cdot 2H_2O \rightarrow K_2SO_4 \cdot CaSO_4 \cdot H_2O + H_2O$$

The formation of potassium gypsum not only forms its own network structure, but also consumes the sulfate in cement, reducing the retarding effect of gypsum, resulting in rapid solidification of cement.

Practice has shown that pseudo-setting rarely occurs in cements mixed with many blending materials. In actual production, in order to prevent pseudo-setting, gypsum with high content of anhydrous calcium sulfate should be used as much as possible to avoid dehydration of gypsum when grinding. When the content of C_3A is more than 8%, in cement, $w(CaSO_4 \cdot 2H_2O)/w(CaSO_4) \geq 0.10$. At the same time, it is cooled down by certain measures during cement grinding process to avoid pseudo-setting caused by gypsum dehydration. In the construction of buildings, extending the mixing time can be performed to eliminate the phenomenon of pseudo-setting.

2.7.4.4 Other setting control admixtures
There are two types of setting control admixtures used in cement: one is retarder and the other is accelerator.

(1) Retarder

As mentioned earlier, gypsum is one of the most commonly used retarders and also with longest history. Internationally, people have studied the use of other substances instead of gypsum as retarders, for example, calcium lignosulfonate – sodium bicarbonate. Adding this retarder instead of gypsum can disperse fine clinker particles, increase liquidity, reduce the required amount of water and also adjust the setting time. It is believed that early hydration and early strength can also be enhanced. In addition, the hydration products will transform from the original elongated C–S–H particles into a C–S–H gel with the roughly same size which occupies a limited space. Observed by scanning electron microscopy, after 2 days of hydration, almost all of the paste structures became massive. At the same time, when it is hydrated at 20 °C, its setting time still meets the requirement.

In addition to tartaric acid, sodium citrate, phosphate, molasses, etc. can also be used as retarders. The effect of different retarders on the setting time of cement paste is shown in Table 2.37.

(2) Accelerator

Another setting control admixture is accelerator. In addition to fluoride, phosphate and Zn, Sn, Pb salts, most soluble inorganic salts can shorten the setting time of cement. The most widely used accelerator is $CaCl_2$. In China, commonly used accelerators are composed of mainly sodium aluminate ($NaAlO_2$), calcium aluminate ($C_{12}A_7$, $C_{11}A_7 \cdot CaF_2$) and silicate (Na_2SiO_3). These accelerators are added into cement from a few to a dozen percentage and can ensure the cement to complete the initial setting, final setting in a few minutes. However, these accelerators generally have shortcomings, like higher alkalinity, reduction of long-term strength of concrete. In response to the shortcomings, researchers have developed an updated nonalkaline accelerator.

During the use of accelerator in actual production, attention should be paid to the dosage, impact on the performance of cement and a series of specific practical problems.

With the social development and people's understanding, accelerators are more and more widely used, and the types will be certainly increased.

2.7.5 Strength

Strength is the most important indicator to judge the quality of cement. It is gradually increasing, and related to the hydration age. Usually before the age of 28 days, the strength is considered as early strength, such as 1, 3 and 7 days strength. At 28 days and later, the strength refers to the late strength. The factors influencing the strength of cement are rather complicated and involve a wide range of aspects. The mechanisms are still lack of firm conclusions so far and need further studies. This section will discuss it through the following aspects.

Table 2.37: Effect of different retarders on the setting time of cement paste.

Retarder	Reference	Tartaric acid		Potassium sodium tartrate	Sodium citrate	Sodium dihydrogen phosphate	Sodium tripolyphosphate	Glucose
Amount (%)	0	0.2	0.3	0.3	0.1	0.3	0.1	0.06
Initial setting (h:min)	3:10	7:40	9:40	10:30	2:10	6:10	5:40	4:20
Final setting (h:min)	5:20	12:00	28:10	13:00	10:30	8:10	11:40	7:30

2.7.5.1 Effect of clinker mineral composition

In Portland cement clinker, the relative contents of four main minerals, C_3S, C_2S, C_3A, C_4AF, have a decisive impact on the cement hydration rate, the morphology and size of hydration products. Undoubtedly, they play a crucial role in the formation and development of cement strength. It can be stated that the mineral composition is the most important factor to the early strength, the strength increase rate and the late strength. Tables 2.38 and 2.39 list the results of the strength of four single minerals in two sets of cement clinker.

Table 2.38: The first group of compressive strength measurement of the four main minerals in Portland cement R_c (MPa).

Mineral	3 days	7 days	28 days	90 days	180 days
C_3S	24.22	30.98	42.16	57.65	57.84
β-C_2S	1.73	2.16	4.51	19.02	28.04
C_3A	7.55	8.14	8.04	9.41	6.47
C_4AF	15.10	16.47	18.24	16.27	19.22

Table 2.39: The second group of compressive strength measurement of the four main minerals in Portland cement R_c (MPa).

Mineral	7 days	28 days	180 days	365 days
C_3S	31.60	45.70	50.20	57.30
β-C_2S	2.35	4.12	18.90	31.90
C_3A	11.60	12.20	0	0
C_4AF	29.40	37.70	48.30	58.30

Although the absolute strengths of single mineral measured under different conditions are different due to the difference of test conditions, the basic law is consistent, that is, the content of silicate minerals is the main factor that determines the strength of cement. Among them, the early strength of C_3S is the highest, and the strength of 28 days basically depends on the content of C_3S. If C_3S content is high, the early strength of cement is high, but the late strength growth is not large. For the cement with a high content of C_2S, early strength is not high, but the late strength increases largely, at 1 year the strength can catch up with or even exceed that of cement with a high C_3S content.

The effect of C_3S and C_2S content on the development of cement compressive strength is shown in Figure 2.67.

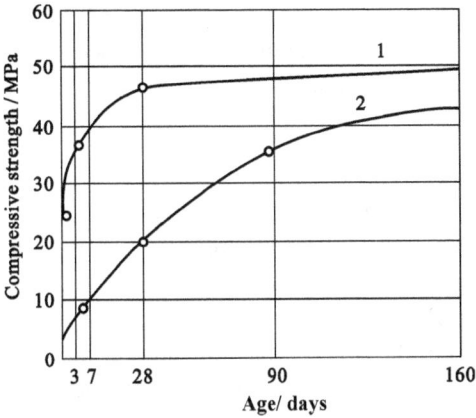

Figure 2.67: Effect of relative content of C_3S and C_2S on strength development. 1, C_3S content is 65.7–71.3%, C_2S content is 6.2–11.8%; 2, C_3S content is 26.0–31.0%, C_2S content is 47.1–59.7%.

The early strength of C_3A increases rapidly. It is generally believed that C_3A is mainly beneficial to early strength, but the absolute value of strength is not high, and the strength growth of C_3A gradually decreases with age, and even shrinks. Experiments show that when the content of C_3A in cement is low, the strength of cement increases with the increase of C_3A. However, the strength of cement decreases with the increase of C_3A after the content of C_3A exceeds a certain optimum value, and the shorter the age, the higher the optimum content of C_3A. The content of C_3A has the greatest impact on the early strength, if it exceeds the optimum content, it will have an adverse impact on the late strength.

With regard to the strength of C_4AF, the experiments show that C_4AF is not only beneficial to early strength, but also contributes to the development of later later strength.

From Tables 2.38 and 2.39, the compressive strengths R_c at 3, 7 and 28 days are higher than that of C_2S at the same age, but the compressive strength of C_2S at later stage increases faster.

Researches from China found that if V^{5+}, Ti^{4+}, Mn^{4+} and other metal ions entered into the iron lattice, and formed solid solutions through nonequivalent substitution with iron ions to further improve the C_4AF hydraulic activity. Therefore, C_4AF is also a kind of clinker with good hydration activity. However, its gelling ability not only depends on the internal causes of the crystal structure, such as the chemical composition of iron-based solid solutions formed under different conditions, crystal defects and the coordination state of atomic groups, but also depends on the hydration environment, hydration products and other factors. As for how to maximize the strength of the iron phase, further study is needed. In addition, the cement contains a trace of the composition of the four major minerals with the formation of solid solutions, the intensity will also have some impacts. For example, calcination of small amounts of MgO, R_2O and TiO_2 together with raw materials can improve the strength of clinker, and the strength of clinker will be reduced by CaF_2 and ZnO.

It should be noted that the strength of cement is not a simple addition of strengths of individual minerals, but also has a certain relationship with the calcination conditions and the structure of a variety of minerals.

2.7.5.2 Effect of cement fineness

Cement fineness has a very important impact on the strength and strength increase rate. The finer the cement, the faster the hydration rate, the easier it is to hydrate completely, then the higher the strength of the cement, especially the early strength. Increasing the fineness of the cement properly can avoid the bleeding, improve workability and cohesion of the cement paste. For the coarse cement, hydration can only happen on the surface, and the unhydrated part only plays as filler. Experiments show different degrees of hydration for different cement fineness. Generally, the relationship between particle size and hydration activity of cement is as follows:

0–30 µm: good activity;
30–60 µm: average activity;
> 60 µm: poor activity;
> 90 µm: very poor activity.

However, when cement is too fine, water demand for hydration is too high, which will increasing the porosity of the hardened cement structure, resulting in decreased strength. Numerous experiments show that when the cement is finer, the early strength increases on 1 day and 3 days, but the strength begins to decline at 7 and 28 days when the particles smaller than 10 µm are more than 50–60%. Therefore, the fineness of cement can only increase the strength of cement within a certain range.

2.7.5.3 Effect of construction condition

The strength of cement paste structure is closely related to the construction process. In the process of construction, the water to cement ratio, aggregate gradation, the degree of stirring and vibration, curing temperature and the admixtures show great impacts on the strength.

(1) Water-to-cement ratio and compaction degree

Many studies have shown that the strength of cement paste structure can be remarkably improved when porosity and the amount of large pores decrease. Figure 2.68 shows the relationship between the compressive strength (R_c) of cement paste and porosity (P), that is, the result carried out by Feldman at high porosity conditions.

The larger the water-to-cement ratio (w/c), the more pores are produced, thereby the strength is reduced, especially the later strength. Figure 2.69 shows the relationship between the comprehensive strength (R_c) and water-to-cement ratio (w/c) of cement paste. It can be seen that there is an approximate linear relationship between

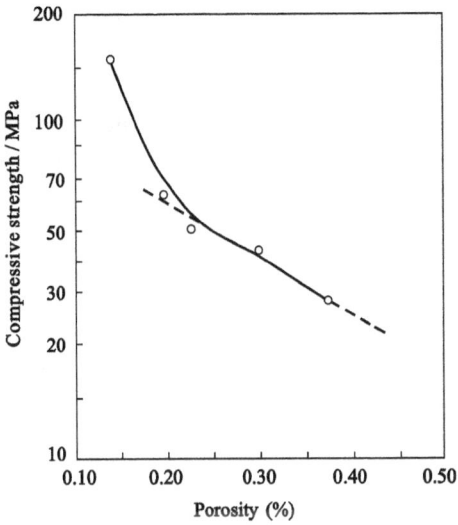

Figure 2.68: Relationship between compressive strength and porosity of cement paste.

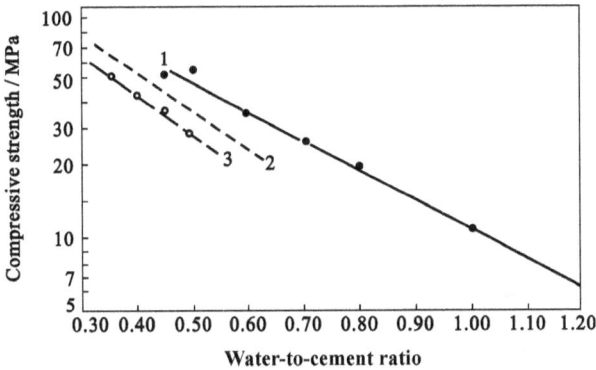

Figure 2.69: Relationship between compressive strength and water-to-cement ratio of cement paste. 1, 2, 3, the data from different sources were used, respectively.

water to cement ratio and compressive strength of cement paste. In fact, the water to cement ratio is a measure of porosity. In the case of the same cement composition and fineness, the relationship between water to cement ratio and strength, and the relationship between porosity and strength are similar.

Whether stirring and vibration during construction are sufficient or not, they have a great impact on the structural strength of the cement paste, especially the flexural strength. Inadequate stirring causes the so-called segregation of components (water, cement and aggregates) in the cement paste, resulting in an uneven distribution of cement paste, and reducing the strength. If there is no enough vibration during construction, the cement paste is not dense enough, bubbles and micro-cracks are produced and have a great impact on the strength.

A lot of practices show that during the construction process, such as vigorous stirring, rolling and press forming, the water-to-cement ratio can be reduced, the porosity of the hardened cement paste can be reduced to less than 20%, and the macro-pores larger than 100 μm occupy not more than 2% of the total volume, even the total porosity with size larger than 15 μm can be controlled within 0.5%, the number and size of microcracks are reduced, the density of cement paste is increased to improve the strength greatly, especially the flexural strength.

(2) Curing temperature

In the cement hydration process, increasing the curing temperature (i.e., the temperature of hydration) can lead to rapid development of early strength, but the late strength, especially the flexural strength will be reduced.

The effect of curing temperature on the strength growth of cement paste is shown in Figure 2.70.

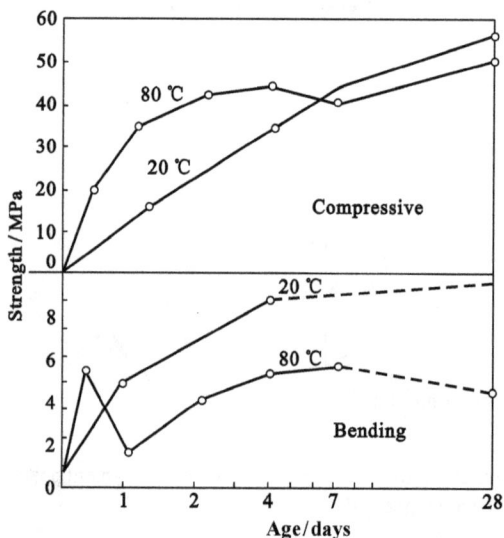

Figure 2.70: Effect of curing temperature on the strength of cement paste.

G. J. Verbeck et al. believed that the main reason for the decrease in strength caused by hydration at high temperature is that the hydration product is rapidly diffused at a high temperature and the resulting hydration products such as gels do not sufficiently diffuse and uniformly precipitate in the space between the cement particles. In this way, the parts of the gel that are sparsely distributed become structural weakness, thereby affecting the increase of strength. At the same time, as the cement particles are wrapped by dense gel layer, later-age hydration is delayed to affect the further development of strength. At room temperature or low temperature, hydration is slow, but hydration products grow and spread well, the structure of the gel evenly distributes and strength is high.

Some other researchers believed that the difference between the thermal expansion coefficient of each component in the paste is the main reason to the damage of the cement paste structure. All components in the cement paste structure, especially saturated air, will expand violently when heated, resulting in a huge internal stress to weaken the bond strength of the cement paste, increase the porosity and even generate microcracks leading to a significantly reduction in flexural strength which is most susceptible to cracks.

It should be noted that, while raising the curing temperature, the paste must be kept wet, otherwise hydration may stop. Generally saturated or steam curing should be used, but according to previous studies, under autoclave conditions, the harmful effects of high temperature on the total strength is more serious than that below 100 °C. From the analysis, it is due to the changes in chemical compositions and physical properties of hydration products, and the increase of porosity of the cement paste. In order to prevent the decrease of the strength, silicon materials, such as fine quartz sand, fly ash and so on, are generally used with an appropriate adding amount in the autoclave.

(3) Admixtures
In modern construction, admixtures are almost used in all concrete and concrete products. Admixtures have a certain effect on the strength of cement paste. If using proper amount of water reducing agent, the water to cement ratio can be significantly reduced to 0.25, which will increase stably strength. Early strength agent can significantly improve the early strength, the late strength can be also increased. The use of other additives such as air-entraining agents, expanders and accelerators may lead to the reduction of later strength, so their dosages should be strictly controlled.

2.7.6 Volume change

Hydration of Portland cement produces different hydration products. This process, coupled with the variation in relative humidity and temperature, leads to volumetric changes in the hardened cement paste. These deformations will negatively affect the physical, mechanical and durability properties, especially when they occur in an intense and uneven volume change. As a result, the volume change of hardened cement paste is an important performance indicator.

2.7.6.1 Chemical shrinkage
When cement hydrates, its constituent compounds are transformed into hydration products undergoing substantial volume increase in the solid phase, while a bulk shrinkage take place in the hardened paste. This bulk shrinkage is termed as chemical shrinkage since it is induced by the chemical reaction between cement and water.

This is further exemplified by the hydration of C_3S. The bulk shrinkage can be expressed by the following equation:

$$2(3CaO \cdot SiO_2) + 6H_2O \rightarrow 3CaO \cdot 2SiO_2 \cdot 3H_2O + 3Ca(OH)_2$$

Density	3.14	1.00	2.44	2.23
Molar mass	228.23	18.02	342.48	74.10
Molar volume	72.71	18.02	140.40	33.23
Volume	145.42	108.12	140.40	99.69

The initial volume of the constituent compounds is calculated to be 145.42+108.12 = 253.54 cm^3, while the volume of hydration products after reaction is 140.40 + 99.69 = 240.09 cm^3, resulting in a volume shrinkage of 253.51–240.09 = 13.45 cm^3. In this case, the chemical shrinkage accounts for 5.30% whereas the solid phase experiences a 65.10% increase with respect to the original volume. This chemical shrinkage occurs to varying extents in other compounds, as tabulated in Table 2.40. It can be seen that the volume of the solid phase after hydration is increased significantly, which fills up the space originally occupied by the mixing water and consequently densifies the hardened paste. Nevertheless, the bulk volume is decreased and this will cause the generation of pores when cement hardening takes place in the air.

Chemical shrinkage, attributed to its strong connection to cement hydration, can be harnessed to indirectly represent the rate and degree of hydration. At a specific age, a higher measured chemical shrinkage denotes a higher rate and the degree of hydration.

Table 2.41 lists the chemical shrinkage for the individual constituents of Portland cement in the order of $C_3A>C_4AF>C_3S>C_2S$. Normally, the amount of chemical shrinkage is linearly proportional to that of C_3A. Based on the compound composition of a typical Portland cement, the total amount of chemical shrinkage upon complete hydration of 100 g cement is 7–9 cm^3, which leads to an impressive shrinkage of $(21–27) \times 10^3$ cm^3 assuming a cement content of 300 kg/m^3 in the concrete. This readily illustrates the considerable porosity as a result of the chemical shrinkage. At the same time, the continuous growth of the solid phase will fill up the residual spaces, which results in a reduction in the total porosity.

2.7.6.2 Wetting expansion and drying shrinkage

The volume of hardened cement paste depends on the water contained therein. When the water content is increased, the gel particles are separated apart due to the molecular adsorption effect leading to volume expansion. When the water content is decreased, volume contraction occurs. The wetting expansion and drying shrinkage is a reversible process to a large extent. Drying shrinkage is related to water depletion, but not in a linear fashion.

Table 2.40: Volume change associated with cement hydration of a few typical cement constituents.

Reaction formula	Mass (g)	Density (g/cm³)	Bulk volume (cm³)		Solid-phase volume (cm³)		Volume change (cm³)	
			Before	After	Before	After	Bulk	Solid phase
$2CaSO_4 \cdot H_2O + 3H_2O =$	290.29	2.62	164.85	148.42	110.80	148.42	-9.97	33.95
$2(CaSO_4 \cdot 2H_2O)$	54.05	1.00						
	344.34	2.32						
$3CaO \cdot Al_2O_3 + 3(CaSO_4 \cdot 2H_2O)$	270.18	3.04						
$+25H_2O=$	516.51	2.32	761.91	715.08	311.51	715.08	-6.15	129.55
$3CaO \cdot Al_2O_3 \cdot 3CaSO_4 \cdot 31H_2O$	450.40	1.00						
	1,237.09	1.73						
$CaO + H_2O = Ca(OH)_2$	56.08	3.32	34.81	33.32	16.79	3.23	-4.54	97.9
	18.02	1.00						
	74.10	2.23						
$2(3CaO \cdot SiO_2)$	456.66	3.11						
$+6H_2O=3CaO \cdot 2SiO_2 \cdot 3H_2O$	108.12	1.00	253.54	240.09	145.42	240.09	-5.30	65.10
$+3Ca(OH)_2$	342.48	2.44						
	222.30	2.23						

(continued)

Table 2.40 (continued)

Reaction formula	Mass (g)	Density (g/cm³)	Bulk volume (cm³)		Solid-phase volume (cm³)		Volume change (cm³)	
			Before	After	Before	After	Bulk	Solid phase
$2(2CaO \cdot SiO_2) + 4H_2O =$	344.50	3.28						
$3CaO \cdot 2SiO_2 \cdot 3H_2O + Ca(OH)_2$	72.08	1.00	177.11	173.63	105.03	173.63	−1.97	65.32
	342.48	2.44						
	74.10	2.23						
$3CaO \cdot Al_2O_3 + 6H_2O =$	270.18	3.04						
$3CaO \cdot Al_2O_3 \cdot 6H_2O$	108.10	1.00	196.98	150.11	88.88	150.11	−23.79	68.89
	378.28	2.52						

Table 2.41: Chemical shrinkage of Portland cement constituents.

Constituent	28 days shrinkage $(\times 10^{-2})$ (cm^3/g)	Ultimate shrinkage $(\times 10^{-2})$ (cm^3/g)
C_3S	5.2	6–7
C_2S	1.2	4
C_3A	17.0	17.5–18
C_4AF	9.0	10–11

There is still a lack of consistent argument over the mechanism of the self-desiccation induced shrinkage. It is generally believed major factors are capillary tension force, surface tension, disjoining force and interlayer water content.

It has been reported the drying shrinkage of hardened cement paste is primarily dependent on the amount of C_3A contained in cement. The drying shrinkage is increased with a higher C_3A content, as shown in Figure 2.71. In the case of similar C_3A content, the amount of gypsum present in cement is the major factor controlling the volume change.

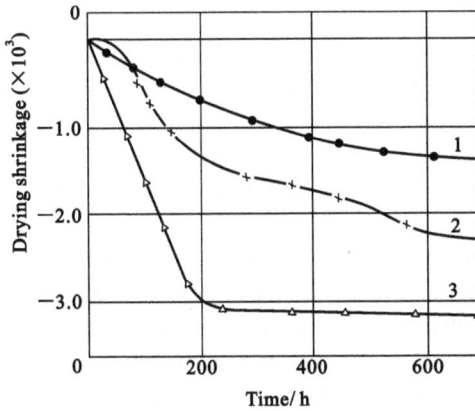

Figure 2.71: Drying shrinkage of cement paste with time. l, C_3A 4%; 2, C_3A 6%;3, C_3A 8%.

The w/c ratio is also another factor and its effect is not readily perceived in the early age when drying shrinkage develops rapidly. After 28 days, pronounced decrease in drying shrinkage will occur at a lower w/c ratio. It is also noted that cement paste at a lower w/c ratio experiences earlier termination in drying shrinkage. As a result, effective drying shrinking mitigation techniques in the field include an appropriate w/c ratio and a well-controlled curing regime.

2.7.6.3 Carbonation shrinkage

When exposed to relative humidity of a specific range, hydration products in the hardened cement paste such as $Ca(OH)_2$, C–S–H will react with CO_2 in the environment and produce $CaCO_3$ and H_2O, which leads to a volume reduction in the hardened paste. This irreversible shrinkage is known as carbonation shrinkage as follows:

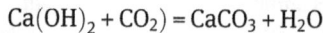

$$Ca(OH)_2 + CO_2) = CaCO_3 + H_2O$$

$$3CaO \cdot 2SiO_2 \cdot 3H_2O + CO_2 = CaCO_3 + 2(CaO \cdot SiO_2 \cdot H_2O) + H_2O$$

The carbonation rate in the atmosphere is normally a very slow process restricted to the near-surface region. This will generate micro cracks on the surface of the hardened cement paste after about 1 year exposure, which only affects its appearance with negligible effect on the strength.

In practice, the most important aspect associated with the volume change in concrete, to be expansion or shrinkage, is whether the change occurs uniformly. When the solid phases formed during hydration generates localized nonuniform expansion, this will lead to unsoundness and damage the hardened cement paste. However, a well-regulated expansion that causes uniform expansion will tend to densify the hardened cement paste, resulting in improved properties such as strength, freeze–thaw resistance and impermeability.

2.7.7 Water retention and bleeding

When mixing mortar or concrete in the laboratory or field, it is a common observation that different types of cement present different phenomena. For instance, some water tends to migrate to the surface in the setting process. This excessive water normally accumulates on the surface of the specimen or a concrete member, or seeps out of the molds. This is known as bleeding while the opposite phenomenon is called water retention.

Bleeding is detrimental to the homogeneity of concrete, because the bleeded water will often gather on the irrigation surface. This will produce a water-rich interlayer between the two concretes, undermining their binding and thus the homogeneity as a whole. Segregation can not only occur between two concretes but also within the interior of concrete. This is because the water separated from the fresh cement mortar will coalesce around coarse aggregate particles and beneath the reinforcing bars and reduces the binding capacity. Moreover, extra pores will be produced upon the evaporation of the bleeded water. This in turn will reduce the strength and water resistance of concrete. The following measures can be carried out to counteract the bleeding of concrete.

(1) Increase the specific surface area of cement
A higher specific surface area of cement will improve the particle packing and its water retention capacity. It is found out cement bleeding is significantly decreased with the increase of specific surface area, when the chemical composition, mix proportion and mixing procedure is maintained the same. A higher specific surface area will shorten the initial setting time and accelerate the

formation of a stable structure, thus reducing water bleeding and an improved 3-day compressive strength. However, attention should be paid to avoid over-grinding cement which will undermine the consistency and increase the water requirement.

(2) Use pozzolanic supplementary materials
Bleeding resistance is improved when pozzolanic supplementary materials (such as diatomite, bentonite, burning clay and coal gangue) are intergrinded with cement and the beneficial effect is normally proportional to their content in cement, which leads to the improved consistency and workability in cement mortar or concrete. However, caution should be executed in the case of excessive use of pozzolanic supplementary materials, which will cause an increase in water demand and a decrease in strength. As a result, the optimized content normally in the range of 5.0–8.0% is recommended.

(3) Addition of limestone powder
A controlled amount of limestone powder adding to cement will improve the water retention capacity and 3 d compressive strength of hardened cement paste. An increasing limestone powder content, coupled with a higher specific surface area of cement, will significantly improve the bleeding resistance of cement. Results indicate that a binary addition of 5% limestone powder and 3% coal gangue into cement reduces bleeding by 1.07% and 1.08%, respectively, compared to the cement with 8% limestone powder and the cement with 8.0% coal gangue with similar specific sur-face area of cement. Thus, coal gangue is more efficient in bleeding mitigation compared with limestone powder.

2.7.8 Hydration heat

Hydration of clinker phases is accompanied by the liberation of latent heat to various degrees, which leads to the accumulation of hydration heat in cement. During winter construction, hydration heat will increase the temperature of cement paste, promoting cement hydration and accelerating construction speed. However, accumulation of hydration heat in mass concrete will not be readily dissipated and a sharp temperature rise will occur in the interior. This causes pronounced temperature gradient and the resultant thermal stresses, which cre-ates cracks due to non-uniform expansion in concrete structures. As a result, hydration heat is an important evaluating indicator for mass concrete projects and the reduction in hydration heat is one of the effective measures in improving the structure quality.

Experimental results indicate the hydration heat of cement is strongly connected to its phase compositions. As listed in Tables 2.42 and 2.43, the amount of hydration

Table 2.42: Hydration heats of clinker phases and cement.

Type	Hydration heat (J/g)	Type	Hydration heat (J/g)
C_3S	500	f-MgO	840
β-C_2S	250	Ordinary Portland cement	375–525
C_3A	1,340	Sulfate-resistant cement and slag cement	355–440
C_4AF	420	Pozzolanic cement	315–420
f-CaO	L,150	Aluminate cement	545–585

f-MgO, free state of MgO.

Table 2.43: Hydration heat of a cement with specific clinker composition (J/g).

Age (days)	C_3S	C_2S	C_3A	C_4AF
3	240	50	880	290
28	377	105	1,378	494

heat is in the decreasing order of $q(C_3A) > q(C_4AF) > q(C_3S) > q(C_2S)$ for different clinker phases. Thus, an effective measure in lowering the hydration heat involves maintaining a high content of C_2S and C_4AF in cement while reducing the amount of C_3A and C_3S.

Figures 2.72 and 2.73 present the effect of C_3A and C_3S on hydration heat of cement, which demonstrates the optimization of clinker composition as an effective measure to reduce hydration heat. Meanwhile, it has to be noted that the state of solid solution for clinker phases of the same composition also has an effect on the hydration heat of cement.

Hydration heat can be roughly calculated as follows:

$$Q_{3\ days} = 240\,w(C_3S) + 50w(C_2S) + 800\,w(C_3A) + 290w(C_4AF) \tag{2.89}$$

$$Q_{28\ days} = 377\,w(C_3S) + 105w(C_2S) + 1378\,w(C_3A) + 494w(C_4AF) \tag{2.90}$$

where the coefficients are the amount of hydration heat for clinker phases, J/g; $w(C_3S)$, $w(C_2S)$, $w(C_3A)$, $w(C_4AF)$ are the mass fractions for clinker phases (%); $Q_{3\ days}$ is the hydration heat after 3-day curing, J/g; $Q_{28\ days}$ is the hydration heat after 28-day curing, J/g.

A multitude of factors play a role in the amount of heat generated during hydration, such as clinker phase composition, state of solid solution, clinker sintering and

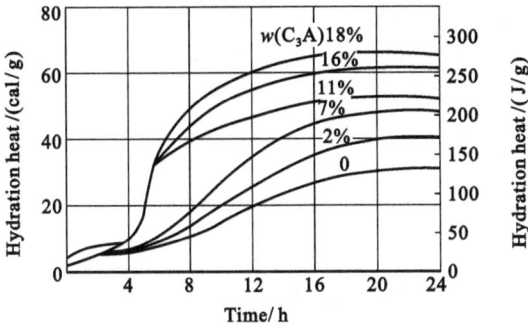

Figure 2.72: Effect of C_3A content on hydration heat of cement. $w(C_3S)$ maintained the same.

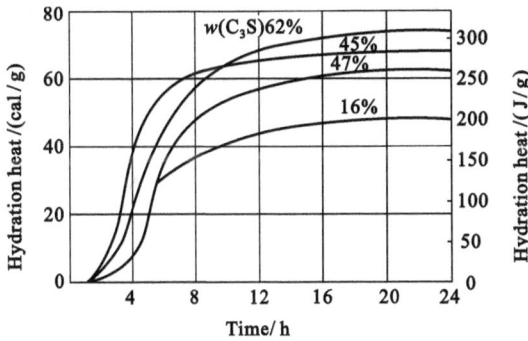

Figure 2.73: Effect of C_3S content on hydration heat of cement. $w(C_3A)$ maintained the same.

cooling regimes, cement fineness, w/c ratio, curing temperature and age. In addition, results obtained under different conditions vary among different researchers. Thus, hydration heat calculated based on the clinker composition is a coarse estimate. Accurate evaluation is more reliable from actual measurements.

2.7.9 Heat resistance

When hardened cement paste is heated, dehydration of the hydration products and the carbonates occurs when the temperature reaches a specific range, which leads to the formation of f-CaO. There will be a secondary hydration when f-CaO is in contact with moisture. This create $Ca(OH)_2$ accompanied by volume expansion, which will undermine the structure of hardened cement paste. The decomposing temperature for different hydration products in hardened cement paste is shown below:

$$C-S-H \text{ gel} : 160-300\,°C$$

$$C-A-H \text{ gel} : 275-370\,°C$$

$$Ca(OH)_2 : 400-590\,°C$$

$$CaCO_3 : 810-870\,°C$$

When the hardened cement paste is heated to 100–250 °C, an increase in its density is observed associated with the gel dehydration and the accelerated crystallization of a part of Ca(OH)$_2$. An improvement in strength is instead resulted. When the temperature is increased to 250–300 °C, the strength of hardened cement paste is reduced when dehydration of C–S–H and C–A–H gels commences. With a further increase in temperature to 400–1,000 °C, decomposition of CaCO$_3$ occurs and the remaining water is evaporated, which significantly reduces its strength and ultimately leads to total failure.

Two successive hydration processes in the heating of hardened cement paste has a pronounced effect on its strength. When a sample prepared using ordinary Portland cement is heated to 500 °C and then cools down in the air, cracks will be generated with a reduction in strength. Samples undergoing 900–1,000 °C exposure will be damaged in 3–4 weeks' exposure in the air. An exposure to the humid air will accelerate the damage of the hardened cement paste.

2.7.10 Permeability resistance

Impermeability refers to the ability of hardened cement or concrete to resist the penetration of various deleterious agents. Most harmful fluids, solutions, gases and other media, are infiltrated from pores and cracks in cement or concrete, so improving impermeability is an effective way to improve durability. For some special projects such as hydraulic concrete, oil tanks, pressure pipes, towers and so on, impermeability has more stringent requirements, so impermeability is an important performance indicator for cement.

Impermeability of hardened cement paste is generally expressed by the permeability coefficient k. According to water penetration test on hardened cement paste, it can be seen that when water infiltrates cement paste, the seepage rate can be expressed as follows:

$$\frac{dq}{dt} = k \cdot A \frac{\Delta h}{L} \tag{2.91}$$

where $\frac{dq}{dt}$ is the seepage rate, cm^3/s; A is the cross-sectional area, cm^2; Δh is the water pressure difference acting on both sides of the specimen, cm H$_2$O; L is the specimen thickness, cm; k is the permeability coefficient, cm/s.

Based on the above formula, seepage rate is proportional to permeability coefficient, for a specific specimen size and water pressure difference. Therefore, permeability coefficient k is commonly used to indicate impermeability level.

Permeability coefficient can be expressed as follows, based on massive experimental results:

$$k = c \cdot \frac{Pr^2}{\eta} \tag{2.92}$$

where ε is the total porosity, %; y is the pore bearing radius (pore volume/pore surface area), m; η is the fluid viscosity, Pa·s; c is the constant.

Thus, the permeability coefficient k is proportional to the square of the radius of the pores and the total porosity ε. Therefore, the size of the pores has a more important effect on impermeability. It is generally believed that the gel pore has little effect on the impermeability, and the permeability coefficient is mainly determined by the size of the capillary pores, as shown in Figure 2.74. When the capillary porosity of the hardened paste is greater than 30%, the permeability coefficient will increase exponentially resulting in a significant decrease in impermeability.

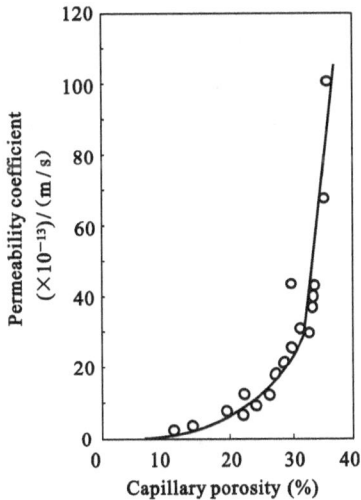

Figure 2.74: Relationship between permeability coefficient and capillary porosity of hardened paste.

As mentioned above, the porosity and pore size of hardened cement paste are related to the w/c ratio, the degree of hydration (or curing age) and the curing conditions. Figure 2.75 shows the relationship between the permeability coefficient of hardened cement paste and concrete and the w/c ratio. It can be seen that the permeability coefficient increases with an increasing w/c ratio. When the w/c ratio is 0.4, the impermeability of the fully hydrated hardened cement paste is similar to dense natural stones such as marble and limestone. When the w/c ratio exceeds 0.6, the increasing rate of the permeability coefficient with w/c ratio is significantly increased. For example, at a w/c ratio of 0.7, the k value exceeds the 0.6 w/c ratio by a few times and exceeds that of a 0.5 w/c ratio by one order of magnitude. According to the analysis, this is the reason that the pore connectivity in the structure has changed. When the w/c ratio is low, the pores are often blocked by the gels, reducing its connectivity and thus leading to a low permeability coefficient; when the w/c ratio increases to a certain extent, not only the total porosity, but also the capillary pore diameter increases. This results in well-connected pores, so the permeability coefficient is greatly increased. Therefore, the appropriate reduction

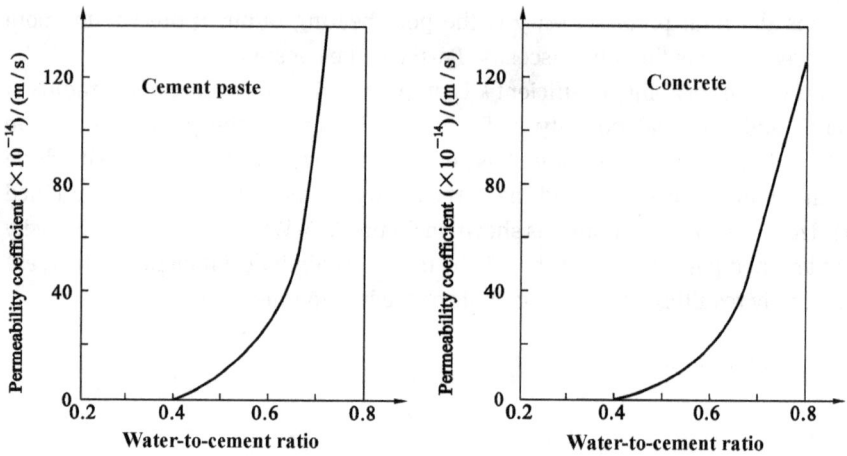

Figure 2.75: Relationship between permeability coefficient and *w/c* ratio for hardened cement paste and concrete.

of *w/c* ratio, pore size refinement and its connectivity are the key measures to improve the impermeability of the hardened cement paste. It is generally believed that the hardened cement paste of a *w/c* ratio below 0.5 has better impermeability.

The curing age and degree of hydration also have an effect on the permeability coefficient, as shown in Table 2.44.

Table 2.44: Relationship between permeability coefficient and curing age (*w/c* = 0.51).

Curing age (days)	Freshly mixed	1	3	7	14	28	100	240
k (m/s)	10^{-5}	10^{-8}	10^{-9}	10^{-10}	10^{-l2}	10^{-13}	10^{-16}	10^{-18}
Comments	Independent of *w/c* ratio		Intern-connected capillary pores				Disconnected capillary pores	

In the course of hydration and hardening of cement, as the age prolongs, the hydration products increase continuously, and the capillary system becomes smaller and tortuous until it is completely partitioned and disconnected, leading to gradual reduction in the permeability coefficient. The influence of age on the permeability coefficient is closely related to the w/c ratio. When the w/c ratio exceeds a certain level (0.7), the impermeability is still low even though the age is longer.

Influencing factors on the impermeability include not only the size of the pore size, the porosity, not also the pore size distribution. Mehta demonstrated this conclusion experimentally. He believes that impermeability depends mainly on the number of large pores, especially those larger than 1,320 μm. B. K. Nyame et al. proposed that the size of the largest connected pores has an impact on impermeability.

In summary, reduction in the w/c ratio and the size of the large pores, especially in the size and proportion of the connected pores are effective measures to improve the impermeability. In the actual construction, because the field conditions are more complicated than the laboratory, additional measures should be adopted, such as selection appropriate aggregate, strengthening the vibration and employing appropriate curing regime.

2.7.11 Frost resistance

Frost resistance is also an important property of hardened cement paste. Durability of Portland cement used in cold areas depends mainly on the ability to resist freeze–thaw cycles. According to a few studies, the destructive effect of freeze–thaw cycles on concrete in cold areas, especially in port, is quite serious.

Water transformed into ice experiences a volume increase by about 9%. Ice formation in the hardened cement paste will create swelling stresses against the capillary wall. When the stress exceeds the tensile strength of the paste, irreversible changes such as micro cracks will occur in the hardened cement paste. After the ice melts, the changes cannot be completely restored. In the next freezing cycle, it will expand the original cracks which will grow with repeated freeze-thaw cycles, eventually leading to serious damage. Therefore, the frost resistance of cement is generally expressed in terms of the maximum number of freeze-thaw cycles the test samples can withstand between −15 °C and 20 °C when the compressive strength loss is less than 25%. The higher the number, the better the frost resistance.

There are several types of water existing in hardened cement paste, that are combined water, adsorbed water (gel water and capillary water) and free water. Combined water does not freeze and gel water can only freeze at very low temperatures (such as −78 °C) due to its extremely small pore size. Under the natural conditions of low temperature, only the capillary water and free water will freeze. The freezing point of capillary water is normally lower than −1 °C due to the dissolved Ca$(OH)_2$ and alkali in the pore solution. At the same time, the capillary water is also subject to surface tension, which further lowers the freezing point. In addition, the smaller the capillary pore size, the lower the freezing point. For example, the freezing point of the water held in 10 nm pore −5 °C, while the water held in the 3.5 nm pore will freeze to −20 °C. However, for general concrete, complete freezing of capillary water can occur at −30 °C. So in the cold area, concrete is more likely subjected to frost cracking.

A lot of practice experiences have proved that the frost resistance of cement is closely connected to its mineral composition, strength, w/c ratio, pore structure and other factors. Generally, an increase in C_3S content or the amount of gypsum incorporated can improve its frost resistance. Under other conditions being equal, the higher the strength of the cement, the higher its capacity to resist expansion stress during ice formation and the better its frost resistance. According to a few studies, when the w/c ratio is controlled below 0.4, the frost resistance of the hardened paste is high, while the frost resistance of the hardened paste is remarkably decreased when the w/c ratio is above 0.55 because of the increasing number in capillary pores of larger size, leading to decreased frost resistance.

In addition, under low-temperature construction, the use of appropriate curing methods to prevent premature freezing, or the incorporation of air-entraining agent in concrete form a large number of dispersed and very fine bubbles are important frost resisting methods.

2.7.12 Erosion resistance

Hardened cement paste in contact with the environment will normally be affected by the agents in the environment. Environmental media harmful to the durability of cement are mainly freshwater, acid and acid water, sulfate solutions and alkali solutions. Under the influence of environmental media, a series of physicochemical changes occur in the hardened cement paste structure, and the strength is reduced or structure is even ruptured.

2.7.12.1 Freshwater erosion

Freshwater erosion, also known as dissolution erosion, refers to the gradual dissolution of the components in hardened cement paste exposed to freshwater, which ultimately leads to paste structure damage.

Among various hydration products, $Ca(OH)_2$ has the highest solubility and is therefore the first to be dissolved. Since the hydration products in cement must be stable in a certain concentration of $Ca(OH)_2$ solution, when $Ca(OH)_2$ is dissolved out, the solution will quickly become saturated if the amount of water is small and in a quiescent state, dissolution will be stopped. However, in the case of flowing water, $Ca(OH)_2$ will be continuously dissolved and carried away, causing other hydration products to be decomposed, especially under the condition of water pressure and low impermeability of concrete. This will further increase the porosity, making water easier to infiltrate into concrete and thus accelerating the dissolution and erosion.

According to the data from Moscoff, the limit concentration of CaO in which all major hydration products of cement can stably exist is as follows:

Hydrates	$c(CaO)$ (g/L)
$CaO \cdot SiO_2 \cdot H_2O$	> 0.05
$2CaO \cdot SiO_2 \cdot H_2O$	> 1.1
$2CaO \cdot Al_2O_3 \cdot H_2O$	0.36 – 0.56
$3CaO \cdot Al_2O_3 \cdot (6-18)H_2O$	0.56 – 1.08
$4CaO \cdot Al_2O_3 \cdot (12-13)H_2O$	> 1.08
$2CaO \cdot Fe_2O_3 \cdot H_2O$	0.64 – 1.06

Thus, under the action of a large amount of flowing water, after CaO in the hardened cement paste is dissolved and taken away, $Ca(OH)_2$ will be the first one to be dissolved. As the CaO concentration in the solution gradually decreases, hydration products of high alkalinity such as calcium silicate hydrates, calcium aluminate hydrate, etc. will be decomposed into low alkaline hydration products. With the continuation of dissolution, low alkaline hydration products will be decomposed, and ultimately ends up as silicate gels of no binding ability, $Al(OH)_3$ and so on, thereby greatly reduce the structural strength. According to the previous studies, when the dissolution of CaO is 5%, the strength is decreased by about 7%, with the CaO dissolution of 24%, the strength will be decreased by 29%.

When the hardened cement paste is exposed to fresh water for a long time, it will be damaged by dissolution and erosion. However, for highly impermeable hardened cement paste or concrete, the dissolution of freshwater is very slow and almost negligible

2.7.12.2 Acid and acidic water erosion

Acid and acidic water erosion, are also known as synergic erosion by leaching and chemical dissolution. It refers to the dissolution of the chemical components or the formation of dissolvable products from their reaction with acid in the hardened cement paste in contact with the acidic solution, leading to structural damage.

The experiments show that the stronger the acid solution, the greater the solubility of the resulting product, the more serious erosion damage. The erosion reaction of the acid and acidic water on the cement paste structure is as follows:

$$H^+ + OH^- \rightleftharpoons H_2O$$

$$Ca^{2+} + 2R^- \rightleftharpoons CaR_2$$

H^+ and acid radical R^- from the dissociation of acid ions will react with OH^- and Ca^{2+} from the ionization of $Ca(OH)_2$ to form water and calcium salts. Thus, it can bee seen that the severity of the acid erosion is determined by the concentration of H^+ or acid in the solution. The stronger the acidity of the solution, the more available H^+, the more

Ca(OH)$_2$ bound and taken away, and the more severe of the erosion. When H$^+$ reaches a sufficiently high concentration, it can also directly destroy the cement structure by reacting with calcium silicate hydrate, calcium aluminate hydrate or even unhydrated calcium silicate and calcium aluminate.

Erosion is closely related to the type of acid anions. Most common acid can react with Ca(OH)$_2$ in the formation of soluble salts. For example, inorganic acid, hydrochloric acid and nitric acid can form soluble CaCl$_2$ and Ca(NO$_3$)$_2$ with Ca(OH)$_2$, which will be carried away by water, resulting in erosion damage. On the other hand, phosphoric acid reacts with Ca(OH)$_2$ to produce almost insoluble Ca$_2$(PO$_4$)$_3$ to block the capillary pores, which slows down the erosion rate. Organic acids are not as strong as inorganic acids, and their erosive capacity is also related to the nature of calcium salts they produce. For example, between acetic acid, formic acid and lactic acid can react with Ca(OH)$_2$ to form soluble salts, while the reaction between oxalic acid and Ca(OH)$_2$ produces insoluble calcium salt, which forms a protective layer on concrete surface to improve its resistance to acid ingress. Under normal circumstances, the higher the concentration of organic acids, the larger the relative molecular mass, the more aggressive the erosion agent is.

The erosion described above normally exists only in chemical plants or industrial waste liquid. In nature, the major erosive agent against cement is carbonate from CO$_2$ dissolved in water.

The presence of carbonate in water will firstly react with Ca(OH)$_2$ in the hardened cement paste to form insoluble CaCO$_3$ on its surface. The newly formed CaCO$_3$ will then react with H$_2$CO$_2$ to form the more soluble Ca(HCO$_3$)$_2$, which dissolves Ca(OH)$_2$ gradually and ultimately leads to the decomposition of calcium silicate hydrate and calcium aluminate hydrate. The reaction formulas are as follows:

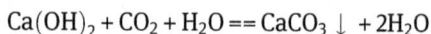

$$Ca(OH)_2 + CO_2 + H_2O == CaCO_3 \downarrow + 2H_2O$$

$$CaCO_3 + CO_2 + H_2O \rightleftharpoons Ca(HCO_3)_2$$

The second reaction of the above equation is reversible, and when the concentration of CO$_2$ and Ca(HCO$_3$)$_2$ in the water reaches equilibrium, the reaction stops. Because natural water itself often contains a small amount of Ca(HCO$_3$)$_2$, which can be balanced with a certain amount of H$_2$CO$_3$, this part of the H$_2$CO$_3$ that does not dissolve CaCO$_3$, has no erosive effect and is known as the balanced carbonate. However, when the water contains excessive H$_2$CO$_3$, it will dissolve the CaCO$_3$ more than the surplus H$_2$CO$_3$ needed for the balance, which will cause erosion to the cement. This part of H$_2$CO$_3$ is called the erosive carbonic acid. Therefore, the higher the content of H$_2$CO$_3$, the stronger the acidity of the solution, the more severe the erosion will be.

The greater the temporary hardness of water, the more amount of balanced carbonic acid is needed, which means the presence of surplus CO$_2$ will not cause erosion. Meanwhile, when the amount of Ca(HCO$_3$)$_2$ or Mg(HCO$_3$)$_2$ is higher, their reaction with Ca(OH)$_2$ in hardened cement paste results in CaCO$_3$ of very low solubility, which will be deposited in the pores and on the surface. This increases the compactness of the structure, hindering the further dissolution of hydration products, thus reducing the erosion effect. In the temporary water, hardness is not high, even if the content of CO_2 is small, it will also have some erosion.

2.7.12.3 Sulfate attack

Sulfate attack, also known as the expansive erosion, is the expansion cracking from the crystallization stress associated with the formation of ettringite by the reaction of sulfate in the pore solution and the components of the hardened cement paste.

Sulfate attack is primarily from the reaction of Na_2SO_4, K_2SO_4 and so on with Ca $(OH)_2$ in the hardened cement paste to from $CaSO_4 \cdot 2H_2O$, as follows:

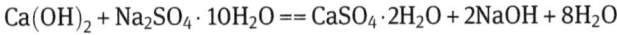

$$Ca(OH)_2 + Na_2SO_4 \cdot 10H_2O == CaSO_4 \cdot 2H_2O + 2NaOH + 8H_2O$$

This reaction causes a volume increase of 114% in the solid phase, generating pronounced crystallization stress in the hardened cement paste which leads to fracture and ultimately failure. However, $CaSO_4 \cdot 2H_2O$ crystallization requires the concentration of SO_4^{2-} reaching above 2,020–2,100 mg/L. When the concentration of SO_4^{2-} is below 1,000 mg/L, $CaSO_4 \cdot 2H_2O$ crystals is restricted from formation due to its high solubility. $CaSO_4 \cdot 2H_2O$ can react with the calcium aluminate hydrate to form ettringite as follows:

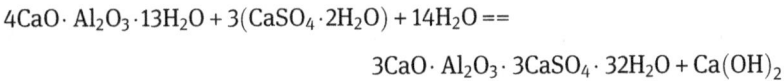

$$4CaO \cdot Al_2O_3 \cdot 13H_2O + 3(CaSO_4 \cdot 2H_2O) + 14H_2O ==$$

$$3CaO \cdot Al_2O_3 \cdot 3CaSO_4 \cdot 32H_2O + Ca(OH)_2$$

Due to the low solubility of Ettringite, crystals can be precipitated when the concentration of SO_4^{2-} is low, creating a volume expansion of the solid phase by 94%. This can also destroy the structure of hardened cement paste. Therefore, sulfate attack is produced at lower sulfate concentrations (250–1,500 mg/L). When its concentration reaches a certain range, it changes into gypsum erosion or a mixed attack of calcium sulfoaluminate with gypsum.

In addition to $BaSO_4$, the vast majority of sulfates have a significant erosion effect on the hardened cement paste. In rivers and lakes, sulfate content is usually low (<60 mg/L), but the content of SO_4^{2-} in seawater often reaches 2,500–2,700 mg/L, and some groundwater flows through intercalations rich in gypsum, thenardite (Na_2SO_4) or other sulfate-rich rock, the dissolution of some of the sulfate into the water can also increase the concentration of SO_4^{2-} in the water causing erosion.

Various cations possess different erosion effect when it comes to sulfate attack. For example, there is more Mg^{2+} in seawater, so the existence of $MgSO_4$ has a greater erosion effect, because the $Ca(OH)_2$ in the paste will reaction with $MgSO_4$ as follows:

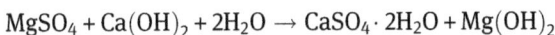

$$MgSO_4 + Ca(OH)_2 + 2H_2O \rightarrow CaSO_4 \cdot 2H_2O + Mg(OH)_2$$

Due to the minimal solubility, $Mg(OH)_2$ easily precipitates out of the solution, so that the reaction proceeds continuously to the right side. Increasing the concentration of $CaSO_4 \cdot 2H_2O$, leading to its crystallization. This produces cracking stresses and destroys the hardened cement paste. Meanwhile, this reaction also promotes the continuous release of CaO to compensate for the consumption, which leads to the continuous decomposition of calcium silicate and calcium aluminate in the cement and destroy the entire paste structure. In fact, $MgSO_4$ causes the decomposition of calcium silicate hydrate by the following formula:

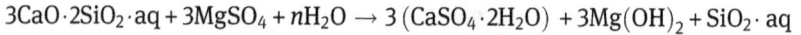

$$3CaO \cdot 2SiO_2 \cdot aq + 3MgSO_4 + nH_2O \rightarrow 3\,(CaSO_4 \cdot 2H_2O) + 3Mg(OH)_2 + SiO_2 \cdot aq$$

In addition, Mg^{2+} will also enter the C–S–H gel to deteriorate its cementing properties. Therefore, in addition to sulfate erosion from $MgSO_4$, Mg^{2+} also causes severe damage, both of which are often referred as "magnesium salt erosion."

Another sulfate attack example is the rapid irreversible reaction of $(NH_4)_2SO_4$ with $Ca(OH)_2$ to generate extremely volatile ammonia. This severe reaction is as follows:

$$(NH_4)_2SO_4 + Ca(OH)_2 \rightarrow CaSO_4 \cdot 2H_2O + 2NH_3 \uparrow$$

2.7.12.4 Erosion of alkali solution

In general, cement and concrete is resistant to alkali attack. However, long-term exposure to an alkali solution with a high concentration (>10%) also causes slow erosion, which is exacerbated at a higher temperature. The erosion consists of mainly chemical and crystallization erosions. Chemical attack is a chemical reaction between the alkaline solution and the components of hardened cement paste to generate products with low cementing property and high solubility:

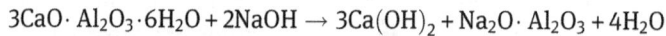

$$2CaO \cdot SiO_2 \cdot nH_2O + 2NaOH \rightarrow 2Ca(OH)_2 + Na_2SiO_3 + (n-1)H_2O$$

$$3CaO \cdot Al_2O_3 \cdot 6H_2O + 2NaOH \rightarrow 3Ca(OH)_2 + Na_2O \cdot Al_2O_3 + 4H_2O$$

Crystal erosion is the evaporation crystallization of the alkali solution that causes crystallization stresses, leading to expansion damage of the hardened cement paste. For example, NaOH in the pore solution reacts with CO_2 in the air to form $Na_2CO_3 \cdot 10H_2O$, leading to volume increases and expansion.

2.7.12.5 Alkali-aggregate reaction

Although cement is an alkaline substance and generally it can resist the alkali erosion, when the alkali content in the cement paste is high, and the aggregates (sand and gravel) contain the active materials, the structure of hardened cement paste will undergo obvious expansion and cracking, and even spalling after a certain period of time. This is known as the alkali-aggregate reaction. Due to the rapid development of modern industry, in order to expand the sources of concrete aggregate and make full use of industrial waste residue, it is becoming more and more important to study the alkali-aggregate reaction when the alkali-containing admixtures are widely used in production to improve the quality of concrete.

Alkali-aggregate reaction is mainly due to the high alkali content of cement (R_2O content is higher than 0.6%), while the aggregate contains active SiO_2, the alkali will react with the active SiO_2 in the aggregate to form alkaline silicate gel. The reaction is as follows:

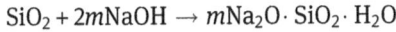

$$SiO_2 + 2mNaOH \rightarrow mNa_2O \cdot SiO_2 \cdot H_2O$$

The alkaline silicate gel formed by the above reaction has a very strong water absorption capacity, which expands during the accumulation of moisture and destroys the structure of the hardened paste.

In general, the alkali-aggregate reaction is usually slow and does not occur significantly over a long period of time. According to the study by T. E. Stanton, there are many factors affecting the alkali-aggregate reaction, which are mainly the alkali content in the cement, the content of the active aggregates and their particle size and water content.

As shown in Figure 2.76, no significant expansion occurs when the alkali content is less than 0.6%, and there is a "critical" content of active aggregates resulting in maximum expansion. For opals, the "critical" content can be as low as 3–5%, while for the less active aggregates, the critical content can reach 10–20%, or even 100%. That is, when the active particles are small, as the content increases, there is more alkaline silicate gel that expands. However, when it is above the "critical" content, the situation is just the opposite, as shown in Figure 2.77. Therefore, adding an appropriate amount of active SiO_2 powder or pozzolan, fly ash and so on can effectively inhibit the alkali-aggregate expansion.

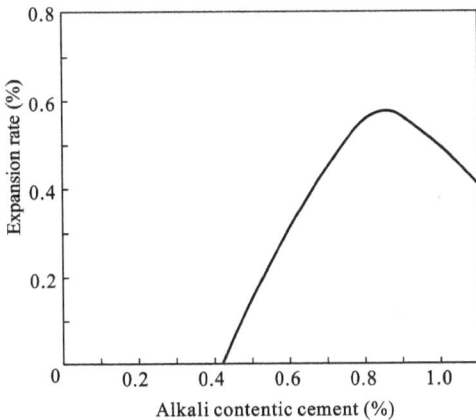

Figure 2.76: Relationship between alkali content in cement and the expansion from alkali-aggregate reaction.

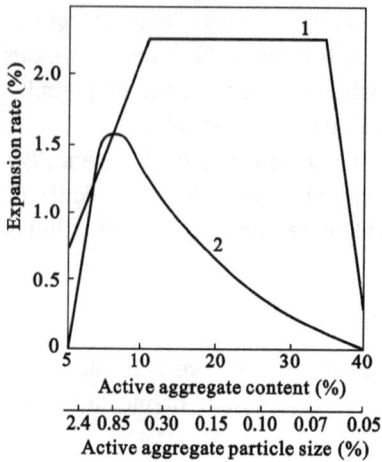

Figure 2.77: Relationship between particle size and the content of active aggregate and the expansion from alkali-aggregate reaction. 1, particle size; 2, content.

In addition, the cement alkali can react with the dolomitic limestone ($CaCO_3 \cdot MgCO_3$) to cause expansion, resulting in concrete damage. This is known as alkali-carbonate reaction, as follows:

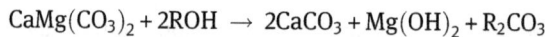

$$CaMg(CO_3)_2 + 2ROH \rightarrow 2CaCO_3 + Mg(OH)_2 + R_2CO_3$$

The above reaction is also called de-dolomite reaction. According to Gilott et al., the above reaction causes the clay minerals in the dolomite to be exposed and swollen to cause damage. Due to the presence of more $Ca(OH)_2$ in the cement paste structure, the following reaction will ensue the regeneration of alkali:

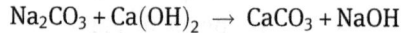

$$Na_2CO_3 + Ca(OH)_2 \rightarrow CaCO_3 + NaOH$$

In this way, the de-dolomite reaction continues in cycles, causing even greater damage.

In summary, to prevent alkali-aggregate reaction and improve the quality of concrete, the following measures can be taken: to minimize the alkali content of cement, to take the appropriate size of the aggregate, reduce the active aggregate content, or to add appropriate amount of reactive silica or volcanic ash, fly ash and so on.

2.7.12.6 Measures to prevent erosion

There are many factors that affect the durability of cement and concrete. In order to improve the durability of concrete, we must firstly consider the environmental conditions for the use of cement. We should carefully design concrete proportioning to minimize w/c ratio, and consider the appropriate construction program to enhance the mixing, vibration and curing, so as to improve the compactness of concrete to enhance its strength, especially the early strength; to improve the pore size distribution to prevent the penetration of erosive media; to consider the use of appropriate additives to improve the performance of concrete, in exceptional circumstances, but also the use of other materials, and surface treatment.

(1) Select the appropriate composition of cement

The quality of cement is the primary concern in relation to the durability of the hardened cement paste. Only by increasing the quality of cement can it fundamentally improve its durability. In the use of cement, the determination of different clinker mineral compositions of cement should be based on the environment.

Resistance to sulfate attack can be improved by reducing the C_3A content and increasing the C_4AF content of the clinker. Figure 2.78 shows the relationship between C_3A content and sulfate resistance.

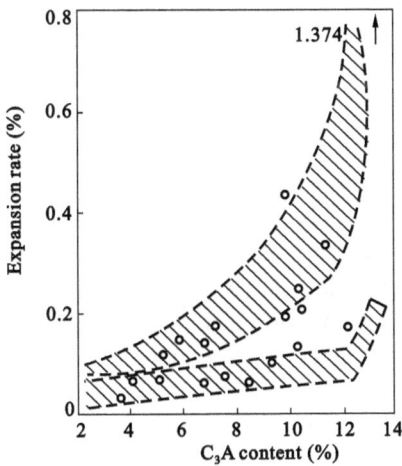

Figure 2.78: Relationship between C_3A content and sulfate resistance in cement.

The results show that under the action of sulfate, calcium ferrite hydrate formed by calcium iron aluminate or its solid solution with calcium sulfoaluminate is cryptocrystalline, precipitated in the form of gel and distributed more evenly. So its expansion is far smaller than Ettringite. Moreover, the rate of sulfate attack decreases with the decrease of $w(Al_2O_3)/w(Fe_2O_3)$. It has been proved by experiments that cement performance is the most stable when $w(Al_2O_3)/w(Fe_2O_3) < 0.7$ under sulfate attack and that the stability of cement is better with $w(Al_2O_3)/w(Fe_2O_3) = 0.7–1.4$; when $w(Al_2O_3)/w(Fe_2O_3) > 1.4$, the cement cannot exist stably.

The engineering practice shown in Figure 2.79 also proves the same conclusion. Since C_3S precipitates more $Ca(OH)_2$ during hydration, and $Ca(OH)_2$ is the main reason for dissolution erosion, it is possible to reduce the content of C_3S and increase the content of C_2S accordingly to improve the erosion and water resistance.

The amount of gypsum in cement also has an effect on the durability of cement. The cement with a reasonable gradation and the best gypsum content has significantly better corrosion resistance than other cements, mainly in the early hydration

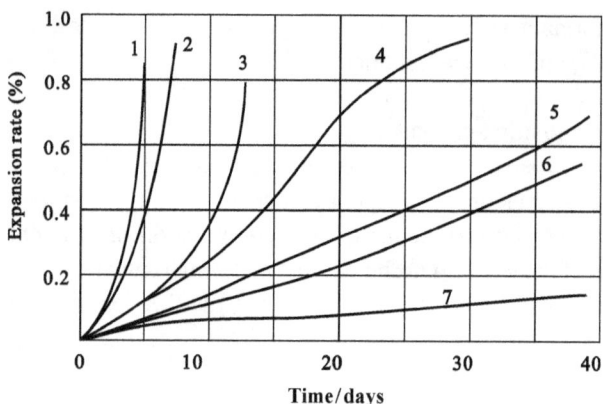

Figure 2.79: Expansion of cement and clinker components in 1.8% magnesium sulfate solution. Samples were water cured for 8 weeks before test.
1, 80% C_3S, 20% C_3A; 2, 80% C_2S, 20% C_3A; 3, ordinary Portland cement; 4, 80% C_3S, 20% C_4AF; 5, C_2S; 6, C_3S; 7, 80% C_2S, 20% C_4AF.

stage where C_3A rapidly dissolves with gypsum to form a large amount of Ettringite while the cement paste still is in a plastic stage and dissipate the expansion stress caused by Ettringite formation. This will not produce damage, but actually make the cement more compact. If the amount of gypsum is not enough to generate a large number of monosulfide calcium aluminate hydrate, it will react with the external corrosive media sulfate to generate secondary ettringite, resulting in expansion and destroy the hardened paste structure. However, it should be noted that the amount of gypsum should not be too high, so as to avoid the latter stage expansion.

(2) Addition of supplementary cementitious materials
The type and amount of SCMs in cement will affect the durability. In general, Portland cement mixed with volcanic material and granulated blast furnace slag can improve its erosion resistance. Because $Ca(OH)_2$ generated from hydration of clinker can combine with active SiO_2 to generate low-alkali hydration products, the reaction is as follows:

$$x Ca(OH)_2 + SiO_2 + H_2O \rightarrow xCaO \cdot SiO_2 \cdot H_2O$$

At a certain content of SCMs, the $n(CaO)/n(SiO_2)$ ratio in the calcium silicate hydrate formed is close to 1, so that the required limit concentration of lime is only 0.05–0.09 g/L, much lower than the average cement required to stabilize hydrated calcium silicate. Therefore, the rate of dissolution in freshwater will be significantly slowed down. At the same time, the concentration of aluminate hydrate can also be reduced, and the low-basicity calcium sulfoaluminate hydrate formed in the liquid phase with a low CaO concentration has a high solubility and a relatively slow crystallization without causing large stress due to expansion. With the addition of SCMs, the proportion of clinker is reduced, along with the content of C_3A and C_3S being correspondingly lower, which will also improve the erosion resistance. In addition, the formation of more gel enhances the compactness of hardened cement paste to prevent intrusion of aggressive media and thus enhances its resistance to erosion. As a result, the erosion resistance of pozzolanic cement and slag cement is better than Portland cement. The erosion resistance of slag cement is related to its slag content, Al_2O_3 content. Reduction in Al_2O_3 content will improve its erosion resistance accordingly.

It must be noted that the pozzolanic cement has poor frost resistance and atmospheric stability. In acidic or magnesium-containing solutions, even cement with pozzolanic materials cannot resist erosion. In addition, when mixed with clay-based pozzolanic materials, due to the high content of active Al_2O_3, its sulfate resistance may be worse, which should be taken seriously. In addition, strict control of the alkali content of cement to prevent or significantly inhibit the alkali-aggregate reaction is an effective way to improve the durability of cement.

Therefore, using different cement under different external environment conditions and selecting cement with higher strength under the same conditions is the fundamental way to improve the durability of cement paste structure.

Aggregates of better quality also help to improve durability. In general, clean aggregates should be selected with high strength and no harmful impurities such as clay, organic matter and sulfate (especially the content of active SiO_2), and the proper grading should be considered according to actual needs.

Mixing water in concrete also should not contain acid, alkali, sugar, oil and sulfate and other harmful ingredients of water. General drinking water such as tap water, river water, well water will suffice.

(3) Improve construction quality

Construction quality is also the key to the durability of concrete. In construction, stirring should be strengthened to prevent segregation. Improve the uniformity and fluidity of the concrete and make the mixture well fill the mold so as to reduce the internal voids. Strengthen the vibration to increase the compactness of the concrete, discharge the internal bubbles as much as possible, and reduce the visible voids and the large pores, especially connected pores to enhance its strength. Thereby enhancing its impermeability, and ultimately achieve the purpose of improving its durability.

In concrete construction, adding appropriate amount of water reducing admixture, air entraining agent and other additives is recommended according to actual needs. The use of water reducing admixture can ensure the ease of workability while greatly reducing the amount of mixing water and reducing the w/c ratio, thereby reducing the internal porosity of concrete and improving its strength. The use of air entraining agent can introduce a large number of 50–125 μm tiny bubbles to isolate the capillary pores within the paste structure. This will hinder the migration of water, reduce bleeding phenomenon. At the same time due to its large deformation capacity, the impermeability, frost resistance, cracking and other capabilities can be significantly improved.

Appropriate curing measures to maintain the appropriate temperature and humidity to ensure the normal hydration and hardening of cement to improve early strength also help to improve the durability of concrete.

(4) Surface treatment

Under special circumstances, the concrete structure is surface-treated to prevent its contact with the erosion medium and thus ensure its durability.

Surface treatment often consists of surface chemical treatment, surface impregnation and coating.

(i) Surface chemical treatment

Surface chemical treatment can increase the density of its surface. Commonly used carbonization method utilizes the reaction of $Ca(OH)_2$ in capillary pores with CO_2 in the air to the formation of $CaCO_3$, which will be deposited on its surface as an insoluble protective shell. This blocks the capillary pores, thereby improving resistance to freshwater leaching and sulfate erosion. However, this method is limited by the actual conditions, and at the same time the effect is not ideal.

Concrete surface treated with an aqueous solution of sodium silicate or fluorosilicate (such as magnesium fluorosilicate, zinc fluorosilicate) will generate extremely insoluble calcium fluoride and silicate gel etc. in the surface pores. This can improve the impermeability, but its effect is limited due to the formation of the thin protective layer. The use of infiltration technique to pressurize silicon tetrafluoride gas into the concrete surface to obtain a thicker protective layer is another effective method, but it is more expensive and not suitable for on-site construction.

(ii) Surface impregnation and coating

In the case of severe erosion environment, the best protective measure is to cut off the concrete from the erosion medium, that is, the concrete surface is coated with impermeable layer, such as asphalt, resin, silicone, asphalt paraffin and so on. In projects where chemical attack is particularly severe, surfacing materials such as ceramic tile, metal, plastic and composite surfacing materials may be used to prevent erosion by direct contact with the concrete and concrete.

In actual production, according to the specific conditions of the project and the needs of objective conditions, for different erosion, types, appropriate preventive measures to improve durability should be taken. If the erosion is weak, cement with a higher strength can be selected, and the w/c ratio can be decreased to below 0.55, and the amount of cement used is increased to not less than $300 \, kg/m^3$, mixing, vibrating should be enhanced. In an environment of strong erosion, you should choose the appropriate cement type, increase the amount of cement (such as $350 \, kg/m^3$ or higher) and further reduce the w/c ratio, or add a certain amount of admixture, or even use coated or impregnated materials for surface treatment. When there are several factors acting at the same time, we should also pay attention to the impact of compounded factors, distinguish the primary and secondary factors and seize the main factors, only in this way can we take effective measures to get better technical and economic results.

Problems

(Section 2.2)

2.1   How to define cement and what are the characteristics of cement?

2.2   How to classify cement and what is the most widely used type of cement?

2.3   What are the chemical and mineral compositions of clinkers in Portland cement?

2.4   What are the properties of the clinker phases in cement?

2.5   Why is it necessary to define the clinker ratios and how are the ratios derived?

2.6   Please briefly describe the physical meaning of KH, SM and IM.

2.7   What is the difference between the two concepts KH and LSF? Please explain why KH can be greater than 1 while LSF can be greater than 100?

2.8   Please explain how the three clinker ratios affect the burning process and the quality of clinkers.

2.9   Please derive the formula for KH when IM < 0.64.

2.10   How will an increasing SiO_2 content affect the burning and quality of clinkers? What is the appropriate content of SiO_2?

2.11   Calculate the value of a KH for a clinker with 20% SiO_2 when the amount of CaO combined with SiO_2 is 50.4%.

2.12   Derive the formula for KH in clinkers with mineralizers $CaSO_4$, CaF_2 and C_3S, C_2S, C_4AF, $C_{11}A_7 \cdot CaF2$, $C3A3 \cdot CaSO_4$.

2.13   Derive the empirical equation of KH:

$$KH = \frac{w(C_3S) + 0.8837w(C_2S)}{w(C_3S) + 1.3256w(C_2S)}$$

2.14   What are the methods to obtain the compound composition of clinker?

2.15   Briefly describe the main factors causing the inconsistency in the calculated values and measured ones in the compound composition of clinker.

(Section 2.3)

2.16 What are the ingredients for cement production and what are their specification?

2.17 What are the requirements for the coals in cement production? What measures to take if the coals fail to meet the requirements?

2.18 Which factors should be considered in the design of clinker composition?

2.19 What are the principles in the design of clinker composition?

2.20 The proportion of raw ingredients is calculated to be w(limestone):w(caly):w(iron powder) = 83.1:14.0:2.9 from Table 2.45. The coal ash content is 2.55% and the fly ash loss in the kiln is 5%. Please the amount of raw ingredients consumed for producing 1 kg clinker.

Table 2.45: Oxide composition of raw materials (mass fraction, %).

Raw ingredients	LOI	SiO_2	Al_2O_3	Fe_2O_3	CaO	MgO
Limestone	41.08	3.84	0.99	0.35	51.58	0.98
Clay	5.72	68.16	14.42	7.32	0.68	1.10
Iron powder	0.46	25.12	8.22	60.02	1.36	1.49

2.21 A precalciner uses four types of raw ingredients to produce clinkers with some of the known data shown in Tables 2.46 and 2.47. The required clinker ratios are KH=0.90, SM=2.6, IM=1.7, the heat consumption per unit clinker is 3,250 kJ/kg. Calculate the proportion of dry-based clinker.

Table 2.46: Chemical composition of raw materials (mass fraction, %).

Raw ingredients	LOI	SiO_2	Al_2O_3	Fe_2O_3	CaO	MgO	Total
Limestone	42.31	0.67	0.53	0.24	55.27	0.39	99.41
Clay	5.22	56.70	28.30	5.27	1.77	0.37	97.63
Iron powder	2.61	32.20	8.50	48.60	3.87	2.06	97.84
Sand stone	1.32	86.78	5.45	1.87	1.41	0.92	97.75
Coal ash	0.00	60.38	28.80	3.23	2.61	0.51	95.53

Table 2.47: Industrial analysis of coal.

w (M_{ad})	w (V_{ad})	w (A_{ad})	w ($F_{c,ad}$)	Q net,ad
0.60%	22.42%	28.56%	49.02%	20,930 kJ/kg

(Section 2.4)

2.22 What are the physical and chemical changes, by which, clinker is formed from raw materials?

2.23 How to accelerate the decomposition of $CaCO_3$?

2.24 What are the characteristics of solid-phase reaction and how to promote the degree and rate of reaction?

2.25 What are the conditions for the formation of C_3S? What are the factors affecting the formation of C_3S?

2.26 What measures can be taken during production to ensure a successful clinker calcination?

2.27 What are the benefits of rapid cooling in clinker?

2.28 Please explain the reason for clinker pulverizing.

2.29 Which cautions should be paid attention to when CaF_2 is used as a mineralizer?

2.30 How to avoid the abnormal setting phenomenon associated with the addition of a single or compound mineralizer?

2.31 What are the effects of alkali content on clinker quality?

2.32 What measures to take in the case of a high MgO content of the clinker in the precalciner?

2.33 What is the main function of a cyclone?

2.34 What is the main function of the pipe connecting the precalciner with cyclone?

2.35 Why can the precalciner substantially increase the production of cement?

2.36 How many reaction stages can be divided in the precalciner and what are the major reactions in each stage?

2.37 What is the clinker cooler and what is its function?

(Section 2.5)

2.38 Why are the blending materials needed in cement and what is the purpose for doing so?

2.39 What are the active and inert blending materials and what are the difference between them?

2.40 What are the specifications for blending materials with hydration activity used in cement? Please give an example.

2.41 What are the factors affecting the hydration activity of blending materials and how to evaluate the hydration activity?

2.42 What are the types of cement blending materials and what are their differences?

2.43 Please specify the cement type for each composition:
 (1) clinker 0%, slag 15%, gypsum 5% (SO_3 content is 3.0%);
 (2) clinker 47%, slag 40%, pozzolan 8%, gypsum 5% (SO_3 content is 3.0%);
 (3) clinker 55%, slag 10%, pozzolan 30%, gypsum 5% (SO_3 content is 3.0%);
 (4) clinker 55%, slag 10%, fly ash 30%, gypsum 5% (SO_3 content is 3.0%).

2.44 Please specify if the cement with the following composition belongs to slag cement
 (1) clinker 65%, slag 20%, pozzolan 10%, gypsum 5% (SO_3 content is 3.0%);
 (2) clinker 53%, slag 30%, pozzolan 5%, limestone 5%, kiln ash 2%, gypsum) 5% (SO_3 content is 3.0%);
 (3) clinker 20%, slag 75%, gypsum 5% (SO_3 content is 3.0%);
 (4) clinker 47%, slag 40%, kiln ash 8%, gypsum 5% (SO_3 content is 3.0%).

2.45 Why is gypsum added during clinker grinding and why should the content of gypsum be restricted at the same time?

2.46 Why is there an optimal gypsum dosage in cement production and how to determine this optimal dosage?

2.47 Why is the national standard needed for cement?

2.48 Why is there a prescribed limit for the SO_3 content in the national standard for cement?

2.49 What are the factors causing the unsoundness in cement and what are the measures to evaluate unsoundness? Why?

2.50 One cement is tested to be unsound and the sample is cracked and even failed. However, strength test on this cement is qualified. Explain why.

2.51 What are the similarity and difference between the open-loop grinding process and the closed-loop grinding process?

2.52 What are the factors affecting the grinding quality of cement?

(Section 2.6)
2.53 How does the C_3S hydration proceed? What are the hydration characteristics of C_3S?

2.54 According to C_3S hydration exothermic curve, what are the stages that can be identified for the hydration process? Briefly describe the reaction characteristics of each stage.

2.55 What are the similarities and differences in the hydration products between C_2S and C_3S?

2.56 How the hydration of C_3S is related to temperature?

2.57 What are the factors affecting the hydration of C_3A?

2.58 What are the hydration products of Portland cement?

2.59 What are the factors affecting the hydration of Portland cement and how?

2.60 What is the difference in hydration rate between four clinker phases and why?

2.61 What are the factors affecting the hydration rate of Portland cement and why?

2.62 What is the difference in hydration products and hydration rate between the pure phase and the corresponding constituent in the clinker and why?

2.63 How to regulate the hydration rate of cement?

2.64 What are the characteristics of slag cement compared to Portland cement and what leads to the difference?What are the techniques to counteract the negative effects associated with the slag cement application?

2.65 Why does slag cement have a lower early age strength but higher later age strength, compared to Portland cement?

2.66 What is the similarity in hydration between Portland cement with and without blending materials?

2.67 What are the processes of setting and hardening?

2.68 Please describe the morphology of C–S–H in hardened cement paste.

2.69 What the types of pores in hardened cement paste and what are their effects on the properties of hardened cement paste?

2.70 What are the types of water existing in hardened cement paste?

2.71 Why does cement sets and hardens and why does cement have strength?

2.72 No diffraction peaks are identified in C–S–H using XRD. Please explain why.

2.73 What are the solid phases produced in the hydration of Portland cement?

2.74 What are the negative effect of overdosing cement with gypsum and why? Why can the boiling method not be used to test the unsoundness associated with excessively high SO_3 content?

2.75 What are the factors affecting the rheological properties of cement paste and how do they affect?

(Section 2.7)

2.76 What are the differences among the true density, apparent density and stacking density of cement?

2.77 Why cannot the cement fineness be too low or too high?

2.78 What are the difference and relationship between standard consistency and fluidity?

2.79 What are the factors affecting the setting time of cement and why?

2.80 Which minerals are the deciding factors on the setting time of cement and why?

2.81 What are the factors affecting the setting rate of cement?

2.82 How does the false setting of cement occur and how to avoid it?

2.83 What is the difference between false setting and flash setting?

2.84 What are the factors affecting the optimum dosage of gypsum and why?

2.85 How do the four major clinker phases affect the strength?

2.86 What are the factors affecting strength of cement?

2.87 How does hardened cement paste gain strength?

2.88 Why is there volume change when cement hydrates?

2.89 What are the factors affecting the volume change in hardened cement paste and how do they affect it?

2.90 What are the hydration products in the mixture of C_3A and $CaSO_4 \cdot 2H_2O$ at three different ratios of $n(SO_3)/n(Al_2O_3)$ (5, 0.8, 0)?

2.91 Assuming C_3S hydrates by the following formula, what is the total porosity upon complete hydration in the hardened cement paste at a 0.5 w/c?

$$2C_3S + 6H = C_3S_2H_3 + 3CH$$

(Note: The densities of C_3S, $C_3S_2H_3$ and CH are 3.14, 2.44 and 2.23 g/cm^3, respectively)

2.92 After the addition of water, the strength of cement is lower after 3 days. Please explain.

2.93 Unsound cement will become sound after stockpiling for a period of time. Why?

2.94 Why does cement bleed? If cement has good water-retaining property, will the bleeding be reduced?

2.95 What are the main factors affecting the hydration rate of cement hydration?

2.96 What are the measures to reduce hydration heat?

2.97 Why does the strength of cement decrease at elevated temperatures?

2.98 How to improve the impermeability of cement?

2.99 Can all water in the hardened cement paste freeze? What types of water will have a negative effect on frost resistance?

2.100 How to improve frost resistance of cement?

2.101 What are the factors affecting cement durability?

2.102 What are the erosion types of cement? Please explain the mechanism.

2.103 What are the appropriate measures against erosion of cement?

2.104 What is alkali-aggregate reaction and how to avoid it during production and application of cement?

2.105 How to improve the durability of hardened cement paste?

3 Characteristic cement and special cement

In recent years, characteristic cements with special performances and purposes have been developed in many countries to meet the requirements of a variety of special projects, such as repair engineering, antiseepage engineering, hydraulic engineering, oil and gas well-cementing engineering, high-temperature engineering, decoration engineering and biological materials. These characteristic cements and special-purpose cements are all called special cements. In other words, special cements refer to the cements other than the general cements.

3.1 Aluminate and high-temperature-resistant cement

The high-temperature resistance concrete can be manufactured with lower energy consumption than sintered refractories and has the advantages including nonstereotypes, noncalcination, high integrity, site grouting, convenient construction and easy to meet the complex requirements of various industrial kilns. Thus, it can prolong the working life of furnace lining, improve furnace operation mechanization degree, speed up the construction progress and reduce labor intensity. Therefore, it has been developed rapidly with increasing varieties in recent years, leading to wide applications in various furnaces for oil, metallurgy, chemical, electric power, building materials and machinery industries. The most widely used cement in kilns and thermal equipment in China are aluminous cement, phosphate cement, refractory castable made of water glass and so on.

3.1.1 Aluminate cement

Aluminate cement is a cementitious material with hydraulicity made by grinding aluminate cement clinker mainly composed of calcium aluminate, as defined in the China standard of *Aluminate cement* (GB/T 201–2015). A proper amount of α-Al_2O_3 powders can be added when grinding the cement with Al_2O_3 content over 68% as required.

Aluminate cement is classified into four groups according to Al_2O_3 content (mass fraction, %):

CA50: $50\% \leq w(Al_2O_3) < 60\%$;
CA60: $60\% \leq w(Al_2O_3) < 68\%$;
CA70: $68\% \leq w(Al_2O_3) < 77\%$;
CA80: $77\% \leq w(Al_2O_3)$.

https://doi.org/10.1515/9783110572100-003

Table 3.1: Main composition of aluminate cement (mass fraction, %).

Type	$w(Al_2O_3)$	$w(SiO_2)$	$w(Fe_2O_3)$	$w(R_2O)$ [$w(Na_2O)+0.658w(K_2O)$]	$w(S)$	$w(Cl)$
CA50	≥50, <60	≤9.0	≤3.0	≤0.50	≤0.20	
CA60	≥60, <68	≤5.0	≤2.0			≤0.06
CA70	≥68, <77	≤1.0	≤0.7	≤0.40	≤0.10	
CA80	≥77	≤0.5	≤0.5			

The chemical composition (mass fraction, %) of aluminate cement should meet the requirements listed in Table 3.1.

The main mineral compositions of cement clinker are CA, CA_2, $C_{12}A_7$, C_2AS, and a minor amount of spinel (MA), perovskite ($CaO \cdot TiO_2$) and ferrite phase (it may be C_2F, or CF, Fe_2O_3 and FeO).

The main raw materials used in aluminate cement production are alumina and limestone. The production of aluminate cement by the melting method is widely used worldwide. In this method, the raw materials need not be grounded and low-grade bauxite can be used, but it has high heat consumption, high clinker hardness and high grinding power consumption. Rotary kilns, equipment for Portland cement production, are widely used in China, with low heat consumption and low grinding power consumption. However, high-quality and uniform raw materials are required by this method. Moreover, its calcining temperature range is narrow with only 50–80 °C and the burning temperature is generally about 1,300–1,380 °C. Low ash fuel is also used in the calcination to prevent ash falling into the material and affect the homogeneity resulting in large blocks and melting. In addition, it is necessary to control the flame temperature of the burning zone. Unlike Portland cement, a retarder such as gypsum is not necessary when grinding aluminate cement.

The alkalinity coefficient (C_m) and a ratio of aluminum to silicon coefficient (N) are mainly controlled for raw material batching of aluminate cement. The mathematical expressions of C_m and N are as follows:
Alkalinity coefficient:

$$C_m = \frac{w(CaO) - 1.87w(SiO_2) - 0.7[w(Fe_2O_3) + w(TiO_2)]}{0.55[w(Al_2O_3) - 1.7w(SO_3) - 2.53w(MgO)]} \tag{3.1}$$

Coefficient of ratio of aluminum to silicon:

$$N = \frac{w(Al_2O_3)}{w(SiO_2)} \tag{3.2}$$

In the formula, $w(CaO)$, $w(SiO_2)$, $w(Fe_2O_3)$, $w(TiO_2)$, $w(Al_2O_3)$, $w(SO_3)$ and $w(MgO)$ are the contents of CaO, SiO_2, Fe_2O_3, TiO_2, Al_2O_3, SO_3 and MgO in clinker (mass fraction, %).

Higher C_m value indicates more CA content, which leads to fast setting and high strength. On contrast, lower C_m value indicates less CA content but more CA_2 content, which leads to slow setting and low strength. C_m value of ordinary aluminate cement is commonly selected as 0.75 when produced by rotary kilns. It should be controlled within 0.8–0.9 and 0.55–0.65 for aluminate cement with high strength and good high-temperature resistance, respectively.

CA is the main mineral in aluminate cement with high hydraulic activity, normal setting and fast hydration hardening rate. CA_2 has low hydration and hardening speed, high late strength and good heat resistance, but low early strength.

Hydration products of CA are related to the ambient temperature. The main product is CAH_{10} when the ambient temperature is below 20 °C, then converted to C_2AH_8 and Al $(OH)_3$ gel when the ambient temperature is 20–30 °C and then transformed into C_3AH_6 and $Al(OH)_3$ gel when the ambient temperature is higher than 30 °C. The hydration of $C_{12}A_7$ is similar to that of CA. The crystallization of C_2AS is very slow, and the hydration of β-C_2S produces C–S–H gel. The metastable phase of CAH_{10} and C_2AH_8 is transformed into a stable C_3AH_6 phase gradually, and the rate is increased with temperature. At the same time, crystal transformation releases a large amount of free water, leading to a sharp increase in porosity. Thus, the long-term strength of aluminate cement (especially in a hot and humid environment) will be significantly decreased, even leading to project failure. Therefore, the use of aluminate cement in structural engineering is limited in many countries.

The setting time of aluminate cement should meet the requirements listed in Table 3.2. The addition of 15–60% Portland cement in aluminate cement would cause flash coagulation. This is because the Portland cement will precipitate $Ca(OH)_2$, and increase the pH value of liquid phase.

Table 3.2: Requirements for setting time of aluminate cement.

Cement type	Initial setting time (min)	Final setting time (h)
CA50, CA60-I, CA70, CA80	≥ 30	≤ 6
CA60-II	≥ 60	≤ 18

Aluminate cement is characterized by the rapid strength development. The strength values of various aluminate cements at different ages must not be lower than the values listed in Table 3.3. Another characteristic is that it can be well hardened at low temperatures (5–10 °C), but the strength will be decreased sharply under higher temperatures (>30 °C). Therefore, the used temperature of aluminate cement should not exceed 30 °C, and steam curing should not be adopted. Aluminate cement has good sulfate resistance because it does not precipitate $Ca(OH)_2$ during hydration. In addition, the hydration products contain $Al(OH)_3$ gel, which densifies the microstructure and improves permeability resistance. Moreover, it has good stability when subjected to carbonated water and dilute acid (pH value is not less than 4), but has a poor corrosion resistance to concentrated acid and alkali. At high temperatures (> 900 °C), solid-phase reaction will occur in aluminate cement with sintering binding, gradually replacing hydration binding. Thus, aluminate cement has a certain resistance to high temperatures. It can still maintain a high strength at high temperatures, especially for low-calcium aluminate cement. It can be used for all kinds of high-temperature furnace lining. At present, aluminate cement is mainly used for the preparation of heat-resistant expansive cement, self-stressing cement and thermal-resistance concrete, with a temperature range of 1,200–1,400 °C.

Table 3.3: The lower limit aluminate cement mortar strength at different hydration time.

Cement type	Compressive strength R_c (MPa)				Flexural strength R_f (MPa)			
	6 h	1 day	3 days	28 days	6 h	1 day	3 days	28 days
CA50	$\geq 20^a$	≥ 40	≥ 50	—	$\geq 3.0^a$	≥ 5.5	≥ 6.5	—
CA60		≥ 20	≥ 45	≥ 85	—	≥ 2.5	≥ 5.0	≥ 10.0
CA70		≥ 30	≥ 40	—	—	≥ 5.0	≥ 6.0	—
CA80		≥ 25	≥ 30	—	—	≥ 4.0	≥ 5.0	—

[a] The producer should provide results when users need.

3.1.2 Refractory castable of phosphoric acid and phosphate

The castable directly produced by mixing phosphoric acid powders and aggregates are called phosphoric acid refractory castable. Phosphate refractory castable is produced by using phosphate solution obtained from the reaction of phosphoric acid with metal oxides as a cementing agent to bind refractory powders and aggregates.

(1) Constituent material

(i) Cementing agent
Orthophosphate (H_3PO_4, phosphoric acid for short) is used for the preparation of refractory castable due to its stable properties. The phosphoric acid on the market usually is an aqueous solution of 85%, and it is often diluted to 40–60% concentrations. In the mass production and construction, the use of phosphoric acid as a cementing agent is more convenient.

Phosphoric acid is a strong acid that is corrosive to humans and clothing so that phosphate is often used as a binder in practice. Of which, two phosphate solutions are widely used: one is aluminum phosphate solution made by the reaction between phosphoric acid and aluminum hydroxide; and the other is magnesium phosphate solution made by the reaction between phosphoric acid and magnesium materials (MgO, Mg(OH)$_2$, periclase and lightly burned magnesia). In addition, there are calcium phosphate, chromium phosphate and other phosphate solutions.

(ii) Refractory powders and aggregates

Refractory powders and aggregates for phosphoric acid and phosphate castable include refractory clay, bauxite clinker, cement clinker, magnesia clinker and siliceous corundum, mullite, zircon, alumina, silicon carbide, chromium slag and so on. Lightweight aggregates include expanded perlite, alumina hollow ball and so on.

The binding properties of phosphate and phosphate solutions are closely related to the cationic species of oxides in aggregates. Reacting with weak alkaline oxides results in high-binding performance; when reacting with strong alkaline oxides, the fast reaction speed results in a porous structure. The reaction with neutral or acidic oxides (SiO$_2$, Al$_2$O$_3$, ZrO$_2$ and Cr$_2$O$_3$) generally does not occur under normal temperature. The heat treatment or adding accelerator is needed to obtain enough strength.

(2) Preparation

The common mixture ratios of phosphoric acid and phosphate refractory castables are 30–40% of refractory powders and 60–70% of refractory aggregates. Besides, 11–14% phosphoric acid or phosphate solution and 2–3% aluminate cement as an accelerator are added. The powders are required to pass 0.08 mm^2 hole sieve, with a residue R less than 20%, and the aggregates sizes d are smaller than 15 mm.

When preparing castable, the powders and aggregates need to be first mixed evenly before adding about 3/5 cementing agent for mixing. The mixture is kept for a period of time for a full reaction between metal impurities such as iron in aggregates and powders with cementing agent to release hydrogen, thus avoiding expansion and loose structure after formation. Then the accelerator is added and mixed evenly, followed by adding the remaining binder and mixing evenly before casting. After 1 day removal of shuttering, curing in the air for 3 days is needed before use.

(3) Product characteristics

Phosphoric acid and phosphate castable combine the advantages of cement, ceramic and refractory materials. It not only has the cement properties such as the plasticity, low-temperature hardening, free to modulate the properties with admixtures, but also has a high-temperature ceramic tendency and high-temperature resistance similar to that of ceramic and castable. It also has its own characteristics, such as high fire resistance (up to 1,500–3,000 °C), good resistance to rapid cold and hot cycle. It can withstand repeated the hot and cold cycle with −30–2,000 °C without damage. It also has good high-temperature wear resistance and a strong binding force. Phosphoric acid and phosphate refractory castable can be used as lining materials for various industrial kilns and thermal equipment. The castable is superior to other types of cement, refractory castable and refractory bricks.

3.1.3 Refractory castable of water glass

Water glass refractory castable is an air-hardening cementitious refractory, which is made of water glass as cementing agent and various refractory aggregates and powders mixed in a certain proportion.

Sodium silicate glass ($Na_2O \cdot nSiO_2$) is used for making water glass castable refractory and its modulus n is generally 2–3. The mix proportion of water glass castable refractory should be determined according to the requirements of the project and raw material conditions. The general ratio is as follows: the dosage of water glass is 13–16% of the total amount of powders and aggregates; the density of castable is generally no higher than 400 kg/m³; the powder dosage is 150–200% of water glass dosage and the finer the powders, the higher the quality (residue from 0.08 mm² hole sieve R should not be more than 30%). Aggregates account for 70–75% of the total amount of powders and aggregates, and the rate of sand S_p is 0.45–0.55. Moreover, the aggregates and their gradation should be reasonably chosen according to the requirements of the specific project.

With different kinds of powders and refractory aggregates, the high-temperature strength and load softening temperature of the castable are also different, so they are used at different temperatures. Table 3.4 lists the loading softening temperatures of castables when different powders are used.

Table 3.4: Softening temperature of castables with different powders.

Powder type	Load softening temperature (°C)		Softening temperature range (°C)
	Start point	4% deformation	
Refractory clay clinker powder	1,010	1,090	80
Quartz powder	820	870	50
Magnesia powder	1,050	1,590	540

Water glass refractory castable is resistant to all kinds of organic acids. Therefore, it is generally used for high-temperature-resistant engineering, which requires resistance to organic acids and organic acid solutions below 1,000 °C.

Using water glass and acid-resistant aggregates, it can be made into an acid-resistant refractory castable, with a poor resistance to dilute acid but a good resistance to concentrated acid. The commonly used acid resistance aggregates include refractory clay bricks, pyrophyllite, andesite and quartz.

3.2 Fast hardening and high-strength cement

Compared with the traditional Portland cement, the fast hardening and high-strength cement are characterized by rapid setting and hardening, high strength, short curing time and short construction period. With the development of modern construction engineering, it is more and more widely used in the engineering such as military repair project, underground engineering, rapid construction project, tunnel engineering and so on. The use of fast hard high-strength cement has a series of advantages:

(i)   When preparing the concrete with the same strength, the cement amount can be saved by 20–50% with the use of fast hardening high-strength cement.
(ii) The high-strength prefabricated parts can be made to reduce the sectional size of components and the amount of materials used and the deadweight and to correspondingly reduce the project cost.
(iii) Because of fast hardening, it can be cured without steam to shorten the time of dismantling formwork, reduce the amount of formwork, shorten the storage time of components and reduce the area of a workshop, so as to reduce the cost.
(iv) The template construction technology could be replaced by anchor and spray technology with the use of fast hardening and high-strength cement so that the construction cost could be greatly reduced.

At present, some of the fast hardening and high-strength cement is able to harden within 5–20 min, and 1 and 2 h compressive strengths R_c are 10–30 and 20–40 MPa, respectively. One day strength can reach up to 75–90% of 28 days strength. The cement strength has reached more than 100 MPa to meet the needs of a variety of special projects. In terms of varieties, more than 10 varieties of fast hard high-strength cement have been developed in China.

3.2.1 Fast hardening Portland cement

Fast hardening Portland cement is a cementitious material with hydraulicity made by grinding Portland cement clinker and a proper amount of gypsum, whose strength grade is expressed by 3 days compressive strength R_c. It is called rapid hardening cement.

The method of produce rapid hardening Portland cement is almost the same as that use for the production of Portland cement. However, it needs stricter control of production process conditions with less harmful ingredients in raw materials and rational design of mineral compositions. Generally, the C_3S and C_3A contents are intended to be higher, and the C_3S and C_3A contents are 50–60% and 8–14%, respectively. According to the raw material situations, an increase of C_3S content without increasing C_3A content is acceptable. If the C_3S and C_3A contents are increased simultaneously, the clinker liquid viscosity is relatively large, which is unfavorable to the formation of C_3S. When using mineralizer, the C_3S content in clinker can be appropriately increased, but not more than 70%.

During the production of fast hardening Portland cement, raw materials are required to have a uniform composition and large surface area. The general requirement of raw materials is that residue of 0.08 mm² hole sieve R is not more than 5%. The clinker requires rapid cooling to avoid C_3S decomposition and crystalline transformation of C_2S.

The specific surface area of cement has a great influence on the cement strength (especially the early strength). Table 3.5 lists the compressive strengths R_c of the same clinker when grinding to different specific surface areas S.

Table 3.5: Effect of specific surface area on compressive strength of cement.

Specific surface area S (m²·kg⁻¹)	Compressive strength R_c (MPa)			
	1 day	3 days	7 days	28 days
298	10.5	27.9	38.6	43.6
464	25.3	43.7	47.8	53.3
630	26.1	44.0	52.6	61.2

When the specific surface area of cement is increased, and the cement size is homogeneous, the cement strength is higher. To accelerate the hardening rate, the specific surface area of the rapid hardening Portland cement is generally controlled at 330–450 m²/kg.

A proper increase of gypsum content is also one of the important measures for the production of fast hardening Portland cement. This ensures enough bauxite formation before cement hardening, which is a benefit for developing the cement strength so that the SO_3 content in the rapid hardening Portland cement is generally controlled with 2.5–3.7%.

Due to the large specific surface area of fast hardening Portland cement, it is easy to be weathering during storage and transportation, and the average storage period is no more than 30 days.

Rapid hardening Portland cement has a high early strength. Its 1 day compressive strength is 30–35% of 28 days compressive strength, and the late strength continues to grow. It has a normal setting time; generally, the initial setting time is 2–3 h and the final setting time is 3–4 h. It has a higher hydration heat and larger early dry-shrinkage. Because its hardened structure is compact, the impermeability and frost resistance is better than that of ordinary cement. It also has a good low-temperature performance. At 10 °C, strengths at different ages are significantly higher than that of ordinary cement. Fast hardening Portland cement is suitable for use under steam curing conditions.

Fast hardening Portland cement is mainly used for rush repair works, military engineering and prestressed reinforced concrete parts. Quick hardening Portland cement is suitable for the preparation of dry hard concrete, and the water–cement ratio (w/c) should be controlled below 0.4.

3.2.2 Fast hardening sulfoaluminate cement

China national standard *Sulfoaluminate cement* (GB 20472–2006) defines that as follows: sulfoaluminate cement is calcined with the proper compositions of raw materials. It is a hydraulic cementitious material with a high early strength made by grinding clinker mainly composed of anhydrous calcium sulfoaluminate and dicalcium silicate, limestone and a proper amount of gypsum. Sulfoaluminate cement is divided into quick hardening sulfoaluminate cement, low-alkalinity sulfoaluminate cement, self-stressing sulfoaluminate cement and so on.

Fast hardening sulfoaluminate cement is a cementitious material with a high early strength prepared by a proper proportion of sulfoaluminate cement clinker, a less amount of limestone and a proper amount of gypsum. Limestone content should be no more than 15% of the cement mass.

Low-alkalinity sulfoaluminate cement is a cementitious material with low alkalinity that is ground by a proper composition of sulfoaluminate cement clinker, a larger amount of limestone and a proper amount of gypsum (see Section 3.10).

Self-stressing sulfoaluminate cement is an expansive cementitious material ground by a proper composition of sulfoaluminate cement clinker and proper amount gypsum (see Section 3.4).

The content of $w(Al_2O_3)$ for manufacturing rapid hardening sulfoaluminate cement clinker should be no less than 30%, the content of $w(SiO_2)$ should be no higher than 10.5% and the 3 days compressive strength of clinker R_c should not be less than 55 MPa.

The strength grade of quick hardening sulfoaluminate cement is expressed by 3 days compressive strength R_c, and divided into four strength grades: 42.5, 52.5, 62.5 and 72.5.

(1) Technical indexes
Specific surface area S of rapid hardening sulfoaluminate cement must not be less than 350 m²/kg. The initial setting time must not be less than 25 min (it can be changed

when the user requires), and the final setting time must be less than 180 min (it can be changed when users require). The strengths at different ages for different strength grades Portland cement must be no less than the values listed in Table 3.6.

Table 3.6: The lower limit of fast hardening sulfoaluminate cement mortar strength at different hydration time.

Strength grade	Compressive strength R_c (MPa)			Flexural strength R_f (MPa)		
	1 day	3 days	28 days	1 days	3 days	28 days
42.5	≥30.0	≥42.5	≥45.0	≥6.0	≥6.5	≥7.0
52.5	≥40.0	≥52.5	≥55.0	≥6.5	>7.0	≥7.5
62.5	≥50.0	≥62.5	≥65.0	≥7.0	≥7.5	≥8.0
72.5	≥55.0	≥72.5	≥75.0	≥7.5	≥8.0	≥8.5

(2) Production process

(i) Raw materials for cement production

The cement raw materials are low-grade alumina materials (alumina), calcareous raw materials (limestone) and gypsum. Their chemical composition requirements are as follows: $w(CaO) >50\%$, $w(MgO) <1.5\%$ in limestone; $w(Al_2O_3) >55\%$, $w(SiO_2) <25\%$ in bauxite; $w(SO_3) >38\%$ in gypsum and $w(A_{ad}) <25\%$ in coal ash.

(ii) Composition and rate of cement clinker

The cement clinker minerals are mainly anhydrous calcium sulfoaluminate $3CaO \cdot 3Al_2O_3 \cdot CaSO_4$ (abbreviated as $C_4A_3\bar{S}$), dicalcium silicate (β-C_2S) and a small amount of $CaSO_4$, perovskite and iron phase. $C_4A_3\bar{S}$ content is 55–75%, β-C_2S content is 15–35% and both sum up to 90% or more in the normal clinker composition. As the main minerals are C_4A_3 and β-C_2S, fast hard sulfoaluminate cement is usually represented by $C_4A_3\bar{S}$-β-C_2S type.

Two rates are mainly controlled for the manufacture of $C_4A_3\bar{S}$-β-C_2S-type cement clinker, that is, the alkalinity coefficient (C_m) and sulfur and aluminum ratio coefficient (P), calculated as follows:

$$C_m = \frac{w(CaO) - 0.7[w(SO_3) + w(Fe_2O_3) + w(TiO_2)]}{1.87w(SiO_2) + 0.55w(Al_2O_3)} \tag{3.3}$$

$$P = \frac{w(Al_2O_3)}{w(SO_3)} \tag{3.4}$$

Alkalinity coefficient (C_m) is generally controlled within 0.9–1.0, and sulfur and aluminum ratio coefficient (P) is usually controlled within 3.5–4.0.

The mineral composition of the cement clinker is calculated as follows:

$$w(C_4A_3\bar{S}) = 1.99w(Al_2O_3) \tag{3.5}$$

$$w(C_2S) = 2.87w(SiO_2) \tag{3.6}$$

$$w(C_2F) = 1.7w(Fe_2O_3) \tag{3.7}$$

$$w(CT) = 1.7w(TiO_2) \tag{3.8}$$

$$w(f-SO_3) = w(SO_3) - 0.131\,w(C_4A_3\bar{S}) \tag{3.9}$$

$$w(f-CaO) = w(CaO) - \left\{ \begin{array}{l} 0.55\,w(Al_2O_3) + 1.87w(SiO_2) + 0.7[w(Fe_2O_3) \\ + w(TiO_2) + w(SO_3)] \end{array} \right\} \tag{3.10}$$

where $w(CaO)$, $w(SiO_2)$, $w(Al_2O_3)$, $w(Fe_2O_3)$, $w(TiO_2)$, $w(SO_3)$ and $w(C_4A_3\,\bar{S})$ are the contents of CaO, SiO$_2$, Al$_2$O$_3$, Fe$_2$O$_3$, TiO$_2$, SO$_3$ and C$_4$A$_3$ $\bar{S}$, respectively (mass fraction, %).

Table 3.7 lists the chemical compositions and rates of sulfoaluminate cement clinker produced by two China special cement plants.

Table 3.7: Chemical composition and rate values of sulfoaluminate cement clinker.

Factory name	Chemical composition (mass fraction, %)								Rate value	
	LOI	SiO$_2$	Al$_2$O$_3$	Fe$_2$O$_3$	CaO	MgO	R$_2$O	SO$_3$	C_m	P
A plant	0.50	9.80	30.50	2.00	43.50	2.50	1.60	8.40	1.02	3.63
B plant	0.15	7.58	35.11	1.57	41.93	1.71	1.82	9.81	0.98	3.58

(iii) Cement clinker calcination

The production of $C_4A_3\,\bar{S}$-β-C$_2$S cement clinker is usually by dry process rotary kiln. The mesh residue of raw material R is usually controlled less than 10% when passing 0.08 mm^2 hole sieve.

The following reactions occur as the material temperature increases during the calcination process:

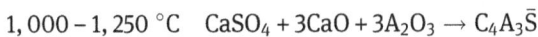

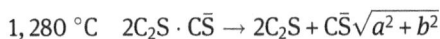

$$900 - 1,000\,°C \quad CaCO_3 \rightarrow CaO + CO_2$$

$$1,000 - 1,250\,°C \quad CaSO_4 + 3CaO + 3A_2O_3 \rightarrow C_4A_3\bar{S}$$

$$CaSO_4 + 4CaO + 2SiO_2 \rightarrow 2C_2S \cdot C\bar{S}$$

$$1,280\,°C \quad 2C_2S \cdot C\bar{S} \rightarrow 2C_2S + C\bar{S}\sqrt{a^2 + b^2}$$

The clinker calcination temperature is 1,250–1,350 °C, and should not exceed 1,400 °C; otherwise, CaSO$_4$ and C$_4$A$_3$ $\bar{S}$ would decompose. During calcination, reducing atmosphere also needs to be prevented. Under reducing atmosphere, CaSO$_4$ would decompose into CaS, CaO and SO$_2$. As the calcining temperature is low, and clinker formation is mainly solid-phase reaction, the liquid-phase amount is little; thus, ring and agglomerate are hard to happen, clinker abrasion is also better and heat consumption is lower. However, the calcination temperature should not be too low. Otherwise, excessive f-CaO would be produced, leading to cement fast setting, unable decomposition of $2C_2S \cdot C\bar{S}$ and less content of C_2S.

(3) Cement hydration

$C_4A_3\bar{S}$–β-C_2S cement hydration would occur with the following reactions:

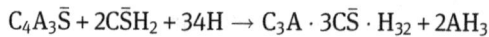

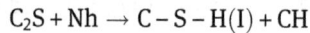

$$C_4A_3\bar{S} + 2C\bar{S}H_2 + 34H \rightarrow C_3A \cdot 3C\bar{S} \cdot H_{32} + 2AH_3$$

$$C_4A_3\bar{S} + 18H \rightarrow C_3A \cdot C\bar{S} \cdot H_{12} + 2AH_3$$

$$C_2S + Nh \rightarrow C - S - H(I) + CH$$

$$3CH + AH_3 + 3\,C\bar{S}H_2 + 20H \rightarrow C_3A \cdot 3C\bar{S} \cdot H_{32}$$

$C_4A_3\bar{S}$ reacts with gypsum and forms Ettringite and $Al(OH)_3$ gel. When the gypsum content is limited, low-sulfur type calcium sulfoaluminate is generated. C_2S sintered at the lower temperature has a fast hydration speed, which generates C–S–H (I) gels; AH_3 and C–S–H (I) fill between the hydrated calcium sulfoaluminate, and densified reinforced cement stone structure, resulting in a high early strength. Fast hardening without shrinkage, microexpansion, expansion and self-stress cement can be prepared by changing the amount of gypsum added in the cement.

(4) Properties and applications

$C_4A_3\bar{S}$–β-C_2S cement has a slow setting time, and the intervals between the initial setting and the final setting time are short. The initial setting time is generally 8–60 min and final setting time is generally 10–90 min. Retarder can be added to slow the cement settling time. This type of cement has a stable late stage strength with gradual increment. It has good low-temperature performance and works well without any special measures when the temperature is higher than –5 °C. However, it has a poor thermal stability, when the temperature is higher than 100 °C, hydration products like Ettringite, AH_3 gel and C–S–H (I) gel begin to dehydrate and strength decreases gradually. Furthermore, the strength decreases sharply when the temperature is higher than 150 °C. This type of cement has a small shrinkage in the air, good frost resistance and impermeability. It is a low-alkali cement with a pH value of 9.8–10.2.

Fast hardening sulfoaluminate cement can be used for emergency repair works, such as seams, plugging, anchor spray, repair aircraft runways, roads and so on. It is suitable for winter construction works, underground works, the preparation of expansion cement, self-stress cement, glass fiber mortar and so on, but are not suitable for use in the environment over 100 °C.

3.2.3 Fast hardening aluminoferrite cement

Fast hardening aluminoferrite cement is defined as the high early strength hydraulic binder, which is ground by cement clinker, limestone and a proper amount of gypsum. In which, cement clinker is calcined by a proper chemical composition of raw materials and its main compositions are anhydrous calcium sulfoaluminate, ferrite phase and dicalcium silicate. The clinker and gypsum contents should not be less than 85% and limestone content should not be more than 15%. Its strength grade is represented by 3 days compressive strength, with four grades of 42.5, 52.5, 62.5 and 72.5.

(1) Chemical compositions and mineral components of clinker
The main chemical compositions of the clinker are Al_2O_3, SiO_2, CaO, Fe_2O_3 and SO_3, the main minerals are $C_4A_3\bar{S}$, C_2S and C_4AF and their content ranges are listed in Table 3.8.

Table 3.8: Chemical composition and mineral composition of fast hardening aluminoferrite cement (mass fraction, %).

Al_2O_3	SiO_2	CaO	Fe_2O_3	SO_3	$C_4A_3\bar{S}$	C_2S	C_4AF
25–30	6–12	43–46	5–12	5–10	35–55	15–35	15–30

(2) Characteristics of the production process

(i) Raw materials and fuel choice
Three kinds of raw materials – limestone, bauxite and gypsum – are used for cement ingredients mix.

CaO and MgO contents in the limestone need to be $w(CaO) > 35\%$ and $w(MgO)$ <1.5%, respectively. $w(Al_2O_3)$ and $w(Fe_2O_3)$ contents need to be 45–55% and 15–20%, respectively. $w(MgO)$ and $w(SO_3)$ contents in gypsum are <3% and >38%, respectively. The following parameters need be controlled for selected fuel coal: $w(V_{ad}) =$ 25–32%, $w(A_{ad}) \leq 12\%$ and $Q_{net, ad} \geq 27{,}200$ kJ/kg.

(ii) Cement clinker calcination
The main calcining equipment for this type of cement is rotary kiln and the burning temperature is generally controlled to 1,250–1,350 °C, which is 100–150 °C lower than that of Portland cement. The cement clinker is mainly formed by solid-phase reaction without worrying about hanging kiln skin, but the reducing atmosphere needs to be avoided to prevent gypsum decomposition. Clinker color is often black, which is similar to that of Portland cement, but different from the sulfoaluminate cement clinker, which is in gray color.

(iii) Cement grinding

In the cement grinding process, limestone addition is limited by 15% and gypsum dosage is generally 15%. Its specific surface area S should be controlled over 350 m²/ kg, typically between 350 and 400 m²/kg.

The cement setting time is generally between 25 and 180 min, that is, the initial setting time is not earlier than 25 min, and the final setting time is not later than 180 min.

(3) Cement strength

Both the early and late strength of fast hardening aluminoferrite cement is high. Its typical mineral composition and strength development rule are listed in Table 3.9.

Table 3.9: Typical clinker minerals composition and cement mortar strength.

Serial number	Main mineral clinker composition (wt%)				Cement compressive strength (MPa)			
	$C_4A_3\bar{S}$	C_4AF	C_2S	Other	12 h	1 day	3 days	28 days
1	32	19	41	8	34.5	39.5	41.7	46.2
2	43	22	23	12	46.9	51.5	55.9	67.2

(4) Main characteristics and uses

(i) High early strength and high strength

Fast hardening aluminoferrite cement has not only a high early strength but also has a growing in late strength with a qualified setting time. The 12h–1d compressive strength is up to 35–50 MPa, and the strength will increase with an increase in the curing period strength and the maximum strength will be up to 100 MPa, which is obtained from results of 5–10 years long-term strength.

(ii) Excellent frost resistance

The early strength is 5–8 times that of the Portland cement when used under low temperatures of 0–10 °C. When used at low temperatures of −20 to0 °C, normal construction can be maintained with concrete mold temperature above 5 °C by adding a few antifreeze agents. About 3–7 days strength can reach up to 70–80% of the designed strength.

(iii) Outstanding corrosion resistance

Fast hardening aluminoferrite cement has an excellent corrosion resistance ability to water, chloride, sulfate, especially their complex salts ($MgSO_4$+NaCl) and on. It is superior to rapid hardening cement, aluminate cement and even high sulfate-resistance cement.

(iv) High impermeability

Due to the high alkalinity in cement hydration liquid (pH > 12), it forms a passivation film on the surface of rebar, which is similar to that of Portland cement-reinforced concrete. Thus, there is no corrosion to steel, and cement has better durability and strong antipermeability performance.

Adding 10% silica fume, 20% fly ash and 2% superplasticizers, high-strength concrete can be prepared, resulting in 28 days compressive strength R_c higher than 100 MPa.

Fast hardening aluminoferrite cement is suitable for use in winter construction projects, fast repair and construction projects, the preparation of shotcrete and the production of prefabricated components.

3.2.4 Rapid hardening fluoroaluminate cement

Rapid hardening fluoroaluminate cement is made by grinding clinker with aluminum fluoride ($C_{11}A_7 \cdot CaF_2$) as the main mineral composition and gypsum. In which, the cement clinker is calcined by proportional bauxite, limestone and fluorite (or plus gypsum). Quick-setting fast hardening cement, molding sand cement in China and ultrafast hardening cement outside China belong to this type.

(1) Mineral compositions

The main minerals in rapid hardening fluoroaluminate cement are calcium fluoroaluminate, Alite phase, Belite phase and tetracalcium aluminoferrite solid solutions. The compositions fluctuate in a wide range and the compositions of several fast hardening aluminoferrite cements are listed in Table 3.10.

Table 3.10: Several rapid hardening fluoroaluminate cements' compositions (mass fraction, %).

Samples	Mineral composition				
	$C_{11}A_7 \cdot CaF_2$	C_3S	C_2S	C_4AF	C_2F
1	20.6	50.4	1.7	4.7	–
2	19.2	52.1	–	5.2	–
3	26.0	55.1	6.7	4.6	–
4	71.3	–	20.0	–	2.2
5	72.4	–	17.6	–	3.0

In Table 3.10, samples 1, 2 and 3 are the type of C_3S-$C_{11}A_7 \cdot CaF_2$ cement, and samples 4 and 5 are the type of β-C_2S–$C_{11}A_7 \cdot CaF_2$ cement. When good quality limestone is used as the raw material, low-grade bauxite can be used to produce

C_3S-$C_{11}A_7$·CaF_2-type cement, but high-grade bauxite or other raw materials with a high aluminate content are required to be used to produce β-C_2S–$C_{11}A_7$·CaF_2-type cement.

(2) Ingredients

The mineral composition of cement clinker is designed first according to the requirements of cement performance. Batching of ingredients is carried out by the method of accumulative cut-and-trial after the chemical composition of clinker is calculated. The amount of CaF_2 needs to be excess of 1% besides satisfying the required amount for $C_{11}A_7$·CaF_2 formation and the reaction with Na_2O, K_2O; since partial fluorine will volatilize during calcination.

The clinker mineral compositions are calculated as follows:

(i) For C_3S-$C_{11}A_7$·CaF_2-type cement:

$$w(C_4AF) = 3.04\,w(Fe_2O_3) \tag{3.11}$$

$$w(C_{11}A_7 \cdot CaF_2) = 1.97[w(Al_2O_3) - 0.64w(Fe_2O_3)] \tag{3.12}$$

$$w(C_3S) = 4.07w(CaO) - 3.47w(Fe_2O_3) - 3.52w(Al_2O_3) - 7.60w(SiO_2) \\ 2.85w(SO_3) + 3.86\,w(Na_2O) + 2.42w(K_2O) \tag{3.13}$$

$$w(C_2S) = 2.87[w(SiO_2) - 0.261\,w(C_3S)] \tag{3.14}$$

The theoretical requirement of CaF_2 is as follows:

$$w(CaF_2) = 0.11[w(Al_2O_3) - 0.64w(Fe_2O_3)] + 0.83\,w(K_2O) + 1.26\,w(Na_2O) \tag{3.15}$$

(ii) For β-C_2S–$C_{11}A_7$·CaF_2-type cement:

$$w(C_{11}A_7 \cdot CaF_2) = 1.97[w(Al_2O_3) - 2.53w(MgO)] \tag{3.16}$$

$$w(C_2S) = 2.87w(SiO_2) \tag{3.17}$$

$$w(C_2F) = 1.70w(Fe_2O_3) \tag{3.18}$$

$$w(MA) = 3.53\,w(MgO) \tag{3.19}$$

$$w(CT) = 1.70w(TiO_2) \tag{3.20}$$

The theoretical requirement of CaF_2 is as follows:

$$w(CaF_2) = 0.11[w(Al_2O_3) - 2.53(Fe_2O_3)] + 0.83\,w(K_2O) + 1.26\,w(Na_2O) \tag{3.21}$$

where $w(CaO)$, $w(Al_2O_3)$, $w(Fe_2O_3)$, $w(SO_3)$, $w(Na_2O)$, $w(K_2O)$, $w(MgO)$ and $w(TiO_2)$ are the amounts of each oxide in the clinker, such as CaO, Al_2O_3, Fe_2O_3, SO_3, Na_2O, K_2O, MgO and TiO_2 (mass fraction,%).

(3) Calcining and grinding of cement clinker

Rapid hardening fluoroaluminate cement clinker is mainly calcined by rotary kilns. Since the raw materials contain fluorine, the calcination temperature is decreased. Raw materials have higher reaction ability when compared with Portland cement. CaO is well absorbed to form calcium silicate minerals in the temperature ranges of solid-phase reactions of Portland cement raw materials. When fluorine content is increased by 2%, the temperature that CaO is completely absorbed is reduced by about 50 °C. The calcination temperature is lightly affected by lime saturation factor KH, silicon rate SM and aluminum rate IM. When the CaF_2 content $w(CaF_2)$ is more than 4%, C_3S can be formed quickly at 1,150–1,200 °C. The reactions are as following:

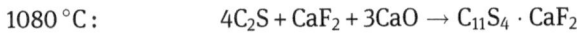

1080 °C: $4C_2S + CaF_2 + 3CaO \rightarrow C_{11}S_4 \cdot CaF_2$

1180 °C: $C_{11}S_4 \cdot CaF_2 \rightarrow 3C_3S + C_2S + CaF_2$

$C_{12}A_7 + CaF_2 \rightarrow C_{11}A_7 \cdot CaF_2 + CaO$

The calcination temperature of cement clinker is generally controlled at 1,250–1,350 °C, and the flame temperature is controlled at 1,350–1,400 °C. Bulk and ring are easily formed under high temperatures. Calcination is uncompleted if the temperature is too low. Clinker must be rapidly cooled. The requirement of cement grinding fineness is high, and generally, the specific surface area S is controlled at 500–600 m^2/kg. Grinding can be carried out with a closed-circuit grinding system by adding a proper amount of gypsum, whose amount must be determined by tests.

(4) Cement hydration

Rapid hardening fluoroaluminate cement hydrates quickly. It can hydrate to form hydrated calcium aluminates including CAH_{10}, C_2AH_8, C_2AH_{13}, C_4AH_{19} and AH_2 $\overline{F}$($\overline{F}$ is CaF_2) in a few seconds. Within several minutes, hydrated calcium aluminates react with $Ca(OH)_2$ and $CaSO_4$ generated from calcium silicate hydration to form low-sulfur type calcium sulfoaluminate and Ettringite. Its main hydration products are the same as Portland cement, which are C–S–H gel and $Ca(OH)_2$. The hardened cement is made of Ettringite as a skeleton, filled with C–S–H gel and alumina. It can quickly reach a high degree of densification.

(5) Properties and applications

Rapid hardening fluoroaluminate cement sets quickly with the initial setting in a few minutes. The initial setting and the final setting time interval is very short, and the final setting time is less than 30 min. Tartaric acid, citric acid and boric acid can be used as retarders to adjust the setting time. It hardens quickly within 5–10 min. The compressive strength R_c after 2–3 h is up to 20 MPa. The 4 h strength of concrete is up to 15 MPa. It also has good low-temperature hardening properties, 6 h strength is up to 10 MPa and 1 day strength is up to 30 MPa.

This type of cement can be used for repair works. It can be made into anchor spray cement. The hydration product of Ettringite will decompose under high temperatures to make sand mold collapse, and it is easy to clear sand without producing harmful gases. Therefore, this type of cement can be made of molding sand cement in the casting industry.

3.2.5 Rapid hardening and high-strength aluminate cement

Rapid hardening and high-strength aluminate cement is made of aluminate cement clinker and has the characteristics of rapid hardening and high strength. The relevant aluminate cement content could be found in Chapter 3.

This type of cement is hydraulic cementitious material ground by clinker with the main composition of calcium aluminate and a proper amount anhydrite. It has fast hardening rate and high strength.

(1) The component materials
(i) Aluminate cement clinker;
(ii) Anhydrite. Anhydrite quality should be consistent with *Natural Gypsum* (GB/T 5483–2008) requirement.

Furthermore, when grinding cement, grinding aids may be added without impairing the cement performance. Its amount should not exceed 1% of the cement mass.

(2) Strength grade
It is determined by the 28 days compressive strength of cement mortar R_c, divided into 625, 725, 825 and 925 classes.

(3) Technical requirements

(i) Specific surface area
S should be no less than 400 m²/kg.

(ii) Setting time
Initial setting time should be no earlier than 25 min, and final setting time should be no later than 3 h. Negotiated by supply and demand sides, the initial setting time can be less than 25 min.

(iii) SO₃ content
$w(SO_3)$ should not exceed 11.0 wt%.

(iv) Strength

Strength values of cement at different ages should not be less than the values listed in Table 3.11. If users need hourly strength, 6 h compressive strength R_c should be more than 20 MPa.

Table 3.11: Strength indicator of rapid hardening high-strength aluminate cement.

Strength grade	Compressive strength R_c (MPa)		Flexural strength R_f (MPa)	
	1 day	28 days	1 days	28 days
625	≥35.0	≥62.5	≥5.5	≥7.8
725	≥40.0	≥72.5	≥6.0	≥8.6
825	≥45.0	≥82.5	≥6.5	≥9.4
925	≥47.5	≥92.5	≥6.7	≥10.2

(4) Cement hydration reaction and products

The hydration reactions can be expressed as follows:

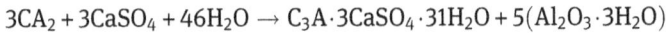

$$3CA + 3CaSO_4 + 37H_2O \rightarrow C_3A \cdot 3CASO_4 \cdot 31H_2O + 2(Al_2O_3 \cdot 3H_2O)$$

$$3CA_2 + 3CaSO_4 + 46H_2O \rightarrow C_3A \cdot 3CASO_4 \cdot 31H_2O + 5(Al_2O_3 \cdot 3H_2O)$$

The main hydration products are Ettringite and hydrated alumina gels, which are stable hydrates under natural conditions. Figure 3.1 shows thermal analysis curves of cement at different ages. As can be seen, a large amount of Ettringite has been formed on 6 h. The amount of Ettringite and hydrated alumina increases with age. No new phase appears when comparing cement paste with hydration ages of one and half

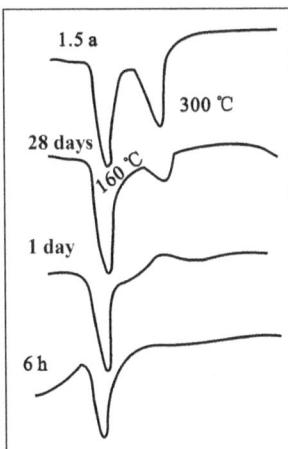

Figure 3.1: Diagram of differential thermal analysis of cement samples at different ages.

year and 28 days. This indicates that the long-term hydration phase of the cement is stable. Concrete placed outdoors for 5 years has increased strength than that of 28 days. Linear expansion of cement paste cured in water is stable. From the phase analysis, strength and expansion properties, it can be concluded that microstructure of hydrated cement is stable. Because the solubility of anhydrite is relatively small (but slightly larger than gypsum), it is more beneficial to dense cement and improves the late strength. Therefore, the combination of microexpansive Ettringite during hydration with filling and densifying of hydrated alumina gel makes cement fast hardening with high strength, good impermeability, frost resistance, corrosion resistance and wear resistance.

(5) Characteristics and applications
Rapid hardening and high-strength aluminate cement have fast hardening speed without rapid setting. It not only has hourly strength characteristics but also has a good growth rate in late strength. Construction requirements can be satisfied without retarders in most situations. It not only retains the advantages of aluminate cement such as high early strength, good impermeability, frost resistance, corrosion resistance and other advantages, but also overcomes the shortcoming of late strength declination.

This type of cement is suitable for modern engineering, military engineering and other special projects that require fast hardening and high strength.

3.2.6 Super-fast hardening and adjustable setting aluminate cement

The China building materials industry standard *Super-fast hardening and adjustable setting aluminate cement* [JC/T 736—1985 (1996)] defines that as follows: Super-fast hardening and adjustable setting aluminate cement is a hydraulic cementitious material and made by grinding the cement clinker mainly composed of calcium aluminate with an addition of a proper amount of gypsum and hardening accelerating agents. With calcium sulfoaluminate as the main hydration product, this type of cement has the properties of adjustable setting and rapidly increasing hourly strength.

(1) Cement strength
It is determined by 2 h strength R_c and has only one grade of 225.

(2) Technical requirements

(i) Specific surface area S
It is controlled above 500 m^2/kg.

(ii) Setting time
Initial setting time should not be earlier than 2 min, and final setting time should not be later than 10 min. When 0.2% (by mass) sodium tartrate is added as a retarder, initial setting time is no earlier than 15 min, and final setting time is no later than 40 min.

(iii) SO$_3$ content
It is no less than 7.0%, and no more than 11.0%.

(iv) Strength
The 2 h compressive strength R_c is 22.06 MPa and flexural strength R_f is 3.43 MPa. The 1 day compressive strength R_c is more than 34.31 MPa, and flexural strength R_f is higher than 5.39 MPa. When 28 days compressive and flexural strength are required to test, the values R_c and R_f should not be lower than 53.92 and 7.35 MPa, respectively.

The main feature of this type of cement is quick setting and hardening. It has a rapid hydration rate on early age, and hydration heat released within 1–2 h occupies 70–80% of 7 days hydration heat; thus, it has a good performance at low temperatures. It can be normally constructed at −10 °C without loss of late strength. Late strength loss is small if precuring measures above 0 °C are taken under −30–15 °C.

(3) Practical applications
Super-fast hardening and adjustable setting aluminate cement are mainly used for rapid repair, rapid construction, plugging and spraying, low-temperature construction and other projects.

The following problems should be paid attention to when applying the cement.
(i) This type of cement cannot be mixed with other varieties of cement, but it can be used in the engineering contacted with hardened Portland cement concrete.
(ii) It must not be used for a long period under the environment temperatures above 50 °C.
(iii) It must be used as soon as possible after mixing to prevent rapid hardening.
(iv) According to the construction conditions and strength requirements, retarders including sodium tartrate and sodium fluorosilicate can be used to adjust setting time.
(v) In the mix proportion of casting and repairing concrete, the w/c ratio should not be higher than 0.42, and the cement amount should be higher than $m > 400$ kg/m^3.

3.2.7 Special high-strength cement

3.2.7.1 Macrodefect-free cement
Macrodefect-free cement was successfully developed and patented by Birchall at Oxford University for the first time, also known as no large defect cement, nonmacroporous

cement or MDF cement. This type of cement has high compression, tensile and bending strengths and excellent crack resistance at low temperatures, good sound absorption and antistatic ability. The compressive strength R_c, tensile strength R_m and bending strength R_f are 200–300 MPa, 30–100 MPa and 100–150 MPa, respectively. It can be made into cement sheet, spring and so on.

Macrodefect-free cement is composed of Portland cement (or aluminate cement), polymer and superplasticizers. When using Portland cement, it is mixed with polyethylene amides, superplasticizers and water (w/c ratio is 0.12–0.16) by shear mixing in a special mixer, and formed by hot pressing and then cured. If aluminate cement is used, the polyvinyl alcohol polymer is appropriate for use.

Late age strength of macrodefect-free cement will decline and water stability is poor. Studies on this type of cement were also reported in China.

3.2.7.2 Ultramicrodense cement

Ultramicrodense cementitious material (DSP, densified system containing homogeneously arranged ultrafine particles) is referred to a dense material containing uniformly distributed ultrafine particles.

Ultramicrodense cement is made by adding 20–25% silica fume in Portland cement and using superplasticizers with a w/c ratio of 0.12–0.15. Its compressive strength R_c can reach 200–270 MPa, which can replace metals, plastics or special-purpose cement products, or be produced into wear-resistant lining materials, pipes or cutting tools for metals.

3.2.7.3 Alinite cement

If adding an appropriate amount of $CaCl_2$ into Portland cement raw materials, the calcination temperature can be lowered to 400–500 °C and a new Alite phase mineral would be produced. The cement produced by this new mineral is referred as Alinite cement.

Alinite cement's formula can be written as $Ca_{11}(Si_{0.75}Al_{0.25})O_{18}Cl$. Its setting time is 2–4 times faster than that of ordinary Portland cement. Its strength development is also fast; in particular, 1 day strength can be higher than that of the rapid hardening Portland cement.

Alinite cement can be made into C60–C80 grade concrete with good workability, or dry-rigid concrete with higher strength.

3.3 Sulfate-resistant cement, moderate-heat and low-heat cement and road cement

In order to meet some special uses, several special types of cement such as sulfate-resistant cement, moderate-heat and low-heat cement and road cement can be produced by selecting the appropriate mineral compositions, changing the fineness or adjusting types and amounts of admixtures according to the properties of the four main minerals in Portland cement clinker. In fact, sulfate-resistant cement, moderate-heat and low-heat cement and road cement are also subsets of Portland cement.

3.3.1 Sulfate-resistant cement

According to the sulfate-resistant ability, it can be divided into moderate sulfate–resistant cement and high sulfate–resistant cement. China national standard *Sulfate resistant Portland cement* (GB 748–2005) defines that as follows: Moderate sulfate–resistant Portland cement is a hydraulic cementitious material ground by appropriate ingredients of Portland cement clinker and proper amount of gypsum, which has sulfate resistance ability to moderate concentrations of sulfate ion erosion. It is called as moderate sulfate–resistant cement in short.

High sulfate–resistant Portland cement is a hydraulic cementitious material ground by appropriate ingredients of Portland cement clinker and the proper amount of gypsum, which has sulfate-resistance ability to high concentrations of sulfate ion erosion. It is short for high sulfate–resistant cement.

3.3.1.1 Design of mineral compositions for sulfate-resistant cement

According to the sulfate corrosion mechanism, the erosion of hardened cement paste by sulfate attack is caused by the reaction between $Ca(OH)_2$ contained in the cement and sulfate ions to form calcium sulfate, and calcium sulfate reacts with hydrated calcium aluminate to destroy the hardened paste. The main approach to improve sulfate resistance is to reduce the amount of $Ca(OH)_2$ produced during hydration and the amount of C_3S and C_3A in hydrated calcium aluminate.

High resistance to sulfate corrosion can be achieved by appropriately reducing C_3A content and correspondingly increasing C_4AF content. Reducing C_3S content and correspondingly increasing C_2S content can also improve the sulfate resistance ability.

In practical production, C_3S and C_3A contents in clinker are controlled as shown in Table 3.12.

Table 3.12: Requirements of sulfate-resistant cement clinker compositions (Mass fraction,%).

Cement type	C_3S	C_3A
Moderate sulfate–resistant cement	≤55.0%	≤5.0%
High sulfate–resistant cement	≤50.0%	≤3.0%

Practical production has proved that mineral composition (by mass, %) should be controlled within the following ranges:

$w(C_3S)$ 40–46%, $w(C_3S+C_2S)$ 77–80%, $w(C_3A)$ 2–4%, $w(C_4AF)$ 15–18% and $w(f\text{-}CaO)\leq0.5\%$.

Correspondingly, three values can be determined according to types of kilns: $KH = 0.80–0.85$, $SM = 2.2–2.5$ and $IM\geq0.70$ for rotary kilns; $KH = 0.84–0.88$, $SM = 1.9–2.1$ and $IM = 0.9–1.0$ for vertical kilns.

3.3.1.2 Production process and technical requirements of sulfate-resistant cement

(1) Characteristics of production process
The production of the sulfate-resistant cement is basically similar to that of Portland cement except for the differences in clinker mineral compositions and values of rate mentioned above. Due to the low KH and IM of sulfate-resistant cement clinker, the clinker formation heat $(\Delta_f H^\theta_m)$ is lower than that of ordinary Portland cement clinker, and it is easily calcined. However, due to the low viscosity of the liquid phase, it is not good for maintenance of rotary kiln skin. The thermal stability must be strengthened with a strictly controlled grain knot situation for rotary kilns. For shaft kilns, the clinker calcination must be strictly controlled to achieve fast burning and fast cooling, pale dark fire operation, reduction of sintering block in the middle of kilns and improving the sintering ability of kilns. In addition, the contents of C_2S and C_4AF in clinker are higher, so the clinker grinding ability is poor, leading to an output reduction of cement mill. In the production, grinding quality can be improved by using a suitable grinding aid or adjusting the gradation of milling balls.

(2) Strength grades of cement
Both types of sulfate-resistant cement have 32.5 and 42.5 grades. The strength at each age should not be lower than the corresponding values listed in Table 3.13.

Table 3.13: Strength indicators of moderate and high sulfate–resistant cement.

Strength grade	Compressive strength (MPa)		Flexural strength (MPa)	
	3 day	28 days	3 days	28 days
32.5	≥10.0	≥32.5	≥2.5	≥6.0
42.5	≥15.0	≥42.5	≥3.0	≥6.5

(3) Technical requirements
In addition to the strict limitation of C_3S and C_3A contents (Table 3.12), other relevant indicators are still required to be limited. The final product should meet the following requirements:

(i) Loss of ignition
It should not exceed 3.0%.

(ii) MgO content
MgO content should not exceed 5.0%. If the cement passes through autoclave stability tests, MgO content is allowed to be 6.0%.

(iii) SO_3 content
It should not exceed 2.5%.

(iv) Insoluble materials content
It should not exceed 1.50%.

(v) Specific surface area
It should be no less than 280 m^2/kg.

(vi) Setting time
Initial setting time is no earlier than 45 min, and final setting time is no later than 10 h.

(vii) Stability
It must be qualified by the boiling test.

(viii) Alkali content
The amount of alkali in the cement is agreed by the supply and demand sides. If using reactive aggregates and users require low-alkali content cement, alkali content calculated by [$w(Na_2O) + 0.658w(K_2O)$] should be not be more than 0.60%.

(ix) Sulfate-resistant properties
The 14 days linear expansion coefficient of moderate and high sulfate–resistant cement should not exceed 0.06% and 0.04%, respectively.

3.3.1.3 Performance and main applications of sulfate-resistant cement
(1) Good resistance to sulfate attack

Moderate sulfate–resistant cement can withstand the pure sulfate solution corrosion with SO_4^{2-} concentration of 2,500 mg/L. High sulfur–resistant cement can withstand the corrosion of pure sulfate solution corrosion, with SO_4^{2-} concentration of 8,000 mg/L and maximum SO_4^{2-} concentration of $(1–2) \times 10^4$ mg/L.

(2) Low hydration heat

(3) Good wear resistance

Swelling, impermeability, frost resistance and other properties are similar to that of Portland cement.

Sulfate-resistant cement is mainly used for sulfate erosion areas such as the harbor, water conservancy, underground, tunnel, road, bridge foundation and other projects. It can also replace ordinary Portland cement for general industrial and civil engineering.

3.3.2 Moderate- and low-heat Portland cement

Moderate-heat Portland cement, for short moderate-heat cement, is a hydraulic cementing material with a moderate hydration heat, made by grinding Portland cement clinker with a proper composition and a proper amount of gypsum.

Low-heat Portland cement, for short low-heat cement, is a hydraulic cementing material with a low hydration heat, made by grinding Portland cement clinker with a proper composition and a proper amount of gypsum.

Because of low thermal conductivity of the concrete, cement hydration heat is hard to release, which is easy to make maximum internal temperature above 60 °C. However, the external surface of concrete has a fast cooling speed, leading to a high-temperature difference of dozens of degrees between concrete surface and internal part. Concrete external cooling leads to shrinkage, and the internal part has not been cooled, resulting in internal stress to produce microcracks causing reduction of water resistance of concrete. The use of cement with low heat amount and low heat rate can reduce the internal temperature rise of mass volume concrete.

To reduce hydration heat amount and heat release rate, it is important to choose reasonable clinker mineral compositions, grinding fineness and adding appropriate blending materials.

Since the hydration heat and rates of C_3A and C_3S are higher than that of C_4AF and C_2S, it is necessary to reduce the contents of C_3A and C_3S, and increase the contents of C_4AF and C_2S, to reduce the hydration heat and heat rate of cement. However, the early strength of C_2S is very low, and so it should not be increased too much and C_3S content should not be too small; otherwise, the strength development is too slow. Thus, when designing the mineral composition of moderate-heat cement and low-heat cement clinker, it is first focused on reducing C_3A content and increasing C_4AF content.

According to China standard of *Moderate-heat Portland cement, low-heat Portland cement* (GB 200—2017), in the clinker of moderate-heat Portland cement, the content of C_3S must not be higher than 55%, the content of C_3A must not be higher than 6% and the content of f-CaO must not be higher than 1.0%. In the clinker of low-heat Portland cement, the content of C_2S must not be less than 40%, the content of C_3A must not be higher than 6% and the content of f-CaO must not be higher than 1.0%.

The content of MgO (mass fraction) must be no greater than 5.0% in moderate- and low-heat Portland cement. The MgO content (mass fraction) in cement is allowed to broaden to 6% if the cement is qualified by autoclaved stability test. The content of SO_3 (mass fraction) must be no greater than 3.5% in moderate- and low-heat Portland cement. The content of loss on ignition (mass fraction) must not be higher than 3%. The content of insoluble substance (mass fraction) must not be higher than 0.75%. The alkali content is calculated according to the value of $w(Na_2O)+0.658w(K_2O)$. When using active aggregates, and low-alkali cement is required to be used by users, the content of alkali must not be higher than 0.60% in moderate- and low-heat Portland cement, or must be agreed by the purchaser and manufacturer. When users make requirements, the content of C_3S must not be higher than 55%, the content of C_3A must not be higher than 6% in moderate-heat Portland cement, the content of C_2S must not be less than 40% and the content of C_3A must not be higher than 6% in low-heat Portland cement.

An increase of the cement grinding fineness will increase hydration heat, especially the early hydration heat; however, if the cement is too coarse, the strength is decreased and the cement amount used in per volume of concrete is increased. Although the cement hydration heat decreases, the heat amount released by the concrete will increase. By regulations in the standard of *Moderate-heat Portland cement and low-heat Portland cement* (GB/T 200−2017), the specific surface area of moderate- and low-heat cement should be no less than 250 m^2/kg. The initial setting time is no less than 60 min, and the final setting time is no more than 720 min. The stability by boiling of medium- and low-heat cement must be qualified.

The strength grades and strengths at different ages for moderate- and low-heat cement are listed in Table 3.14. The upper limits of the hydration heat at different ages are listed in Table 3.15.

Table 3.14: Strength grades and strengths at different ages.

Type	Strength grade	Compressive strength (MPa)				Flexural strength (MPa)		
		3 days	7 days	28 days	90 days	3 days	7 days	28 days
Medium-heat cement	42.5	≥12.0	≥22.0	≥42.5		≥3.0	≥4.5	≥6.5
Low-heat cement	32.5		≥10.0	≥32.5	≥62.5		≥3.0	≥5.5
	42.5		≥13.0	≥42.5	≥62.5		≥3.5	≥6.5

Moderate-heat cement is mainly applied in the surface layer of the overflow surface of dam and fluctuating water level zone required higher wear resistance and frost resistance. Low-heat cement is mainly applied in the dam, internal part of large volume structures and underwater engineering.

Table 3.15: Hydration heat of cement for different strength grades.

Type	Strength grade	Hydration heat (kJ/kg)		
		3 days	7 days	28 days
Medium-heat cement	42.5	≤251	≤293	
	32.5	≤197	≤230	≤290
Low-heat cement	42.5	≤230	≤260	≤310

3.3.3 Low-heat and microexpanding cement

Low heat and micro-expansion cement (GB 2938—2008) standard is defined as follows: Low-heat and microexpanding cement is a hydraulic cementitious material with low hydration heat and low expansion mainly composed of granulated blast furnace slags and produced by adding an appropriate amount of gypsum and Portland cement clinker to grind. The technical indicators are as follows:

(1) Mineral content of cement clinker
Calcium silicate mineral content in Portland cement clinker (mass fraction, %) should be no less than 66%; CaO content/SiO$_2$ content ratio [$w(CaO)/w(SiO_2)$] is no less than 2.0; clinker strength grade must be over 42.5 and contents of f-CaO and MgO (mass fraction, %) must not exceed 1.5% and 6.0%, respectively.

(2) Grinding aids
Cement grinding aids are allowed to add, and the amount should not exceed 0.5% by mass of cement. Grinding aids must meet the requirements of *Cement grinding aids* (GB/ T 26748—2011).

(3) Additive materials
It is agreed by both the supply and demand to add a few additive materials to improve the expansion performance of the cement.

(4) Strength grades
The strength grade of low-heat small-expansion cement is 32.5. The strength at each age should not be less than the values list in Table 3.16.

(5) SO$_3$ content
It should be 4.0–7.0%.

(6) Specific surface area
It should not be less than 300 m^2/ kg.

Table 3.16: Cement grade and strengths at each age.

Strength grade	Flexural strength (MPa)		Compressive strength (MPa)	
	7 days	28 days	7 days	28 days
32.5	≥5.0	≥7.0	≥18.0	≥32.5

(7) Setting time

The initial setting time should not be earlier than 45 min, and the final setting time should not be later than 12 h, which may also be agreed by the supplier and users.

(8) Stability

It must be qualified with boiling cooking method.

(9) Hydration heat

Cement hydration heat at each age should not be more than the values listed in Table 3.17.

Table 3.17: Hydration heat of cement at different ages.

Strength grade	Hydration heat (kJ/kg)	
	3 days	7 days
32.5	≤185	≤220

(10) Linear expansion coefficient

The linear expansion coefficient of 1, 7 and 28 days should be ≥0.05%, ≥0.10% and ≤0.60%, respectively.

(11) Cl⁻ content

It should not be more than 0.06 % (mass fraction,%).

(12) Alkali content

The content of cement alkali is agreed by both supply and demand.

The low-heat and microexpanding cement are mainly composed of granulated blast furnace slag (70 %), Portland cement clinker and gypsum. The quality coefficient of slag is generally more than 1.80, and the controlled cement specific surface area is 350–450 m^2/kg during grinding. The addition of appropriate admixture (such as molasses) can be added to improve the grinding efficiency.

Ettringite and hydrated calcium silicate gel are the main hydration products of cement. Its early strength is 30–40% higher than that of moderate heat Portland cement and low-heat Portland cement. Its hydration heat of 7 days is lower than that of low-heat slag Portland cement. Moreover, it has microexpansion performance. Linear expansion rates of cement paste are 0.14–0.24% for 7 days and 0.15–0.35% for 28 days. These characteristics are important to prevent cracking of large volume concrete engineering. In addition, this type of cement has almost the same corrosion resistance as sulfite-resistant cement. It is mainly used for dam projects.

3.3.4 Low-heat slag Portland cement

China standard of *Low heat slag Portland cement* (JC/T 1082–2008) is defined as follows: Low-heat slag Portland cement is referred as a hydraulic cementitious material, finely ground by a converter slag or an electric furnace steel slag (short for slag), Portland cement clinker, a proper amount of granulated blast furnace slag and gypsum. Slag must be consistent with *slag used in the cement* (YB/T 022–2008) and its dosage (mass fraction, %) should be no less than 30%. Portland cement clinker should meet the requirements of *Portland cement clinker* (GB/T 21372–2008), and 28 days compressive strength should not be less than 55 MPa. Low-heat slag Portland cement is mainly applied in the dam or mass volume concrete; its technical specifications and requirements are as follows:

(1) Grinding aids
The addition of the grinding agents should not exceed 0.5% of the cement mass. The grinding agents should conform to the provisions of GB/T 26748–2011.

(2) Strength grades
Cement strength grades are divided into 32.5 and 42.5. Strengths at different ages should be greater than the values listed in Table 3.18.

Table 3.18: Strength grade of cement and its strength on each age.

Strength grade	Flexural strength (MPa)		Compressive strength (MPa)	
	7 days	28 days	7 days	28 days
32.5	≥3.0	≥5.5	≥12.0	≥32.5
42.5	≥3.5	≥6.5	≥13.0	≥42.5

(3) SO_3 content
SO_3 content should not exceed 4.0%.

(4) The specific surface area
The cement surface area should not be less than 350 m²/kg.

(5) Setting time
Cement initial setting time should not be earlier than 60 min, and final setting time should not be later than 12 h.

(6) Stability
It must be qualified with boiling cooking method. Cement made with slag consisting of over 13% MgO must be qualified by the pressure and steam test.

(7) Alkali content
Alkali content of cement should be decided by both parties of supply and demand. When the reaction between cement in concrete and aggregates is harmful and users require low-alkali content, alkali content in low-heat slag Portland cement is represented by a calculated value of [$w(Na_2O) + 0.658w(K_2O)$], and it should be less than 1.0%.

(8) Hydration heat
Hydration heat of this type of cement at different ages should not exceed the values listed in Table 3.19.

Table 3.19: Cement hydration heat at each age.

Strength grade	Hydration heat (kJ/kg)	
	3 days	7 days
32.5	≤197	≤230
42.5	≤230	≤260

3.3.5 Road Portland cement

Cement concrete pavement is often subjected to the effects including friction from high-speed running vehicles, shocks from load-carrying vehicles, alternating freezing and thawing and expansion stresses caused by temperature and wetting/drying between pavement and embankment. These effects result in durability reduction of concrete pavement constructed by ordinary cement. To meet the requirements for cement concrete pavement, road Portland cement is required to have good wear resistance, low shrinkage, good frost resistance and impact resistance, high flexural strength and good durability.

(1) Definition of road Portland cement
Road Portland cement is a hydraulic cementitious material produced by grinding road Portland cement clinker, gypsum with a proper amount and blending materials, regulated by *Road Portland cement* (GB 13693—2017). It is road cement for short. The content of clinker and gypsum (mass fraction) in road Portland cement is 90–100%, and the content of active blending materials (mass fraction) is 0–10%.

(2) Constituents in Portland cement for pavement

(i) Portland cement clinker
The content of C_3A must not be higher than 5.0%; C_4AF content must not be less than 15.0% and f-CaO content must not be higher than 1.0%.

(ii) Blending materials
Blending materials are class F fly ash qualified with GB/T 1596—2017, granulated blast furnace slag qualified with GB/T 203—2008, granulated phosphorus slag from electrical furnace qualified with GB/T 6645—2008, slag qualified with YB/T 022—2008, ground slag qualified with GB/T 18046—2017 or steel slag powder qualified with GB/T 20491—2017.

(iii) Gypsum
Natural gypsum is class G or M class natural gypsum or mixed gypsum at second or higher levels that must be qualified with GB/T 5483—2008. Desulfurized gypsum must be qualified with GB/T 21371—2008.

(iv) Grinding aids
Grinding aids are allowed to add during cement grinding, and the amount should not exceed 0.5% of the cement mass. Grinding aids should be consistent with the provisions of GB/T 26748—2011.

(3) Design of mineral composition of road Portland cement clinker
The common requirements for clinker mineral composition of Portland cement for pavement are high C_3S and C_4AF contents but low C_2S and C_3A contents. This is because the increase of C_3S and C_4AF contents helps in improving the early strength of cement, cohesion and dynamic pressure, wear resistance and enhance the strain capacity; a decrease of C_2S and C_3A contents helps in reducing the impact of shrinkage deformation. The clinker mineral contents (mass fraction, %) are generally controlled within: 50–70% of $w(C_3S)$, 15–20% of $w(C_2S)$, $w(C_3A) \leq 5.0\%$, $w(C_4AF) \geq 16.0\%$ and $w(f\text{-}CaO) \leq 1.0\%$.

The main factors that influence shrinkage and wear resistance of Portland cement concrete pavement are w/c ratio and the cement amount used in concrete of unit volume. That is to say, measures to reduce the w/c ratio and the cement amount are helpful to improve the abrasion resistance of concrete and reduce shrinkage. Therefore, the production of high-strength grade cement can reduce the amount of cement in concrete and reduce the w/c ratio, improving the wear resistance of concrete. In this sense, C_3S content in the clinker should be increased. When C_3S content is high, the flexural strength, compressive strength and early strength of concrete are high, and wear resistance is also improved, but brittleness and shrinkage are increased, and so hydration heat increases. Thus, it is not appropriate to raise C_3S much higher. In practice, it is mainly through the appropriate increase in C_4AF content and controlling in C_3A content that we achieve wear-resistant and dry shrinkage-resistant purposes.

(4) Preparation of Portland cement for pavement

(i) Grinding fineness
An increase in grinding fineness of Portland cement for pavement can improve the strength. But too fine cement will lead to excessive shrinkage of hardened cement, which is prone to fine cracks, and the pavement is easy to be damaged. Studies have shown that when the specific surface area is increased from 272 to 325 m^2/kg, the shrinkage slightly increases, but it has a great impact on the strength, especially the early strength. Therefore, the cement surface area can be generally controlled within 300–320 m^2/kg, and the residue of 0.08 mm^2 hole sieve should be controlled between 5% and 10%.

(ii) Gypsum amount
Appropriately increasing the amount of gypsum can improve the strength of Portland cement for pavement and reduce shrinkage. This is because when the gypsum content is increased, part of gypsum continues to react with C_3A during early hydration period, generating calcium sulfoaluminate hydrate and thereby reducing the cement shrinkage. However, the SO_3 content in road cement should not exceed 3.5%, and the gypsum dosage must meet the requirements that the initial setting time must not be earlier than 90 min and the final setting time must not be later than 720 min when adding gypsum.

In addition, appropriate amount of blending materials can be selectively incorporated (such as steel slag) to improve the wear resistance of road cement.

(5) Technical indicators of road cement

(i) MgO content
The MgO content should be no more than 5.0%. The MgO content in cement is allowed to be up to 6% if the cement autoclaving test is qualified.

(ii) SO₃ content
SO$_3$ content should not exceed 3.5%.

(iii) Loss on ignition
The ignition loss should not exceed 3.0%.

(iv) Chloride ion
The content of chloride ion in cement should be no more than 0.06%.

(v) Alkali content
The alkali content in cement is calculated according to the value of $[w(Na_2O)+0.658w(K_2O)]$. When using active aggregates and when low-alkali cement is required to be provided by users, the content of alkali in cement must not be higher than 0.60% or must be agreed by the purchaser and manufacturer.

(vi) Specific surface area
The specific surface area of cement should be 300–450 m^2/kg.

(vii) Setting time
The initial setting time should not be earlier than 90 min, and the final setting time should not be later than 720 min.

(viii) Stability by boiling
It must be qualified with the Rayleigh clip.

(ix) Dry shrinkage rate
Dry shrinkage rate at 28 days should not be higher than 0.10%.

(x) Wear resistance
Abrasion amount at 28 days should not be higher than 3.0 kg/m^2.

The code for road Portland cement is P.R. It is divided into two grades as 7.5 grade and 8.5 grade based on flexural strength at 28 days, for example, P.R. 7.5. The strengths at different ages for all grades must be no less than the values listed in Table 3.20.

Table 3.20: Road cement grades and strengths at different ages.

Strength grade	Flexural strength (MPa)		Compressive strength (MPa)	
	3 days	28 days	3 days	28 days
32.5	≥3.5	≥6.5	≥16.0	≥32.5
42.5	≥4.0	≥7.0	≥21.0	≥42.5
52.5	≥5.0	≥7.5	≥26.0	≥52.5

(6) Properties and uses of Portland cement for pavement

(i) Good wear resistance
The wear rate is the only 20–40% when compared with the same level of Portland cement.

(ii) High strength
The Portland cement for the pavement has a high early strength and a high flexural strength. The early strength enhancing rate is no less than the same level of R-type silicate cement; moreover, flexural strength growth rate is higher than that of compressive strength. The 28 days flexural strength index is higher than the same level of R-type Portland cement.

(iii) Small shrinkage
Shrinkage rate is obviously better than that of Portland cement (about 10%) with a short period of shrinkage stabilizing. The number of reserved road construction joints can be reduced to improve the flatness of the road surface and driving comfort.

(iv) Low hydration heat and good durability
Hydration heat of road cement may achieve the requirements of moderate-heat Portland cement. It has good durability under the alternating freezing and thawing environment.

Portland cement for pavement suits various concrete pavements and also for other projects that have high-performance requirements on wear resistance and antishrinkage.

3.3.6 Steel slag cement for pavement

China national standard *Steel slag cement for pavement* (GB 25029–2010) stipulates that steel slag cement for the pavement is a hydraulic cementitious material made by grinding converter steel slag or electric furnace slag (short for steel slag), road Portland cement clinker, granulated blast furnace slag and an appropriate amount of gypsum. The technical indicators are as follows:

(1) Material composition
The amounts of components (mass fraction, %) should comply with requirements listed in Table 3.21.

(2) Technical requirements
The slag used should be consistent with *Slag used in cement* (YB/T022–2008); the steel slag powder should conform to *Steel slag powder for cement and concrete* (GB/

Table 3.21: Requirements of components in steel slag cement for pavement (mass fraction, %).

Clinker + gypsum	Steel slag or slag powder	Granulated blast furnace slag or granulated blast furnace slag powder
> 50 and < 90	≥ 10 and ≤ 40	≤ 10

T20491—2017); the granulated blast furnace slag should conform to GB/T 203—2008; the granulated blast furnace slag powder should conform to *Granulated blast furnace slag powder for cement and concrete* (GB/T 18046—2017); cement clinker of Portland cement for pavement should meet GB 13693—2017; natural gypsum should conform to requirements of G-type or M-type second-grade (including and above) or blended gypsum and industrial by-product gypsum should meet *Industrial by-product gypsum for the cement* (GB/T 21371—2008).

(3) Grinding aids
When it is allowed to be added during cement grinding, the amount of grinding aids should not exceed 0.5% of cement mass. Grinding aids should meet GB/T 20748—2011 standard requirements.

(4) SO_3 content
SO_3 content (mass fraction, %) in cement should not be more than 4.0%.

(5) Setting time
The initial setting time should not be less than 90 min, and the final setting time should not be more than 600 min.

(6) Stability
Tested by steam test method, expansion rate under pressure and steam should be less than 0.50%.

(7) Shrinkage rate
The shrinkage rate at 28 days should not be more than 0.10%.

(8) Wear resistance
The abrasion amount at 28 days should not be higher than 3.0 kg/m^2.

(9) Grinding fineness
Fineness is represented by specific surface area and it should be not less than 350 m^2 /kg.

(10) Cl⁻ content

Cl⁻ content in cement (mass fraction, %) should be no more than 0.06%.

(11) Alkali content

Alkali content is represented by the calculated value of $[w(Na_2O) + 0.658w(K_2O)]$ in the steel slag cement for road use. If the active aggregates are used and users request to provide low-alkali content cement, the alkali content in the cement should not exceed 0.60% or negotiated value.

(12) Strength

Slag cement for pavement is divided into 32.5 and 42.5 levels. The compressive strength and flexural strength at each age should be in accordance with requirements listed in Table 3.22.

Table 3.22: Strength index of slag cement for road.

Strength grade	Compressive strength (MPa)		Flexural strength (MPa)	
	3 days	28 days	3 days	28 days
32.5	≥16.0	≥32.5	≥3.5	≥6.5
42.5	≥21.0	≥42.5	≥4.0	≥7.0

3.4 Expansive and self-stressing cements

Ordinary Portland cement usually presents shrinkage when hardening in the air. The shrinkage value depends on varieties of cement, clinker mineral composition, the fineness of cement, gypsum dosage and w/c ratio. Generally, the average shrinkage rate of 28 days is 0.02–0.035%, and that of 180 days is 0.04–0.06%. After concrete is produced, the shrinkage rate of 7–60 days is large, and tends to decrease after 60 days. As a result of shrinkage, microcracks are generated inside the concrete so that the properties of concrete can be deteriorated. For example, the decrease of strength, impermeability and frost resistance result in penetrating of the external corrosive mediums (corrosive gas and water vapor) into concrete by direct contacting with steel to cause corrosion and reduce concrete durability. Due to the drying shrinkage of the cement, the desired effect can not be reached when casting the joint of prefabricated unit or the connection point of buildings, or blocking and repairing the cracks.

When using expansive concrete, because there is a certain bond force between steel and concrete during hardening, the steel will expand together with concrete, resulting in the elongation of the steel due to tensile stress, while the expansion of concrete is limited by steel under the corresponding compressive stress. Afterward, even the drying shrinkage still cannot offset all of the expanded size. The remaining expansion not only can reduce the cracking phenomenon, but also, more importantly, the tensile stress produced by external factors can be offset by the compressive stress in advance. This will decrease the actual tensile stress of concrete to extremely low values, effectively compensate for the poor tensile strength defects of concrete. This precompressive stress is produced by cement itself, and it is called self-stress. The stress generated in concrete is expressed by the self-stress value (σ value).

The prestressing of expansive cement concrete is shown in Figure 3.2.

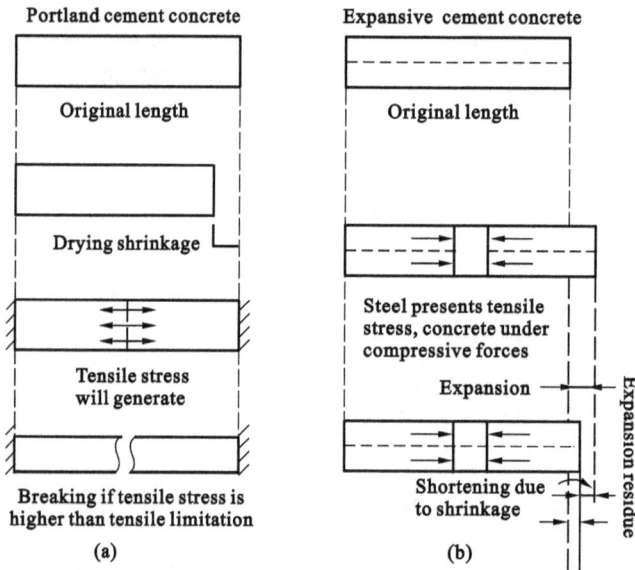

Figure 3.2: Schematic diagram of prestressing of expansive cement concrete: (a) Portland cement concrete and (b) expansive cement concrete.

During the process of hydration, this type of cement has a considerable amount of energy being used to expand and transferred to expansion energy. Generally, the larger the expansion energy, the higher is the potential expansion value. According to the development law of expansion, the expansion rate is usually fast in the early period, and then slows down to a steady state. After reaching the inflation stabilization period, the expansion will basically be stopped. In addition, in the absence of any restriction, the resulting expansion is generally referred as free expansion, which does not produce stress at this time. When subjected to unidirectional, bidirectional or three-directional restraints, the resulting expansion is called limit expansion, which produces stress at this time, and the higher the limit, the higher is the self-stress value (σ).

Depending on the expansion values and applications, the expansion cement is mainly used to compensate the contraction expansion and produce the self-stress, and so expansion cement can be divided into the expansion cement and self-stressing cement. The former has a low expansion energy, and the compressive stress generated by the expansion can approximately offset the tensile stress caused by the drying shrinkage, which is mainly used to reduce or prevent the dry shrinkage in concrete, while the latter has a high expansion energy, and it is sufficient to make the concrete still have a great self-stress after drying shrinkage, which is used for the preparation of various self-stressing reinforced concretes.

3.4.1 Basic principle of expansive cement production

Although there are many chemical reactions that cause expansion in concrete, suitable methods to produce expansive cement are the following three main ways.
(1) Mixing a certain amount of CaO (calcined lime) with cement, and the hydration of CaO will cause a volume expansion. The calcination temperature of CaO usually is controlled at 1,150–1,250 °C.
(2) Adding a certain amount of MgO (magnesia) in cement, and the hydration of MgO will cause a volume expansion. The calcination temperature of MgO usually is controlled at 900–950 °C.
(3) The formation of high sulfur calcium sulfoaluminate hydrates (Ettringite) will produce a volume expansion.

The expansion performance of MgO and CaO is usually not stable during actual production process due to the sensitivity of MgO and CaO hydration speeds under ambient temperature, and the hydration and expansion speeds depend on calcination temperature and particle size. Apart from these reasons, the poor binding performance (low strength) of MgO and CaO also limits their applications in actual engineering, while expansive cements with the Ettringite component are usually used.

The formation of Ettringite can cause destruction of cement stone, which has been regarded as a factor in the destruction of concrete for a long time. Lossier (in France) was the first researcher to recognize that the expansion of Ettringite in concrete could be used to counteract shrinkage and chemical stress. In the mid-1930s, he proposed a plan to make expanded cement, which was made of expansive agent, silicate cement and slag. The expansive agent was produced by calcining the bauxite, gypsum and chalk mixture, and the expansion rate could be controlled by adding a proper amount of slag. In a solution, Ettringite can be formed with a proper ions concentration of calcium, aluminate and sulfate. In the general case, dihydrate gypsum or anhydrite provides sulfate ion, calcium aluminate (C_3A, $C_{12}A_7$, CA and CA_2) provides aluminate ion and the hydration of calcium silicate minerals (C_3S and C_2S) provides calcium ion, while anhydrous calcium sulfoaluminate can provide all of the aforementioned ions.

There are different theories about the expansion mechanism of Ettringite formation.

Lafuma believed that the volume expansion could be caused by the solid-phase reaction of Ettringite in high concentrations of $Ca(OH)_2$ solutions, which could buildup of large amounts of water from other parts. But in fact, the liquid-phase reaction is faster than the solid-phase reaction as Ettringite is formed from aluminate, and its formation also will produce the volume expansion even in a low $Ca(OH)_2$ concentration. B. B. Mikhailov thought that the expansion was due to the crystal growth of Ettringite, which produced the crystal pressure. But Mehta found that the crystal size of formed Ettringite in expansion component do not grow up in 6–7 days; however, the expansion value reaches the maximum within 7 days. He thought that colloidal Ettringite absorbed water to expand in a saturated $Ca(OH)_2$ solution due to its high-specific surface area and special microstructure. According to researches in Chinese building materials research institute, it was believed that the primary reason of Ettringite expansion in hardened cement paste is due to the increase of Ettringite solid volume, which is caused by increasing the supply of external water, and also due to the mutual repulsion, which is caused by the cross-growing of crystals to generate crystallization pressure. The expansion value is related to the morphology, quantity and formation time of Ettringite. The structure, strength, self-stress and other properties of hardened cement slurry are related to properties of the main strength phase such as the forming characteristics, quantity and restriction conditions of binding phase.

Although it is explained in different ways, the expansion of Ettringite has been confirmed by a large number of experiments. During the process of cement hydration, the strength of cement stone can be increased by the filling of initial formed Ettringite in inner space hydrates. However, with the continuous hydration of cement, the quantity of Ettringite will increase and its crystal size will grow, resulting in an expansion stress to weaken and destroy the structure of cement stone and reduce the strength. At the same time, strengthening components will repair and enhance the structure of the cement to increase the strength and impede the expansion. It can be concluded that expansion can weaken and destroy the connection between the hydrated particles, while the strength can strengthen the internal relations between them to obtain mechanical characteristics. Therefore, the strength and expansion of self-stress cements are mutually dependent on each other and contradictory dialectical unity. Only strength without expansion cannot produce expansion energy, while expansion without sufficient strength does not generate enough expansion to tension the steel. Expansion is the premise, while the strength is the guarantee; both are indispensable. For example, the increase of strengthening components increases the strength of cement but decreases the expansion. On the contrary, increasing the expansion component increases the cement expansion and decreases the strength. This requires that the expansive cement and self-stressing cement should be guaranteed to meet the desired expansion value, and also to ensure high strength and enhancement rate.

A reasonable proportion of components for expansion cements is necessary to control influence factors of expansion and strength. If the amount of aluminate cement is high, and the quantity of gypsum is large, both the expansion value and expansion rate are high. And the more silicate cement with the higher strength will result in the more obvious restriction of the expansion, and also, the expansion value and expansion speed are low. Cement should be suitably finely ground, the fineness for self-stress cement is usually 400–500 m^2/kg. Large particles result in a low initial strength and a large expansion to decrease self-stress value. Fine particles lead to a high early strength to limit the expansion and more Ettringite is produced at the early stage to decrease the self-stress value. When preparing concretes, a large expansion is generated with a large dosage of cement. A high w/c ratio leads to a low concrete strength and a premature expansion, which is harmful to the self-stress value. It is proved by practice that enough Ettringite should be formed when the cement stone reaches a certain strength to produce sufficient self-stress. Therefore, the self-stressing cement must be placed in the air by static or steam curing to produce a certain strength (10–15 MPa), and then placed in water for further curing. Therefore, a high self-stress with a high expansion and a high strength is achieved by reasonable proportioning, fineness and curing system to control expansion rate and strength growth.

3.4.2 Types of expansive cement

Expansive cement is different from ordinary cements that contract in air. In China, the compensate shrinkage cement is called expansive cement. The cement that is used for producing self-stressing concrete is called self-stressing cement. Because there are steels in concrete, the steel reinforcement is stretched and subjected to tensile stress when the cement is expanding, while the concrete is subjected to a compressive stress, and this is called the self-stress value.

In order to form stable Ettringite, the liquid phase must have the corresponding concentration of Ca^{2+}, Al^{3+} and SO_4^{2-}, and these ions come from different sources and can form different types of expansive cements. Ca^{2+} usually comes from silicate cement and aluminate cement or lime; Al^{3+} comes from calcium aluminate or hydrated calcium aluminate, and $C_4A_3\bar{S}$ or alunite as well and SO_4^{2-} comes from lime or $C_4A_3\bar{S}$ and alunite.

Expansive cements and self-stressing cements have the following types:

(1) The main component is Portland cement clinker
(i)   Portland expansive cements and self-stressing cements are produced by Portland cement clinker or adding aluminate and gypsum.
(ii) K-type expansive cements are produced by Portland cement clinker or by adding expansive agents. The expansion agents are made of C_4A_3, lime and gypsum.

(iii) S-type expansive cements are made of silicate cement clinker with a high content of C_3A.

(iv) Alunite expansive cements and self-stressing cements are made of silicate cement clinker with calcined or uncalcined alunite and gypsum.

(v) Expansive oil well cements are made of silicate cement clinker and a certain amount of MgO calcined at a fixed temperature.

(vi) Casting cement is made of silicate cement clinker and a certain amount of CaO calcined at a fixed temperature.

(vii) Aerated cement is made of silicate cement clinker and aluminum powder or iron powder and oxidation agent (NH_4Cl).

(2) The main component is aluminate cement clinker

(i) Expansibility impervious cement is produced by aluminate cement and by adding hydrated calcium aluminate and hemihydrate gypsum or high-strength construction gypsum.

(ii) Impervious cement without shrinkage is produced by aluminate cement and adding hemihydrate gypsum and slaked lime.

(iii) Gypsum bauxite expansive cement, rapid hardening expansive cement and self-stress aluminate cement are made of aluminate cement and gypsum.

(3) The main component is sulfoaluminate cement clinker

Early strength microexpansive sulfoaluminate cement and sulfur aluminate cement are made of sulfoaluminate cement and dihydrate gypsum.

3.4.3 Production processes and properties

3.4.3.1 Nonshrinking and rapid hardening Portland cement

The cementitious material hardening in water with rapid hardening and without shrinkage produced by grinding silicate cement clinker, with a certain amount of dihydrate gypsum and expansive agents, is called nonshrinkage rapid hardening silicate cement clinker (casting cement).

The MgO content in nonshrinkage rapid hardening silicate cement clinker must be less than 5.0%, SO_3 content must be no more than 3.5%, the sieve residue from a 0.08 mm sieve must be less than 10%, the initial setting time is less than 30 min and the final setting time is more than 6 h. The free expansion rate of cement paste sample in water curing is more than 0.02% after 1 day, and less than 0.3% after 28 days.

Nonshrinkage rapid hardening silicate cement is made of high-quality silicate cement clinker, gypsum and calcined lime (–1,300 °C) expansive agent. To enhance the long-term strength, a suitable amount of alkali slag particles can be added into the aforementioned cement. This type of cement has a lot of advantages including fast hardening, high early and long-term strength, microexpansion and nonshrinkage properties. Therefore, it can be extensively used to various building engineering, such as the casting concrete in joint point of fabricated framework, joint mortar of steel concrete, concrete project, joint casting of concrete project, concrete grouting of mechanical instrument and other instant engineering.

3.4.3.2 Self-stressing Portland cement
The levigated expansive hydrate binding material that is made of a certain amount of ordinary Portland cement or aluminate cement, and natural dihydrate gypsum, is called self-stressing Portland cement.

The self-stressing Portland cement can be divided into four types based on the 28 days self-stressing values (σ), marked as S_1, S_2, S_3 and S_4. The self-stressing value of each type should comply the requirements listed in Table 3.23. The specific surface area (S) of self-stressing Portland cement should be larger than 340 m^2/kg, the initial setting time is less than 30 min and the final setting time is longer than 390 min. The free expansion rate (ε) of 28 days should be less than 3%, and the stable period should be more than 28 days.

Table 3.23: Self-stressing values and types of self-stressing Portland cement.

Type	S_1	S_2	S_3	S_4
Self-stressing value (MPa)	$1.0 \leq S_1 < 2.0$	$2.0 \leq S_2 < 3.0$	$3.0 \leq S_3 < 4.0$	$4.0 \leq S_4 < 5.0$

The proportion of self-stressing Portland cement is usually as follows: 67–73% Portland cement, 12–15% aluminate cement, and 15–18% dihydrate gypsum. The free expansive rate (ε) of cement paste is 1–3%, and it reaches the maximum value within 7 days. The expansion mainly comes from Ettringite, which is the hydration product of CA, CA$_2$ and gypsum of aluminate cement. Because of the high alkalinity of liquid phase and short stability of cement slurry, the expansion property is not easy to control, resulting in the instable quality, poor impervious performance and low self-stressing value of the product. Self-stressing Portland cement can be used to produce water transport pipes of self-stressing steel concrete and sprinkling irrigation of self-stressing ferrocement tube, and is not suitable for producing high-pressure water or gas transport pipes with large diameters.

3.4.3.3 Self-stressing aluminate cement
Self-stressing aluminate cement is a binding material with large expansion produced by a certain amount of aluminate cement and dihydrate gypsum. Self-stressing

aluminate cement can be divided into 3.0, 4.5 and 6.0 levels according to 28 days self-stressing values (σ) of 1:2 standard mortars. It is mainly used to produce steel (steel wires) concrete (mortar) self-stress pressure pipes.

Generally, the ratio of self-stressing aluminate cement is as follows: 60–66% aluminate cement and 30–40% dihydrate gypsum (ca.16%±0.5% SO_3 content). The main hydrates of self-stressing aluminate are Ettringite and $Al(OH)_3$, and the reaction equations are as follows:

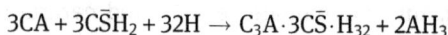

$$3CA_2 + 3C\bar{S}H_2 + 41H \rightarrow C_3A \cdot 3C\bar{S} \cdot H_{32} + 5AH_3$$

$$3CA + 3C\bar{S}H_2 + 32H \rightarrow C_3A \cdot 3C\bar{S} \cdot H_{32} + 2AH_3$$

Ettringite will be produced by the reaction of CA, CA_2 and gypsum of aluminate cement during the hydration process with a low CaO concentration of liquid phase. At the same time, the relative high amount of $Al(OH)_3$ gel will be generated as well. These can not only effectively enhance the density of cement stone but also act as an important plastic buffer effect during growing and expanding of Ettringite crystals, which can continuously improve the strength of cement stone due to the high deformability. Moreover, because of the low supersaturated degree during the generation of Ettringite, the generated Ettringite has well dispersed with uniform distribution, resulting in a disperse crystal pressure and low destroy risk of cement structure. Therefore, it can be considered that Ettringite and $Al(OH)_3$ gel build up the strength and expansion factors in cement stone.

Aluminate self-stressing cement has good antipermeability and a high self-stressing value, which can reach 5 MPa. Moreover, its production process can be easily controlled, and the quality is also stable. However, it has a long expansion stable stage with a relative high production price.

3.4.3.4 Expansive and self-stressing sulfoaluminate cement

The adjustable expansive hydrate binding material, produced by using anhydrous calcium sulfoaluminate and dicalcium silicate as the main minerals and adding a certain amount of dihydrate gypsum, is called expansive sulfoaluminate cement. It can be divided into microexpansive and expansive sulfoaluminate cements based on free expansive rates of cements.

According to the Chinese standard *Sulfoaluminate cement* (GB 20472–2006), the expansive hydrate binding material, which is made of a certain sulfoaluminate cement clinker and gypsum, is called self-stressing sulfoaluminate cement.

The components of self-stressing sulfoaluminate cement are as follows: $w(Al_2O_3) \geq 30\%$ (mass fraction, %), $w(SiO_2) \leq 10.5\%$ (mass fraction, %), and $w(Al_2O_3)/w(SiO_2) \leq 6$. The 3 days compressive strength of clinker must be higher than 55 MPa.

Self-stressing sulfoaluminate cement can be divided into four levels (3.0, 3.5, 4.0 and 4.5) according to 28 days self-stressing values. For all of self-stressing sulfoaluminate, the compressive strengths must be higher than 32.5 and 42.5 MPa after 7 and 28 days, respectively.

The physical and chemical properties and alkali content of self-stressing sulfoaluminate must meet the requirements listed in Table 3.24. The self-stressing values of each level self-stressing sulfoaluminate in different hydration stages should meet the requirements listed in Table 3.25.

Table 3.24: Physical and chemical requirements of self-stressing sulfoaluminate cement.

Program		Value
Specific surface areas (m²/kg)		≥ 370
Setting time[a] (min)	Initial setting time	≥ 40
	Final setting time	≤ 240
Free expansive rate (%)	7 days	≤ 1.30
	28 days	≤ 1.75
Alkali content [$w(Na_2O)+0.658w(K_2O)$] (%)		≤ 0.50
28 days self-stressing enhancement (MPa/day)		≤ 0.010

[a] According to real situation.

Table 3.25: Self-stressing value requirements of each level self-stressing sulfoaluminate cement in different hydration stages.

Level	7 days self-stressing value (MPa)	28 days self-stressing value (MPa)	
	≥	≥	≤
3.0	2.0	3.0	4.0
3.5	2.5	3.5	4.5
4.0	3.0	4.0	5.0
4.5	3.5	4.5	5.5

The expansion in cement is caused by the formation of Ettringite. The main hydrates are Ettringite and Al(OH)₃ gel. The formed Ettringite acts as a framework in the initial stage of hydration, Al(OH)₃ gel and C–S–H gel acts as a buffer layer to expansion. Therefore, the mild expansive characteristic combined with the denser cement stone results in a good compactness and antipermeability. The expansive extent and self-stressing values depend on the dosage of gypsum, the higher dosage and the bigger self-stressing value. Generally, the self-stressing value can reach 2–7 MPa. It can be used to produce high-pressure water, gas and oil transport pipes with large diameters.

3.4.3.5 Expansive and self-stressing aluminoferriate cement

The adjustable expansive hydrate binding material that is produced by using iron phase, anhydrous calcium sulfoaluminate and dicalcium silicate as the main minerals, and adding a certain amount of limestone and gypsum, is called expansive aluminoferriate cement. It can be divided into microexpansive and expansive aluminoferriate cement according to free expansive rates of cements.

According to the Chinese standard of *Self-stressing aluminoferriate cement* (JC/T437−2010), the expansive hydrate binding material, which is made of a certain aluminoferriate cement clinker and gypsum, is called self-stressing aluminoferriate cement.

The components of self-stressing aluminoferriate cement are as follows: $w(Al_2O_3) \geq 28\%$, $w(SiO_2) \leq 10.5\%$, $w(Fe_2O_3) \geq 3.5\%$ (mass fraction, %) and $w(Al_2O_3)/w(SiO_2) \leq 6.0$. The alkali content $w(R_2O)$ in cement is calculated and regulated with $[w(Na_2O)+0.658w(K_2O)] \leq 0.50\%$.

The physical properties of self-stressing aluminoferriate cement should meet the requirements index listed in Table 3.26.

Table 3.26: Physical properties of self-stressing aluminoferriate cement and requirements index.

Program		Technical index
Specific surface areas, $S(m^2/kg)$		≥ 370
Setting time[a] (min)	Initial setting time	≥ 40
	Final setting time	≤ 240
Free expansive rate, $\varepsilon(\%)$	7 days	≤ 1.30
	28 days	≤ 1.75
Compressive strength, R_c (MPa)	7 days	≥ 32.5
	28 days	≥ 42.5
28 days self-stressing enhancement (MPa/day)		≤ 0.01

[a] According to real situation.

Self-stressing aluminoferriate cement can be divided into four levels (3.0, 3.5, 4.0 and 4.5) according to 28 days self-stressing values, σ. The self-stressing values of each level self-stressing aluminoferriate cement in different hydration stages should meet the requirements listed in Table 3.27.

Table 3.27: Self-stressing value requirement of each level self-stressing aluminoferriate cement in different stage.

Strength level	7 days self-stressing value (MPa)	28 days self-stressing value (MPa)	
3.0	≥2.0	≥3.0	≤4.0
3.5	≥2.5	≥3.5	≤4.5
4.0	≥3.0	≥4.0	≤5.0
4.5	≥3.5	≥4.5	≤5.5

3.4.3.6 Alunite expansive cement

According to the Chinese standard of *Alunite expansion cement* (JC/T 311–2004), the expansive hydrate binding material, which is produced by using silicate cement clinker, aluminum clinker, gypsum and granulated blast furnace slag or fly ash in a certain proportion, is called alunite expansive cement.

After calcination under a certain temperature, the active material with a content of Al_2O_3 more than 25% is known as aluminum clinker.

Silicate cement clinker must meet the Chinese standard GB/T 21372–2008, and clinkers over 42.5MPa grades are suitable to be used.

It is allowed to add grinding agents in accordance with the standard GB/T 20748–2011, which must not be more than 1% of the cement mass.

The sulfate content (SO_3) in expansive alunite cement should be less than 8.0%. The specific surface area (S) must not be less than 400 $m^2/$ kg. The initial setting time is no earlier than 45 min, and the final setting time is no later than 6 h.

The strength of each strength grade cement should not be less than the values listed in Table 3.28, which is divided into three grades: 32.5, 42.5 and 52.5 according to 28 days self-stressing values.

Table 3.28: Level and strength of alunite expansive cement in specific age.

Strength level	Compressive strength (MPa)			Flexural strength (MPa)		
	3 days	7 days	28 days	3 days	7 days	28 days
32.5	≥13.0	≥21.0	≥32.5	≥3.0	≥4.0	≥6.0
42.5	≥17.0	≥27.0	≥42.5	≥3.5	≥5.0	≥7.5
52.5	≥23.0	≥33.0	≥52.5	≥4.0	≥5.5	≥8.5

The limiting expansion rate of 3 days shall not be less than 0.015% and 28 days shall not be higher than 0.10%. According to the standard of *Alunite expansion cement* (JC/T 311–2004) impermeability testing method (see Appendix), the 3 days impermeability must be qualified. However, if the cement is not used in seepage prevention engineering, this test is not necessary.

Alkali content is decided by both of supply and demand; in case the aggregate harmful reactions may occur in cement and concrete, and the low alkali is required by users, the alkali content, $w(R_2O)$, of alunite expansion cement with [w (Na$_2$O) + 0.658 w (K$_2$O)] values should not be higher than 0.60% according to Na$_2$O content.

Generally, alunite expansive cement will reach expansion stable in 14–28 days. The self-stress value can reach 0.2–1.0 MPa, meeting the requirements of compensating contraction to reduce or prevent shrinkage cracking of concrete when cement amount is 380 kg/m^3 and w/c ratio is 0.50–0.55 with the reinforcement ratio (ρ) of 0.5–1.6% to the reinforced concrete. The antiwater pressure of 1:2 mortar with a 3 cm thickness can reach 2.9–3.9 MPa. The antiwater pressure of the expansive concrete with a thickness of 15 cm and cement content of 350 kg/m^3 is higher than 2.9 MPa. Under the 9.8 MPa constant pressure of gasoline for 7 days and nights, the penetration thickness is only 1–2 cm. This type of cement can present good durability and water resistance as a waterproof concrete. The strength of alunite expansive cement is high, especially in the later period; the strength is increased rapidly with a stable long-term strength enhancement. The concrete has good stability in air, cold resistance, antisulfate property and no rust effect on steel bars.

Alunite expansive cement is mainly used for compensating shrinkage concrete structure engineering, antiseepage concrete seepage control engineering or plaster, in situ concrete engineering after pouring seam joint of concrete and bricks, precast beam–column joint, slurry mortar anchor and pipe joints, the machine base and the secondary grouting anchor bolt material, reinforcement and repair of engineering and so on.

3.5 Oil well cement

The oil well cement is used for well cementation engineering of oil and gas wells. In the exploration and exploitation of oil or natural gas, a steel casing is inserted into the well and then injected with cement to seal the casing with the surrounding strata. During cementing operations, layers of oil, gas and water in strata are isolated and sealed to prevent each other from contacting so that a well-isolated oil flow channel from the oil layer to the ground is formed in well layers, as shown in Figure 3.3.

The temperature and pressure at the bottom of the oil well increase with an increase in depth of oil well. For every 100 mm depth, the temperature is increased by about 3 °C, and pressure is increased by 1.0–2.0 MPa. As the depth of oil well is over 7,000 m, the bottom temperature can reach 200 °C, and the pressure can reach 125.0 MPa. Therefore, the most important problems in the production and use of oil well cement are caused by high-temperature and high-pressure effects, especially the effects of high temperature on the properties of cement.

The application characteristics of oil well cement are as follows: the cement paste (w/c ratio is 0.5) is pumped into the gap between the pipe wall and the rock. It is

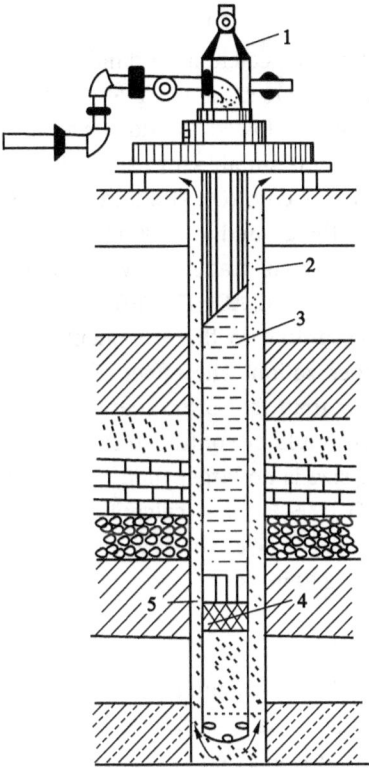

Figure 3.3: Well cementation schematic of oil well: 1, through cement pump; 2, cement paste; 3, slurry; 4, tie plug and 5, well wall.

required that the cement slurry has a good fluidity and a proper density in the cementing process. The cement paste can quickly harden after completing the casing, the interval between the initial and final setting times is short. In addition, it is required that the casted cement has a sufficient strength in the short term to prevent the sediment and destroying by water flow. Oil well cement is mainly subjected to the bending stress, and its hardening process is carried out under high temperatures and pressures. In special cases, there are special requirements for oil well cement, such as antisulfate corrosion, plugging gap and so on.

Under the influence of temperature and pressure on cement hydration, the temperature is the main and the pressure is secondary factor. High temperature significantly reduces the strength of Portland cement. Therefore, the requirements on performance of cement are different according to the oil well depth.

According to the standard *Oil well cement* (GB 10238–2015), oil well cement is divided into eight levels, including common type (O), sulfite (MSR) and high-resistant sulfate (HSR) in China. The oil well cement and use range for all levels are as follows:

(1) Grade A oil well cement

Grade A oil well cement is mainly composed of hydraulic calcium Portland cement clinker and by adding a suitable amount of gypsum (meet the GB/T 5483—2008 standard). In the production of grade A cement, it is allowed to add a grinding agent that conforms to GB/T 26748—2011. The product is suitable for using with no specific performance requirements, only the common type (O). Grade A oil well cement is suitable for the oil well with a depth from the ground to 1,830 m deep.

The suitable mineral compositions (mass fraction, %) of grade A oil well cement are $w(C_3S)$=53–60%, $w(C_2S)$=13–20%, $w(C_3A)$=7–9%, $w(C_4AF)$=12–14% and $w(f\text{-}CaO)$≤1.0%. Raw materials should be prepared according to the aforementioned requirements and should be processed in homogenization. The effect of cement grinding process on oil well cement is much larger than that of ordinary cement. Therefore, the temperature of the grinding material should be controlled so that it is close to the ambient temperature and the particle size is no larger than 15 mm. In grade A oil well cement, $w(MgO)$≤6%, $w(SO_3)$≤3.5%, the calcination loss amount≤3.0% and the content of insoluble substance≤0.75%. The specific surface area (S) is controlled to 350–370 m^2/kg, no less than 280 m^2/kg.

(2) Grade B oil well cement

Grade B oil well cement is mainly composed of hydraulic calcium silicate cement clinker, and adding a suitable amount of gypsum (meet the GB/T 5483—2008 standard). In the production of grade B oil well cement, it is allowed to add a grinding agent that conforms to GB/T 26748—2011. The product is suitable for the use of antisulfates in underground conditions or high resistance to sulfates, and there are two types: medium antisulfate (MSR) and high antisulfate (HSR). Grade B oil well cement is suitable for the oil well with a depth from the ground to 1,830 m. The $w(C_3A)$ mineral of medium antisulfate clinker is less than 8%. The $w(C_3A)$ mineral of high antisulfate clinker is less than 3%, and $[w(C_4AF)+2w(C_3A)]$ is less than 24%. In grade B oil well cement, $w(MgO)$≤6%, $w(SO_3)$≤3.0%, the calcination loss amount, $w(LOI)$ ≤3.0%, the content of insoluble substance and w(insoluble substance)≤0.75%.

(3) Grade C oil well cement

Grade C oil well cement is mainly composed of hydraulic calcium silicate cement clinker and a suitable amount of added gypsum (meet the GB/T 5483—2008 standard). In the production of grade C cement, it is allowed to add a grinding agent that conforms to GB/T 26748—2011. The product is suitable for the use of high initial strength in underground conditions, and there are three types: ordinary (O), medium antisulfate (MSR) and high antisulfate (HSR). Grade C oil well cement is suitable for the oil well with a depth from the ground to 1,830 m. In grade C oil well cement clinker, the ratio value (KH) =0.9–0.92, $w(C_3A)$ = 2–8% and in the production, $w(SO_3)$ = 2.5–3.0%; the specific surface area (S) is controlled to 440–

460 m^2/ kg, no less than 400 m^2/kg. In grade C oil well cement, $w(MgO)$ ≤6%, the calcination loss amount, w (LOI)≤3.0%, the content of insoluble substance, w (insoluble substance) ≤0.75%. In which, the ordinary type oil well cement requires $w(C_3A)$ ≤15% in the clinker and $w(SO_3)$≤4.5% in the cement. The medium sulfate–resistant type requires $w(C_3A)$ ≤8% in the clinker and $w(SO_3)$≤3.5% in the cement. The high sulfate-resistant type requires $w(C_3A)$ ≤3.0%, $[w(C_4AF)+2w(C_3A)]$≤24% in the clinker and $w(SO_3)$≤3.5% in the cement.

(4) Grade D oil well cement

Grade D oil well cement is mainly composed of hydraulic calcium silicate cement clinker and by adding a suitable amount of gypsum (meet the GB/T 5483–2008 standard). In the production of grade D cement, it is allowed to add a grinding agent that conforms to GB/T 26748–2011. In addition, the suitable set-controlling admixture can be used for the common mill or mixing in production. The product is suitable for using under medium-temperature and pressure conditions, and there are two types: medium antisulfate (MSR) and high antisulfate (HSR). Grade D oil well cement is suitable for a well depth from 1,830 to 3,050 m.

(5) Grade E oil well cement

Grade E oil well cement is mainly composed of hydraulic calcium silicate cement clinker and by adding a suitable amount of gypsum (meet the GB/T 5483–2008 standard). In the production of grade E cement, it is allowed to add a grinding agent that conforms to GB/T 26748–2011. In addition, the suitable set-controlling admixture can be used for the common mill or mixing in production. The product is suitable for using under high-temperature and pressure conditions, and there are two types: medium antisulfate (MSR) and high antisulfate (HSR). Grade E oil well cement is suitable for a well depth from 3,050 to 4,270 m.

(6) Grade F oil well cement

Grade F oil well cement is mainly composed of hydraulic calcium silicate cement clinker and by adding a suitable amount of gypsum (meet the GB/T 5483–2008 standard). In the production of Grade F cement, it is allowed to add a grinding agent that conforms to GB/T 26748–2011. In addition, the suitable set-controlling admixture can be used for the common mill or mixing in production. The product is suitable for using under high-temperature and pressure conditions, and there are two types: medium antisulfate (MSR) and high antisulfate (HSR). Grade F oil well cement is suitable for the well depth from 3,050 to 4,880 m.

In addition to compressive strength and thickening time, the technical indexes requirements of grades D, E and F are the same. In grade F oil well cement, w (MgO)≤6%, $w(SO_3)$≤3.0%, the calcination loss amount, $w(LOI)$≤3.0%, the content of insoluble substance, w(insoluble substance)≤0.75%. The $w(C_3A)$ mineral of medium antisulfate clinker is less than 8%. The $w(C_3A)$ mineral is less than 3%, and $[w(C_4AF) +2w(C_3A)]$ is less than 24% in high antisulfate clinker.

(7) Grade G and H oil well cement

Grade G and H oil well cements are mainly composed of hydraulic calcium silicate cement clinker and by adding a suitable amount of gypsum (meet the GB/T 5483–2008 standard). In the production of grades G and H oil well cements, it is allowed to add gypsum or water or the two together, but not to add other admixtures. The product is a basic oil well cement with two types: medium antisulfate (MSR) and high antisulfate (HSR). Grades G and H oil well cements are suitable for the well depth from ground to 2,440 m. The application depth and temperature range can be extended as using coagulant or retardants. The mineral composition, quality standard and technical requirements of grades G and H oil well cements are exactly the same. There is a difference in w/c ratio: grade G is 0.44 and grade H is 0.38. Therefore, grade H has a smaller specific surface area (S), only 270–300 m^2/kg. In terms of chemical composition requirements, for medium antisulfate cement, the requirements are as follows: w (MgO)≤6.0%, $w(SO_3)$≤3.0%, the calcination loss amount, w (LOI)≤3.0%, the content of insoluble substance, w (insoluble substance) ≤0.75%, $w(C_3S)$=48–58%, $w(C_3A)$≤8% and the total alkali amount $[w(Na_2O)+0.658w(K_2O)]$≤0.75%. For high antisulfate cement, apart from $w(C_3A)$≤3%, $[2w(C_3A)+w(C_4AF)]$≤24%, $w(C_3S)$ = 48–65%, others are the same as that of medium antisulfate oil well cement.

Physical performance requirements of oil well cement include w/c ratio, cement specific surface area, the initial consistency within 15–30 min, thickening time under certain temperature and pressure, and compressive strength of under certain temperature, pressure and curing age and so on. The physical mechanical properties of aforementioned oil well cements can be seen in GB 10238–2015.

There are two methods to produce the oil well cements: one is to produce a special mineral clinker to meet the chemical and physical requirements of a certain grade of cement and the other is to use the basic oil well cement (G and H grade cement) and by adding the corresponding admixtures to meet the technical requirements of cement grade. The former method usually brings more difficulties to cement plants, and so the latter method is usually used now. The situation of oil wells and gas wells is very complicated. In order to adapt to the specific conditions of different oil wells and gas wells, it is sometimes necessary to add some admixtures to cement such as weight enhancers, extenders or retardants.

3.6 Decoration cement

Decorative cement refers to white cement and colored cement, mainly used in building decoration works, which can be made into colored mortar or made of various colors and white concrete. Compared with natural decorative materials, decorative cement has a lot of advantages including convenience, color adjustment, low price and so on.

3.6.1 White Portland cement

3.6.1.1 The main technical requirements of white Portland cement

According to Chinese standard *White Portland cement* (GB/T 2015–2017), white Portland cement, for short, white cement, is a hydraulic cementing material made by grinding clinker of white Portland cement, gypsum with a proper amount and blending materials. Clinker of white Portland cement is the clinker produced by firing raw materials with a proper composition to partially melt, in which calcium silicates are the main components with a small content of iron oxide. The content of MgO in the clinker is no more than 5.0%.

In white Portland cement, the content of clinker and gypsum is 70–100%, and the content of natural minerals including limestone, dolomitic limestone and quartz sand is 0–30%. Grinding aids qualified with GB/T 26748–2011 can be added during cement grinding, and the addition amount shall not be higher than 0.5% of the cement mass.

In white Portland cement, the content of SO_3 must not be higher than 3.5%, the content of water-soluble six valent chromium should not be greater than 10 mg/kg and the content of chloride ions should not be greater than 0.06%. The alkali content in white cement is calculated according to the value of [$w(Na_2O)$ +0.658$w(K_2O)$]. When using active aggregates and when low-alkali cement is required to provide by users, the content of alkali in cement must not be higher than 0.60% or must be agreed by the purchaser and manufacturer. The sieving residue from 45 μm^2 hole sieve of cement must not be more than 30%. The initial setting time should not be less than 45 min, and the final setting time should not be longer than 600 min. Stability must be qualified with boiling test. The whiteness of cement should be no less than 89 for grade 1, and be no less than 87 for grade 2. The internal radiation index I_{Ra} of cement must be no more than 1, and the external radiation index I_y must not be greater than 1. White Portland cement is divided into three strength grades, 32.5, 42.5 and 52.5, and strengths at different ages for each grade must not be lower than the values listed in Table 3.29.

3.6.1.2 Production process of white Portland cement

(1) Raw and fuel materials

The production process of white Portland cement is similar to that of Portland cement. The color of Portland cement is mainly caused by iron oxide. The cement

Table 3.29: Strengths at different ages for each grade of white Portland cement.

Strength grade	Compressive strength (MPa)		Flexural strength (MPa)	
	3 days	28 days	3 days	28 days
32.5	≥12.0	≥32.5	≥3.0	≥6.0
42.5	≥17.0	≥42.5	≥3.5	≥6.5
52.5	≥22.0	≥52.5	≥4.0	≥7.0

clinker displays different colors as the content of Fe_2O_3 is changed in the clinker, as shown in Table 3.30. Therefore, the production of white cement is mainly to reduce the content of Fe_2O_3 in the clinker, and Table 3.31 lists the chemical composition and rate values in white Portland cement clinker.

Table 3.30: The color of white Portland cement clinker.

Fe_2O_3 content in cement clinker	Color of cement clinker
3–4	Dark gray
0.45–0.70	Light green
0.35–0.45	White (greenish)

Table 3.31: Chemical composition and ratio rate of white Portland cement clinker.

No.	Chemical composition (mass fraction, %)				Rate value		
	SiO_2	Al_2O_3	Fe_2O_3	CaO	KH	SM	IM
1	22–27	5–6	<0.5	65–67	0.90–0.95	4–4.9	>10
2	23.40	6.11	0 33	67.52	0.85	3.85	18.5
3	23.58	5.66	0.32	64.12	0.83	3.95	17.1
4	25.10	5.50	0.43	65.68	0.80	4.23	12.7

The raw materials for producing white cement are required to have high purity, good quality and low-fuel ash, especially the low iron content. The main raw materials used in the production of white cement are limestone raw materials and clay materials. The lime material is usually made of pure limestone and chalk. The clayey material chooses kaolin, pyrophyllite or sandy clay with low iron content. Correction of raw materials is commonly made by using porcelain and quartz sand. The quality requirements for raw materials are listed in Table 3.32.

Table 3.32: Quality requirements of raw materials.

Name of raw materials	Chemical composition (mass fraction, %)					
	SiO_2	Fe_2O_3	CaO	MgO	Al_2O_3	R_2O
Limestone	—	<0.05	>55	<3	—	—
White clay	60–70	<1.0	—	<1	13–25	<2.5
Chinastone	78–82	<0.5	—	<1	<1	<3.0
Quartz sand	>97	<1.0	—	<1	<1	—

As a retardant, the whiteness of gypsum should be higher than 90 and the impurities and color parts should be removed when used in cement. When fluorite is used as a mineralizer, the content of CaF_2 in fluorite is higher than 85%, and the appearance should be clean and clear. The fuel for burning white cement clinker is best to use ash-free fuel–gaseous fuel (such as natural gas) and liquid fuel (such as heavy oil). If using solid fuels, it is necessary to use the high-quality coal; generally, the volatiles are around 25–40%, the calorific value is greater than 27,170 kJ/kg, ash content is less than 10% and Fe_2O_3 content of ash content is less than 15%; otherwise, whiteness of clinker will be significantly affected. Under the same conditions, the whiteness for the use of solid fuels is reduced by 5–10% compared with when gas and liquid fuels are used.

(2) Preparation of raw materials

The ingredient calculation and ingredient scheme of cement raw materials are mainly based on the composition design of clinker, which is determined by the rate values of cement clinker. In order to select the rate values, the influences of clinker rate values and the mineral compositions on whiteness of the cement should be taken into account in addition to considering the influence of Fe_2O_3 of the raw materials.

Research results all over the world show that the whiteness of cement increases with the increase of C_3S content and the ratio of C_3S to C_2S [$w(C_3S)/w(C_2S)$] in cement clinker and vise versa. This is due to the fact that the Alite mineral is whiter than Belite, and color oxides can be easily to be fixed in Belite. A high lime saturation coefficient (KH) of cement clinker leads to a high white degree (W) of clinker. However, too high lime saturation coefficient results in difficult calcining of cement clinker and an increase of free calcium oxide to decreases the strength of cement. Generally, KH is chosen between 0.88 and 0.95, and the influence of different C_3S contents and C_3S to C_2S ratios on the whiteness of cement clinker is listed in Table 3.33.

Silicon rate (SM) is also an important factor to be considered when preparing raw materials. Due to the low content of Fe_2O_3 in white cement clinker, the silicon rate is much higher than that of Portland cemen t clinker. Generally, silicon rate is between 3.5 and 5.0. If silicon rate is too small, the relative content of Al_2O_3 will be high,

Table 3.33: The whiteness of cement clinker with different mineral compositions.

No.	Mineral composition (mass fraction, %)				$w(C_3S)/w$ (C_2S)	Whiteness of clinker (compared with $BaSO_4$, %)
	C_3S	C_2S	C_3A	C_4AF		
1	83.5	0.0	15.0	1.5	∞	84.3
2	50.1	33.4	15.0	1.5	1.5	73.0
3	25.1	58.4	15.0	1.5	0.43	70.3
4	68.0	17.0	14.35	0.65	0.4	85.0
5	34.0	51.0	14.35	0.65	0.67	79.0

resulting in an increase of C_3A content and rapid setting of cement. If the silicon rate is too high, the mineral content of melt in the cement clinker will be low, making calcination difficult to reduce white degree of the clinker due to higher C_2S content.

Because of the extremely low content of Fe_2O_3 in white cement clinker, the aluminum rate (*IM*) is generally not considered as a production control index. It is normally higher than 12.

The preparation of raw materials of white silicate cement is similar to that of Portland cement. Different raw materials are usually kept in storage after two-stage crushing.

Due to different Fe_2O_3 contents in limestone with different particle sizes, some white cement plants apply a rich selection process for crushed limestone to remove particles with a particle size smaller than 5 mm (outside China). And this can obviously decrease the Fe_2O_3 content in raw materials. For grinding process of raw materials, various raw materials are ground by roller mills in white cement plants in China, and then blended with mixers as required for proportion to reduce mixed metals.

(3) Calcining and cooling of cement clinker

The sintering temperature for white silicate cement is generally higher than that of Portland cement due to low flux mineral content, high lime saturation coefficient and high silicon rate. It can reach 1,500–1550 °C. In order to reduce the sintering temperature, it is better to add a small amount of mineralizer in raw materials. Fluorite mineralization agent (CaF_2) is the most commonly used one, and it should not be added more than 0.5% (to the mass fraction of clinker). In this case, the whiteness of clinker can be improved, while the whiteness will decrease when the addition is more than 0.5%.

When the ordinary Portland cement is calcined, the atmosphere in the kiln is oxidizing atmosphere. The atmosphere of kiln can significantly affect the whiteness of cement clinker.

Table 3.34 lists the whiteness of raw materials with the same composition that is calcined in different atmosphere and then fast cooled in the water. In the

Table 3.34: The cement clinker whiteness in different calcination atmospheres with the same cement raw material compositions.

Calcination atmosphere	Cement clinker whiteness W (%)
Oxidation	81.9
Neutralization	85.0
Reduction	87.7

nonoxidizing atmosphere, the mechanism to increase clinker whiteness is to change its phase composition and structure. In the reduction atmosphere, the strongly colored trivalent iron (Fe^{3+}) is reduced to a low-colorimetric divalent iron (Fe^{2+}); thus, the whiteness is improved. In order to obtain a good reduction atmosphere, in addition to considering the atmosphere of the kiln, it can also add a small amount of petroleum pitch or coke into the raw materials to replenish the reduction atmosphere. In addition, the whiteness of clinker can also be improved by using refractory lining with low chromium and low iron in the calcining zone and transition zone.

The whiteness differs with different cooling methods and speeds. Generally, a higher initial temperature of cement clinker with a faster cooling speed leads to higher whiteness. Under the condition of rapid cooling, the aluminate cannot be massively dissolved into iron phase in time. Therefore, a small amount of C_6AF_2 iron aluminate with lighter color is generated. At the same time, by quenching, a large number of minerals in the cement clinker are fixed in the form of fine crystals, which increase the diffuse scattering of light to enhance whiteness. The most used method is to directly put the high-temperature cement clinker into cold water or to spray a lot of water on the surface of high-temperature cement clinker. Table 3.35 demonstrates the effect of cooling methods on cement clinker whiteness.

Table 3.35: The effect of cooling method on clinker whiteness.

Cooling method of cement clinker	Cement clinker whiteness W (%)
1,450 °C, Cement clinker quenching in water	83
1,350 °C, Cement clinker quenching in water	81
1,250 °C, Cement clinker quenching in water	77
1,200 °C, Cement clinker quenching in water	73
Cement clinker cooling slowly in kiln	63

The whiteness of cement clinker can also be improved by bleaching clinker in a special bleaching device because the cement clinker is affected by different media. Table 3.36 lists the whiteness of cement clinker cooled in different media. In reduction media of

Table 3.36: The effect of cooling media on cement clinker whiteness.

Cooling media of cement clinker	Cement clinker whiteness W (%)
Oxidation state (air)cooling	75
Neutralization state (nitrogen)cooling	80
Reduction state (hydrogen)cooling	84
Quenching in water	85.5
Quenching in water after conversion by natural gas	90.3

hydrogen, cracking gas, natural gas and coal gas, the strongly colored Fe_2O_3 can be reduced to low-colorimetric FeO. The white cement clinker is treated by natural gas for about 1 min, and then put into cool water, the corresponding whiteness is relatively high. This type of device must be sealed tightly when in use and cannot let oxygen get mixed in bleached gas; otherwise the bleaching effect will be reduced.

It is also a measure to improve the whiteness by contacting high-temperature cement clinker out of kiln with the powder cooling material. Because of the contact, the cooling material is wrapped in the outer surface of high-temperature clinker, which does not allow the clinker to come in contact with the air, and bleach the clinker by cold and heat exchange. In China, some plants use kiln ash to cool the produced white cement clinker or mix the kiln ash with the produced white cement clinker. These methods not only can help in clinker cooling, but also can reduce production costs of white cement.

(4) Drying and grinding of cement clinker

The cement clinker usually contains a lot of moisture residue after bleaching, and should be dried before grinding with gypsum for producing the cement. The drying equipment generally adopts rotary dryer, and the drying temperature has a certain influence on the whiteness of clinker. Cement whiteness at different drying temperatures in lab conditions are listed in Table 3.37.

Table 3.37: Cement whiteness at different drying temperatures in lab conditions.

Drying temperature /°C	200	250	300	400	500	700
Cement whiteness (%)	91.9	91.1	89.3	88.4	87.1	86.0

From Table 3.37, it can be seen that the drying temperature must not be higher than 300 °C; otherwise, the whiteness is reduced.

Cement grinding is carried out in a pipe mill. In order to prevent pollution of iron and its oxides, the mill liner is made of granite, ceramic or high-quality wear-resisting steel. The grinding machine is made of siliceous pebbles or high chromium cast iron. The iron conveyance equipment needs to be carefully painted to prevent iron dust from mixing.

During grinding, the whiteness of used gypsum must be higher than that of clinker to guarantee the whiteness of cement. In this case, high-quality fibrous gypsum is usually used after impurities are washed out.

It is allowed to add 5% limestone or kiln ash according to the Chinese standards. As a blending material, the whiteness of limestone must be higher than 85%, and the content of Al_2O_3 should not exceed 2.5%. The whiteness of kiln ash must exceed 70%. It is also allowed to incorporate suitable grinding agents with an amount less than 0.5% to cement mass that does not affect cement performance. The use of sunflower fatty acid residues as grinding agents for white cement is more effective in increasing the yield, reducing energy consumption and improving whiteness than other similar residues (such as seed, hydrogenated fats and fish oil fatty acid).

3.6.2 Color cement

The production methods of color cement can be divided into indirect method and direct method.

(1) Indirect method
Indirect method means that the color cement is made by grinding white Portland cement or ordinary Portland cement together with color pigments (or used in the field). The pigments are required to be strong coloring, insoluble in water, easy to disperse, durable under light and atmosphere, alkali resistant and not damaging for cement. The commonly used pigments are iron oxide (red, yellow and brown red), manganese oxide (black and brown), chromium oxide (green), ochre (ochre), group cyan (blue), carbon black (black) and so on. Red, brown and dark-colored cements are produced by using ordinary Portland cement clinker. Light-colored cements are produced by using white Portland cement clinker.

(2) Direct method
The direct method is to add coloring matters to cement raw materials. Colored cement is made by calcining color cement clinker and grinded. For example, adding Cr_2O_3 can get green color, adding CoO can get light blue in reduction flame and get rose red in oxidation flame. Adding Mn_2O_3 can get light yellow color in reduction flame and get light purple color in oxidation flame.

The intensity of color varies with the dosage of colorant (0.1–2.0%). The shortcoming of this method is that the color agent is little in clinker, and not easy to control accurately and evenly. Moreover, the change of atmosphere in the kiln will also cause uneven color of clinker. In addition, when preparing products, the color will be diluted due to hydration of color clinker minerals.

In raw materials of aluminate or sulfur aluminate cements, adding various color agents can get color aluminate or sulfur aluminate cement clinkers after calcination, and various color cements are then made by grinding. This type of cement has bright color and high early strength. Since the hydration of this type of cement does not precipitate $Ca(OH)_2$, there is no color fading phenomenon compared to common color cement.

Color cement can be used to produce color cement paint, and make color cement slurry and mortar for producing color cement products.

3.7 Masonry cement

In the current residential buildings in China, the brick-concrete structure still accounts for a large proportion, and accordingly masonry mortar is a building material in great demand. Therefore, it is of great practical significance to save cement and energy, and reduce costs in brick-concrete constructions.

Masonry mortar made for building constructions in China is usually produced by 32.5 and 42.5 strength grade cement, and the commonly used mortar is M5 grade (5.0 MPa). The requirement on strength is not high; the strength ratio of cement to mortar is significantly higher than the value required by the general technical economy principle (4–5 times). However, in order to meet the workability requirements of masonry mortar, it is often necessary to use more cement, resulting in high strengths of masonry mortar and waste of cement. It is necessary to produce masonry cement with low strength grades.

The production method of masonry cement is the same as that of ordinary cement but with low dosage of clinker and high dosage of blending materials. In order to produce masonry cement, clinker and blending materials can be ground separately and then mixed together, or ground separately and then mixed ground, or directly mixed ground. The production manner of masonry cement should be in accordance with the properties of components and grinding equipment. When producing fly ash masonry cement, it is reasonable to apply the two-stage grinding process. That is, cement clinker and gypsum powders are ground and sieved by the 0.08 mm^2 hole sieve with a residue of about 35%, and then ground together with fly ash to produce products.

Masonry cement is suitable for the masonry mortar of industrial and civil buildings, the mortar of inner walls and foundation cushion and so on. It is allowed to produce block, tile and so on. It is not used for the preparation of concrete; it is allowed to be used for low-strength grades concrete by tests, but not for load-bearing structures such as reinforced concrete.

Currently, masonry cement has two standards to apply: one is Chinese standard *Masonry cement* (GB/T 3183–2017), and the other is industry standard of building materials *Steel slag masonry cement* (JC/T 1090–2008).

3.7.1 Chinese standards for masonry cement

Regulated by Chinese standard *Masonry cement* (GB/T 3183—2017), masonry cement, with the code of M, is a hydraulic cementing material with a good water-retaining property made by grinding Portland cement clinker mixed with specified blending materials and gypsum with a proper amount.

Masonry cement is made up of clinker qualified with GB/T 21372—2008, natural gypsum qualified with GB/T 5483—2008 or industrial by-product gypsum qualified with GB/T 21371—2008, active or (and) inactive blending materials or (and) kiln ashes qualified with JC/T 742—2009.

Active blending materials are granulated blast furnace slag qualified with GB/T 203—2008, fly ash qualified with GB/T 1596—2017, pozzolanic materials qualified with GB/T 2847—2005, granulated phosphorous slag from electric furnace qualified with GB/T 6645—2008 and granulated blast furnace titanium slag qualified with JC/T 418—2009 (2015).

Inactive blending materials are granulated blast furnace slag with an activity below the regulation in GB/T 203—2008, fly ash with an activity below the regulation in GB/T 1596—2017, pozzolanic materials with an activity below the regulation in GB/T 2847—2005, granulated phosphorous slag from electric furnace with an activity below the regulation in GB/T 6645—2008, granulated blast furnace titanium slag qualified with JC/T 418—2009 (2015) and limestone powder qualified with GB/T 35164—2017.

Grinding aids qualified with GB/T 26748—2011 can be added during cement grinding, and the addition amount shall not be higher than 0.5% of the cement mass.

Masonry cement is divided into three strength grades, 12.5, 22.5 and 32.5. The strengths at each age must not be lower than the data listed in Table 3.38.

Table 3.38: The required strengths of masonry cement at each age.

Strength grade	Compressive strength (MPa)			Flexural strength (MPa)		
	3 days	7 days	28 days	3 days	7 days	28 days
12.5	–	≥7.0	≥12.5	–	≥1.5	≥3.0
22.5	–	≥10.0	≥22.5	–	≥2.0	≥4.0
32.5	≥10.0	–	≥32.5	≥2.5	–	≥5.5

The technical requirements for masonry cement are as follows:
(1) The content of SO_3 in cement should not be more than 3.5%.
(2) The content of chloride ions in cement should not be higher than 0.06%.
(3) The content of water-soluble chromium (VI) in cement should not be greater than 10 mg/kg.

(4) The sieving residual amount from 80 μm^2 hole sieve is no higher than 10.0%.
(5) The initial setting time is no less than 60 min, and the final setting time is no longer than 720 min.
(6) The stability of cement must be qualified by boiling test.
(7) The water retention rate should not be less than 80%.
(8) The internal radiation index I_{Ra} of cement must not be higher than 1, and the external radiation index I_γ must not be greater than 1.

3.7.2 Industrial standards for steel slag masonry cement

Chinese industry standard of building materials *Steel slag masonry cement* (JC/T 1090 2008) regulates the following: a hydraulic binding material with good workability made by converter steel slag or electric furnace steel slag, granulated blast furnace slag as the main components and adding suitable amount of Portland cement clinker and gypsum, is called steel slag masonry cement.

The steel slag in steel slag masonry cement should conform to the provisions of YB/T 022—2008. Granulated blast furnace slag should conform to the provisions of Chinese standard GB/T 203—2008. The Portland cement clinker should conform to the provisions of Chinese standard GB/T 21372—2008. Gypsum should conform to GB/T 5483—2008.

SO_3 content in steel slag masonry cement must not exceed 4.0%. If the water leaching stability is qualified, SO_3 content can be relaxed to 6.0%. The specific surface area of steel slag masonry cement should not be less than 350 m^2/ kg. The initial setting time must not be earlier than 60 min, and the final setting time must not be later than 12 h. The stability of steel slag masonry cement is tested by boiling method to be qualified. The cement made by steel slag with MgO content over 5% must pass the pressure and steam stability test. When the content of MgO in steel slag is 5–13% and the content of granulated blast furnace slag is more than 40%, the produced cement need not perform pressure and steam test. However, if the content of SO_3 in cement is more than 4.0%, the water leaching stability test is necessary. Cement retention rate should not be less than 80%. Steel slag masonry cement can be divided into 17.5, 22.5 and 27.5 strength grades, and the strength of each grade cement in different hydration stages must not be less than the values listed in Table 3.39.

3.8 Radiation-proof cement

With the rapid development of atomic energy and wide applications of radio-active isotopes in various fields, the problem of physical protection is becoming more and more prominent. The protection consists of three aspects: slowing fast neutrons, capturing slow or initial slow neutrons and absorbing all kinds of

Table 3.39: The strength of each grade steel slag masonry cement.

Strength grade	Compressive strength R_c (MPa)		Bending strength R_f (MPa)	
	7 days	28 days	7 days	28 days
17.5	≥7.0	≥17.5	≥1.5	≥3.0
22.5	≥10.0	≥22.5	≥2.0	≥4.0
27.5	≥12.5	≥27.5	≥2.5	≥5.0

γ-rays and X-rays. The ability of a moderated neutron is inversely proportional to its own mass number; therefore, hydrogen is a good neutron-retarding material, while the ability to absorb γ-rays and X-rays increases with the increase in the atomic weight.

3.8.1 Barium cement

Barium cement is a binding material made by grinding clinker with $3BaO \cdot SiO_2$ as the main mineral and suitable amount of gypsum. The clinker is obtained by calcination of raw materials with proper compositions ground by raw materials of barite and clay with addition of a small amount of coke for helping decomposition of barite. The production process of barium cement is the same as that of ordinary Portland cement. The barite w ($BaSO_4$) is required to be higher than 88%, $w(SiO_2)$ is less than 9% and the quality of barite must be uniform. The barite and clay are ground to the specified fineness, and then mixed evenly with a suitable ratio because the uniformity of raw materials can directly affect the calcination in kilns. The preparation of raw materials should meet the range of the following clinker ingredients (mass fraction, %): $w(BaO)$ 79–81%, $w(SiO_2)$ 14–16%, $w(Al_2O_3) < 4\%$, $w(Fe_2O_3) < 3\%$, $w(BaO)/w(SiO_2) = 5.40$–5.56. The raw materials are calcined in the kiln to produce clinker at the calcination temperature of 1,500–1,600 °C, and then the clinker is ground to the specific surface area of 350–450 m²/kg by adding a suitable amount of gypsum powders to make barium cement.

The density of barium cement is higher than that of Portland cement, reaching 4.5–5.0 g/cm³. The initial setting time of barium cement is 20–30 min and the final setting time is 40–60 min. The early strength of barium cement is high, and the 3 days compressive strength can generally reach over 65% of 28 days. Prepared barium cement concrete has the strong absorption effect to γ-rays, and it is mainly used in the production of radioactive isotope protection facilities, such as protective shield walls, source storage rooms, activity protection and so on and the low-temperature parts of atomic reactor shells. It must not be used as a heat radiation protection wall. Table 3.40 lists the required minimum thickness of materials including barium cement concrete to prevent X-rays.

Table 3.40: Required minimum thickness of materials to prevent X-rays.

Protective material type	Minimum thickness under the following voltage (mm)	
	60 kV	200 kV
Lead plate	0.9	4
Barium cement	7	29
Lead glass	8	34
Ordinary cement concrete	65	270

3.8.2 Boron-containing cement

Boron-containing cement is a cementitious material made by grinding aluminate cement clinker and calcined camsellite and natural anhydrite in proper amounts to required fineness.

The 3 days strength of aluminate cement clinker used in boron-containing cement must be above 42.5 MPa. The calcination loss of calcined camsellite should be less than 2%, and the amount of B_2O_3 is higher than 20%. It is recommend to control the rotary kiln temperature to 800–1,000 °C when calcining camsellite and anhydrite. For example, the main chemical compositions in boron containing cement (mass fraction, %) are $w(B_2O_3)$ 8.61%, $w(Al_2O_3)$ 30.63%, $w(SiO_2)$ 6.44%, $w(Fe_2O_3)$ 5.21 %, $w(CaO)$ 25.45%, $w(SO_3)$ 8.03%, $w(MgO)$ 13.02%.

The high content of crystal water and a certain boron content in boron-containing cement have a significant effect to slow down the fast neutron, absorb and reduce thermal neutron and prevent γ-ray radiation. Table 3.41 lists a comparison between crystal water content and the boron element in several types of cement concrete.

Table 3.41: The contents of crystal water and boron in several types of cement concrete.

Concrete type	Mix proportion (cement: fine aggregate: coarse aggregate)	w/c ratio	Boron content (kg/m³)	Crystal water content (kg/m³)
Boron cement, gravel, concrete	1:2.1:4.5	0.47	7.91	81.0
Boron cement, boron magnesium sand, barite	1:1.9:4.2	0.43	37.60	181.2
Ordinary cement, gravel concrete	1:2.0:4.6	0.50		64.8

The early strength increasing rate of boron cement is high. Because of the certain content of B_2O_3 and high content of chemically bonding water, the boron elements contained in the former can absorb thermal neutron to decrease and capture the radiation and heat shield. The hydrogen in bonded water can slow down fast neutrons. Concrete with higher volume density and more boron is made of this boron-containing cement together with boron aggregates and heavy material aggregates to prevent mixed radiation (γ-rays and neutrons). The concrete is suitable for fast neutron and thermal protective shield engineering, such as nuclear reactors, particle accelerators and neutron applications laboratory biological shield, the atomic radiation shield defense engineering and so on.

3.8.3 Strontium cement

By replacing the raw material of limestone in Portland cement with strontium carbonate, strontium cement is produced by grinding the clinker with tristrontium silicate as the main mineral and a suitable amount of gypsum. Its chemical compositions (mass fraction, %) are as follows: $w(SrO)$ 71–76%, $w(SiO_2)$ 10–15%, $w(Al_2O_3)$ 4–7%, $w(Fe_2O_3)$ 3–6% and $w(MgO)$ 0–2%. Its sintering temperature is higher than that of Portland cement, around 1,550 °C, and the relative density is higher than that of ordinary Portland cement. It can be made into a uniform and solid anti-radiation concrete with heavy materials, but the radiation performance is slightly worse than that of barium cement.

3.8.4 Conducting and magnetic cement

Cement as a hydraulic binding material can bind particulates (such as sand and stone) or block materials (such as brick, tile, etc.) into a whole material in construction, which can be used for building projects in ground, underground and underwater. Portland cement itself does not have the conductive and permeability function. Cement can be made with conductivity and magnetic permeability by filling conductive and magnetic materials during manufacturing. In addition to the general cement characteristics, the conductive and magnetic cement has the characteristics of conduction and magnetic conductivity.

There are several specific manufacturing methods to produce conducting and magnetic cement.

(1) Conducting and magnetic Portland cement or ordinary Portland cement with different strength grades are made by no more than 3–10% gypsum, 3–20% conductive graphite and 100–500% iron powders (to cement mass).

(2) Conducting and magnetic sulfoaluminate cement are produced by grinding clinker, limestone, a certain amount of gypsum, no more than 3–20% conductive

graphite and 100–500% iron powders (to cement mass). The main components in clinker are anhydrous calcium aluminate and dicalcium silicate, obtained from the calcination of raw materials.

(3) The iron sulfide slag iron powders and cement are mixed in a suitable proportion with the addition of appropriate amount of rare earths. According to the iron content of sulfur slag, the proportion of iron powder is 20–80%, and the rare earths ratio is 0.01–5%. If the proportion of iron powders is increased appropriately, the conductive cement can be made.

Conductive magnetic cement can be used for electromagnetic shielding, electromagnetic wave absorption and heating. By using its electrical conductivity, the positive and negative electrodes are installed on the corresponding side, and the heat can be generated by the resistance, which can be used to melt ice for facilities as well as can be used in antimoisture and anticold areas.

3.8.5 Nonmagnetic cement

In defense engineering and precision engineering, the concept of low-magnetic concrete is proposed in order to prevent magnetic interference and magnetic exposure. Since cement is the main component of concrete performance, it is necessary to obtain low-magnetic cement to make low-magnetic concrete. Therefore, the concept of nonmagnetic cement is presented.

Nonmagnetic cement is a kind of cement with an extremely low magnetism. The cement is mainly used for the storage of precision instruments, positioning, navigation and other buildings that have less demand on magnetic properties of cement, and the buildings against geomagnetic and electromagnetic interference.

There is a certain relationship between magnetism and compositions of cement, such as Ca, Si and Al are typical weak magnetic materials, and so the magnetism of cement mainly depends on the content of Fe. That is to say, the Fe_2O_3 component in cement is the main reason for magnetic property. Ordinary cement and compound cement have certain ferromagnetism, while white cement is a typical weak magnetic material. White cement is made of high-purity limestone and white clay (do not contain iron oxide and manganese) with low ash fuel. During the grinding process, hard stones are used instead of liner and steel balls in ball mills to prevent iron incorporation. The main minerals in the clinker are C_3S, C_2S and C_3A. It cannot be widely applied in engineering due to limitation in strengths.

In addition to the influence of Fe content, magnetism of silicate-type cement is mainly affected by iron filings coming from abrasion of liner and steel balls in ball mills when mixing cement products during production. The main measure to decrease magnetism of silicate-type cement is to decrease the mixing amount of iron filings from abrasion as much as possible by using high-strength stone materials to replace steel liner and steel balls in ball mills during cement production.

In order to facilitate constructions, the residual magnetism can be used as an index to identify the magnetic properties of Portland cement. The index value is 2.0 Gs (magnetic field strength unit, Gauss). It is generally considered that if the value is less than or equal to this value, the cement can be considered with nonmagnetic properties and vice versa.

3.9 Acid-proof cement

Acid-proof cement is the earliest used cement in chemical industry to resist acid corrosion. In the early twentieth century, the air-hardening cement including water glass acid-proof cement and sulfur cement were produced. In the mid of twentieth century, with the rapid development in polymer industry, the acid-proof binding materials, which use synthetized resin as a main binding component, were proposed.

3.9.1 Water glass acid-proof cement

Water glass acid-proof cement can harden in air and resist corrosions of most inorganic and organic acids. It is made by grinding acid-proof fillers and hardening agents in a proper ratio together or separately before mixing uniformly into powders, and then by mixing with a proper amount of water glass solution.

Water glass is the binding material in water glass acid-proof cement; it is expressed by $R_2O \cdot nSiO_2$. R_2O represents Na_2O or K_2O; n represents the molar ratio of SiO_2 to R_2O, and it is called modulus. Soluble silicate is divided into sodium silicate and potassium silicate. Sodium silicate has a lot of advantages including low price and extensive source, leading to wide applications. The two technical indexes of soluble silicate are modulus and density. When producing the acid-proof cement, the modulus and density of soluble silicate are 2.6–2.8 and 1.38–1.45 g/cm^3, respectively.

There are many types of fillers. Crushed natural acid-proof materials, such as neutral feldspar, quartz keratophyre, quartzite, dense feldspar, pyroxite, powdery quartz and ceramic powders, and artificial stone materials such as cast stones can be used as fillers in soluble silicate acid-proof cement. The requirements of fillers in acid-proof cement are as follows: the acid-proof degree should be larger than 93%, fillers should have good adsorption ability to soluble silicate, the water retention rate should not exceed 0.5% and the fineness requires the 0.08 mm^2 hole sieving residues less than 15%. While the extremely fine powders are also detrimental because they will increase the amount of soluble silicate to cause shrinkage and decrease cement durability. The amount of fillers depends on the required consistence of cement. Usually, the consistence is within the range of 7–15 cm.

The commonly used hardening agent is fluorine sodium silicate in soluble silicate acid-proof cement. Fluorine sodium silicate has a small solubility in water (0.56% at 17 °C), and its water solution presents acidic property due to hydrolysis. The reaction rate will not be too fast due to low solubility to make the produced cement paste lose its workability. On the other hand, it can react with soluble silicate due to its hydrolysis. Therefore, fluorine sodium silicate can be considered as a setting and hardening promoting agent for soluble silicate. Due to the low solubility, it is necessary to grind the fluorine sodium silicate powders fine enough. The required fineness is that residues of 0.08 mm² hole sieving are less than 10%. The amount of fluorine sodium silicate is about 14–18% of soluble silicate mass when producing the acid-proof cement.

The production process of soluble silicate acid-proof cement is as follows: fillers and hardening agents are filled into mixing equipment according to proportioning (generally, 940–960 g powder fillers, 40–60 g hardening agents and 250–350 g soluble silicate in every kilogram dry mixture). After stirring for 2 min of the dry mixture, a certain amount of soluble silicate is added, and they are stirred wetly for 2 min to produce homogeneous slurry.

The setting time of soluble silicate acid-proof cement depends on proportion, properties of soluble silicate and the ambient temperature. Decreasing the modulus and increasing the density can increase the setting time, and vice versa. The setting time can be reduced when the amount of fluorine sodium silicate is increased or ambient temperature is increased. The setting time of soluble silicate acid-proof cement is about 2–3 h in summer, and will be prolonged to 5–6 h or longer in winter.

The strength of soluble silicate acid-proof cement depends on the quality, type, proportion of raw materials, as well as curing temperature and humidity. The strength and durability can reach the optimum when the curing temperature and relative humidity are 18–30 °C and 80–90%, respectively, in engineering. Low curing temperatures will result in a low early strength. Cured by heating at early ages will result in fast water loss of soluble silicate and decrease strength. The strength can be improved by properly increasing the density of soluble silicate. The compressive strength of soluble silicate acid-proof cement can reach 50–60 MPa after 10 days of curing.

Soluble silicate acid-proof cement has a good ability to resist concentrate acid, and poor ability to dilute acid. This is because although the acid has high concentration, the viscosity is also high and leads to low destroying ability. Furthermore, the acid solution can react with unreacted soluble silicate to precipitate silicate gels to help improve the strength. Because of the gradual leaching of soluble salts (products of reaction between soluble silicate and fluorine sodium silicate) under water condition, pores are generated in the inner space of cement. At the same time, the dehydration and condensation of silicate gels with evaporation of water also result in the formation of micropores in the inner space of acid-proof concrete. Therefore, the compactness is not high, and the cement has poor antipermeability with high shrinkage. These characteristics enable the cement to resist corrosion of organic solvents and acidic gases, but it has poor water resistance. This type of cement also cannot resist the corrosions of hot phosphate (over 300 °C), hydrogen fluorinate and higher aliphatic acid. The soluble silicate acid-proof cement is able to endure over 300 °C of high temperature, and endure less than 1,000 °C of high temperature when using firebrick powder as filler.

The soluble silicate acid-proof cement has high mechanical strength and good acid-proof ability; it can be made into concrete bricks and acid-proof components. It also can be made as acid-resistant plaster, mortar and packing in the chemical industry, metallurgy, papermaking, sugar and textile industries and the general acid resistance engineering. However, in food industry, the toxicity of sodium fluorosilicate must be considered.

3.9.2 Sulfurous acid-proof cement

Sulfur powders are melted (heating to 130–150 °C), and fillers preheated to 110 °C are gradually added in batches within 30 min with stirring before adding polysulfide rubber with 20 min continuously stirring till bubbles disappear. Sulfurous acid-proof cement is then produced when temperature is raised to 170 °C and kept for 1 h with stirring.

Sulfur is the binding material in sulfurous acid-proof cement, and toughening agents are polysulfide rubber or polyvinyl chloride. Sometimes, epoxy resin can be added to decrease shrinkage cracks and improve bonding force. Sulfurous acid-proof cement has the same fillers with soluble silicate acid-proof cement. In China, the relative ratios of producing sulfurous acid-proof cement are 58% sulfur, 40% acid proof filler and 2% polysulfide rubber.

Sulfurous acid proof cement has a poor thermal stability, and the use temperature generally should not exceed 90 °C. But it can endure corrosions caused by any concentration of sulfuric acid, hydrochloric acid, phosphoric acid, as well as nitric acid within 40% concentration and acetic acid within 50% concentration. When used in engineering with hydrofluoric acid or fluorosilicic acid, carbon powder (such as graphite) is used as filler.

Sulfuric acid-resistant cement is used as an acid-proof bonding material, which is used in chemical plants as an adhesion agent of acid-resisting vessel bricks and anticorrosion floor and floor joints, and also used in fixing equipment foundation embedded parts, repairing concrete pipes and pools and other components. It is also used for bonding electric porcelain bottles in electrical industry and bonding concrete sleeper components in railway departments.

3.9.3 Polymer acid-proof cement

The polymer acid-proof cement uses synthetic resin as a binding material, and it is made by adding curing agent, fillers and toughening agent or thinner with warm or properly heating curing.

Epoxy, phenolic and unsaturated polyester polymerized cement are mainly used in domestic construction anticorrosion engineering. The corrosion resistance of polymer cement mainly depends on the corrosion resistance of used polymer.

The corrosion resistance of polymeric acid proof cement is excellent, but the price is relative high. It is mainly used as bonding layer, filleting and filling material, binding material, sealing material and equipment foundation covering and so on.

3.10 Low-alkalinity sulfoaluminate cement

Glass fiber–reinforced cement is a promising high-strength lightweight composite, which is a cement-based, glass fiber–incorporated material with high tensile and compressive strengths. This high-quality composite is difficult to produce by compositing ordinary Portland cement and common fiberglass. For ordinary Portland cement, its hydration will produce a lot of $Ca(OH)_2$; this precipitated product has a high alkalinity (pH = 12.5), which will react with SiO_2 of ordinary glass fiber to form hydrated calcium silicate. This type of reaction is irreversible, until the reactant runs out. In this case, the glass fiber will be corroded to become brittle, lose strength and fail to the desired quality. To prevent corrosion of fibers, three ways are proposed.
(i)   Using low-alkalinity cement with the generation of a small amount or without Ca $(OH)_2$ during hydration;
(ii) Improving the chemical stability of framework of glass fibers;
(iii) Coating and isolating glass fibers.

In addition to the research and development of high alkali–resistant glass fibers in China, low-alkali sulfur-aluminate cement has been successfully developed, which is suitable for glass fiber–reinforced cement products produced with various glass fibers.

Regulated in Chinese standard GB 20472–2006, low-alkalinity sulfoaluminate cement is a hydraulic hardening binding material with low alkalinity made by grinding sulfoaluminate cement clinker with proper composition and limestone in a large amount, and an appropriate amount of gypsum. The limestone content must not be less than 15%, and no more than 35% of cement mass.

In low-alkalinity sulfoaluminate cement clinker, the Al_2O_3 content (mass fraction, %) should not be less than 30% and SiO_2 content (mass fraction, %) should not be higher than 10.5%. The 3 days compressive strength of the clinker must be not less than 55.0 MPa.

The strength grade of low-alkalinity sulfoaluminate cement is indicated by its 7 days compressive strength, and strength grades are 32.5, 42.5 and 52.5. The physical properties of low-alkalinity sulfoaluminate cement must meet the requirements listed in Table 3.42, and strengths for each strength grade at different hydration ages must not be less than the values listed in Table 3.43.

Table 3.42: Requirements on physical properties of low-alkalinity sulfoaluminate cement.

Material properties		Index requirements
Specific surface area, $S(m^2/kg)$		≥ 400
Setting time[a] (min)	Initial setting time	≥ 25
	Final setting time	≤ 180
Alkalinity (pH)		≤ 10.5
28 days free expansive degree, $\varepsilon(\%)$		0–0.15

[a] According to real situation.

Table 3.43: Lower limiting strengths of low -sulfoaluminate cement for each strength grade at different hydration ages.

Strength grade	Compressive strength R_c (MPa)		Bending strength R_f (MPa)	
	1 day	7 days	1 day	7 days
32.5	≥ 25.0	≥ 32.5	≥ 3.5	≥ 5.0
42.5	≥ 30.0	≥ 42.5	≥ 4.0	≥ 5.5
52.5	≥ 40.0	≥ 52.5	≥ 4.5	≥ 6.0

The main hydration products of low alkalinity are dehydrated calcium sulfoaluminate ($C_4A_3\bar{S}$), dicalcium silicate (β-C_2S) and calcium sulfate ($CaSO_4$). During proportioning, the concentration of CaO in the liquid phase can be decreased to less than 83 mg/L and the pH can be reduced to less than 10 as long as gypsum is greatly overdosed. In this case, the hydration products are mainly Ettringite, dihydrate gypsum and hydrated alumina gels. The fineness of cement is determined by the production process of glass fiber–reinforced cement products.

The experiment results of accelerating at 50 °C show that the corrosion effect of low-alkalinity sulfoaluminate cement on all types of glass fibers is lower than that of any current hydraulic hardening cement. When composited with alkali-resistant glass fibers, the fiber strength retention rate is more than 90% after accelerating test in one year.

The low-alkalinity sulfoaluminate cement has strong carbonation resistance but poor heat resistance. In actual applications, the ambient temperature should not exceed 80 °C. Its antifreezing property is directly influenced by curing conditions and w/c ratio. The frost resistance can be enhanced by carefully curing and using water-reducing agents. Low-alkalinity sulfoaluminate cement is mainly used to produce glass fiber–reinforced cement products by compositing with various glass fibers to make various sheets, small wave tiles, composite outer wall boards, air duct and activity rooms. It is not suitable for concrete products and structures such as steel fibers, steel reinforcement, wire mesh and steel-embedded parts.

Problems

3.1   What are the main ways to produce special cements?

3.2   According to the properties of aluminate cement, what are the problems to be paid attention to while using it?

3.3   Why aluminate cement has properties of high early strength and good high-temperature resistance?

3.4   What are the common refractory castables?

3.5   Try to compare the similarities and differences of rapid hardening Portland cement, rapid hardening sulfoaluminate cement, rapid hardening fluoroalumi-nate cement, rapid hardening iron aluminate cement and rapid hardening high-strength aluminate cement.

3.6   What are the similarities and differences in production methods between rapid hardening Portland cement and ordinary Portland cement?

3.7   What are the characteristics of mineral compositions in rapid hardening fluor-aluminate cement clinker and rapid hardening sulfoaluminate cement clinker?

3.8   What is ultrarapid hardening aluminate cement with regulated setting? What are the point that should be paid attention to while using it?

3.9   What are the technical requirements and properties of sulfate-resistant cement? What are the main functions?

3.10 What are medium- and low-heat Portland cement and low-heat slag cement? What are the characteristics of its production process?

3.11 What are the similarities and differences between low-heat microexpansive cement and low-heat steel slag Portland cement?

3.12 What points should be paid attention in the production process of road cement?

3.13 What are the technical requirements of cement concrete pavements for roads? Why?

3.14 What are the differences between steel slag road cement and road Portland cement?

3.15 How to produce expansive cement? What are the main types of expansive cement? What are its main applications?

3.16 What is self-stress? How is it generated? What are the types of self-stress cement?

3.17 What are the grades of oil well cement? What are the requirements for oil well cement? What are the technical properties for each grade of oil well cement?

3.18 What are the main controlling factors to increase whiteness of white Portland cement?

3.19 How to produce color cement?

3.20 What is masonry cement? What is the significance of producing masonry cement?

3.21 What are the functions of antiradiation cement? What are the types?

3.22 What are the common acid-resistant cement varieties? What are the applications?

3.23 What components are mixed to make soluble silicate acid proof cement? What is the approximate proportion?

3.24 What is low-alkalinity cement? Why is low-alkalinity cement produced?

4 Gypsum

Gypsum is an air-hardening cementitious material with calcium sulfate as the main component, which is important for building materials. Gypsum and its products have excellent properties, such as light weight, fire resistance, sound insulation, thermal insulation and so on. Owing to its simple production process with low-energy consumption, gypsum is a perfect energy-efficient material.

4.1 Raw materials for the production of gypsum

Natural gypsum is the main raw material for the production of gypsum, which is the gypsum stone embodied in nature and is an important nonmetallic mineral resource with wide range of uses. The discovery and usage of gypsum have a long history. For example, the Mawangdui tomb was constructed of plaster as cementitious material in Changsha, China, 2,000 years ago. Gypsum was abounded near Paris in France and plaster had been widely used in masonry and wall plastering of building from the seventh century to the end of the Middle Age. Thus, the plaster is also known as the plaster of Paris. In everyday life, gypsum is used as the coagulant for making tofu and chalk, sculpture, art and craft goods and so on; it can be said that gypsum exists everywhere.

The resources of gypsum are abundant in China. The proven reserves of natural gypsum of 4.715×10^{10} t ranks the first in the world, mainly including the gypsum mines in Nanjing, Dawenkou, Pingyi and so on. With the development of economy, gypsum has been listed one of the top development in nonmetallic minerals industry in China. In addition to the raw material used in cement industry, architectural gypsum products play an important role as new interior wall materials in China.

In addition to natural raw materials, chemical by-products and wastes (also known as chemical gypsum), consisting of $CaSO_4 \cdot 2H_2O$ or the mixture of $CaSO_4 \cdot 2H_2O$ and $CaSO_4$, can also be served as the raw materials for producing gypsum. The urgency of environmental protection, that is, disposal of the large amount of chemical gypsum (phosphogypsum [PG] and desulfuration gypsum), has become the top priority. For example, annual emissions of PG have reached about 2×10^7 t and the total amount of fertilizer by-products gypsum has reached 4×10^7 t. Hence, from the perspective of ecological balance, environmental protection and rational utilization of resources, comprehensive utilization of industrial by-product gypsum has become the inevitable development trend.

https://doi.org/10.1515/9783110572100-004

4.1.1 Natural gypsum

(1) Crystal structure of dihydrate gypsum

Gypsum exists in two stable forms in the nature: one is natural anhydrous gypsum (AH), called anhydrite, with the chemical formula of $CaSO_4$. This form can only be used for producing the anhydrite cement (alunite expansive cement) because of its dense crystallization and hardness.

Another form is the natural dihydrate gypsum (DH) containing two molecules of crystal water, also known as soft gypsum, gypsum or raw gypsum, with the chemical formula of $CaSO_4 \cdot 2H_2O$. It is the main raw material for producing gypsum. The theoretically chemical compositions are that the CaO content, $w(CaO)$, is 32.57%, the SO_3 content, $w(SO_3)$, is 46.50% and the H_2O content, $w(H_2O)$, is 20.93%.

DH belongs to monoclinic system and its crystal structure is shown in Figure 4.1. A bilayer structure is formed by tetrahedral connecting between Ca^{2+} and $[SO_4]^{2-}$ with H_2O molecules distributed between the layer structures.

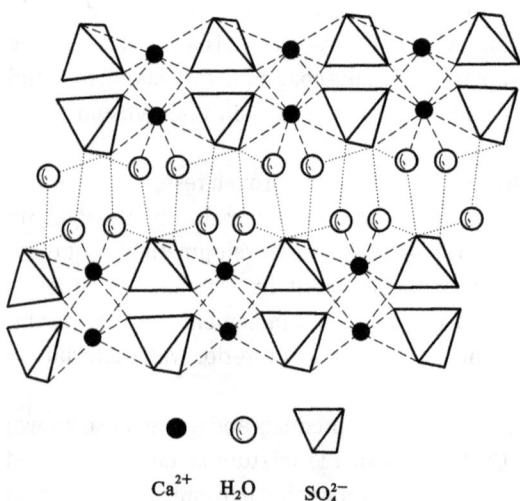

Ca^{2+} H_2O SO_4^{2-}

Figure 4.1: Crystal structure of DH.

Natural DH is white or colorless and transparent. It has a Mohs hardness of 1.2–2.0 and a density ρ of 2.2–2.4 g/cm³. The dissolubility of that in water (calculated by $CaSO_4$) is 2.05 g/L at room temperature.

DH can be divided into five categories based on the physical properties. Selenite is colorless and transparent, sometimes with slightly light and glass luster. Satin spar is fibrous aggregates with silky luster. Alabaster is grainy bulk and is white and transparent. Common gypsum is a compact bulk with dim burnish. Gypsite is earthy mixture with clay and other impurities.

(2) Quality inspection of gypsum

Stipulated by *Methods for chemical analysis of gypsum* in China national standard GB/T 5484–2012, quality inspection of gypsum is based on the mass fractions of $CaSO_4 \cdot 2H_2O$ and $CaSO_4$. Different applications of gypsum require different contents of $CaSO_4 \cdot 2H_2O$ in gypsum ores.

The methods to evaluate the quality of gypsum can be divided into two: semi-analytical method and total analytical method.

(i) Semianalytical method for gypsum

Semianalysis is also called simple analysis or basic analysis of gypsum. It refers to the analysis of the main components in gypsum, that is, the determination of the contents of crystal water, CaO and SO_3.

When gypsum is used as a cement retarder, mass fractions of $CaSO_4 \cdot 2H_2O$ and $CaSO_4$ are the criteria for its quality evaluation. It can be analyzed in accordance to China National standard GB/T 5484–2012. When it is used for producing building gypsum, normal consistency, setting time and compressive strength can be tested according to China National standard *Calcined gypsum* (GB/T 9776–2008).

(ii) Total analytical method for gypsum

Total analysis is also called the comprehensive test, including the determination of the contents of crystal water, adhesive water, CaO, SO_3, SiO_2, Al_2O_3, Fe_2O_3, MgO and so on.

4.1.2 Industrial by-product gypsums

Industrial by-product gypsums, commonly known as chemical gypsums, are by-products or industrial solid wastes (also known as industrial residues) mainly composed of $CaSO_4$ (containing 0–2 molecules of crystal water) by chemical reactions in industrial operations. The production of phosphorous chemical fertilizers and compound fertilizers is the major industry for the production of industrial by-product gypsums. By-product gypsums can also be produced during wet desulfurization of flue gas using limestone or lime, preparation of hydrogen fluoride through fluorite decomposed with sulfuric acid and preparation of citric acid with fermentation method.

(1) Phosphogypsum

PG is an industrial solid waste produced in factories of synthetic laundry powders and plants of phosphate fertilizer when preparing phosphoric acid. The main chemical compositions are CaO, sulfuric acid (expressed in SO_3), SiO_2, Al_2O_3, Fe_2O_3, P_2O_5 and F. China is the most populated country in the world, and agricultural production is important for this country. With the development of agriculture, the requirement of fertilizer is increasing. The demands of fertilizer of farmland (excluding forestry, meadowland and aquaculture) in 2005 and 2010 were 4.37×10^7 and 5×10^7 t, respectively. The components of plant nutrition are N, P_2O_5 and K_2O, in which the demand for phosphorous fertilizers is about $(1-1.2) \times 10^7$ t.

Phosphorous fertilizers are mainly prepared by natural phosphorite, in which the major component is $[3Ca_3(PO_4)_2 \cdot CaF_2]$. Phosphorus acid solution is extracted from phosphate using sulfuric acid and processed into ammonium phosphate and triple superphosphate producing high concentration phosphate fertilizer and compound fertilizer.

PG is one of the product in the reaction between apatite (or $[Ca_5F(PO_4)_3]$) and sulfuric acid:

$$Ca_5F(PO_4)_3 + 5H_2SO_4 + 10H_2O \rightarrow 3H_3PO_4 + 5(CaSO_4 \cdot 2H_2O) + HF$$

A slurry of the mixture can be formed by the reaction between phosphate ore and sulfuric acid. The liquid phosphoric acid and solid calcium sulfate residue in the mixture will be filtered and washed to separate phosphate acid and calcium sulfate, and the resultant calcium sulfate residue is PG. It has a content about 64–69% of $CaSO_4 \cdot 2H_2O$, which is the main component. In addition, there are 2–5% phosphoric acid, 1.5% fluorine, as well as free water and insoluble acidic powder residues. It will produce 5 t phosphorgypsum along with 1 t phosphoric acid. At present, the annual output of PG in China has been more than 5×10^7 t. PG can replace natural gypsum to produce ammonium sulfate and be used as agricultural fertilizer. PG qualified with the Chinese national standard *Phosphogypsum* (GB/T 23456—2009) can be used as a cement retarder and can be used to produce gypsum cementitious materials and products.

Worldwide, only 15% of the PG has been recycled in construction materials, agricultural soil improvement, retarders in cement production and other fields. The remaining 85% is stacked as a solid waste. Stacking of unprocessed PG not only takes up a lot of land, but also results in pollution in surrounding ecological environment such as groundwater, atmosphere and soil.

When sulfuric acid is decomposed with sulfuric acid, the compounds containing Fe, Al, Mg and other impurities in the phosphate rock are decomposed and dissolved in the phosphoric acid solution. A small amount of undecomposed phosphate rock, fluorine compound, acid insoluble matter and carbonized organic matter are present as small particles and precipitate out together with calcium sulfate generated in the reaction. After the phosphate solution is separated from the calcium sulfate precipitate, calcium sulfate filter cake always contains a small amount of unwashed acid because of balanced water consumption allowed to reach a certain phosphoric acid concentration.

The impurities in PG will slow down setting of plaster during hydration and will reduce product strength. Impurities can be soluble or insoluble. Soluble impurities mainly include the following.

(i) Free phosphoric acid
Hemihydrated gypsum (HH) is also known as plaster or calcined gypsum with the chemical formula $CaSO_4 \cdot 0.5H_2O$. When it is used in buildings, free phosphoric acid in HH is corrosive to structural materials, or corrosive to models and equipment used for prefabricated gypsum parts.

(ii) Calcium phosphate, dicalcium phosphate and fluorosilicate
Calcium phosphate, dicalcium phosphate and fluorosilicate will slow down setting rate of gypsum.

(iii) Potassium and sodium salts
Potassium and sodium salts result in crystal flowers on the surface of dried gypsum products.

The types of insoluble impurities are as follows.

The first type includes silica sand that exists in phosphate rocks and does not change when phosphate rocks decompose by acidolysis, unreacted minerals and organics.

The second type includes dicalcium phosphate crystallized together with calcium sulfate during acidolysis of phosphorite rocks, other insoluble phosphates and fluoride.

Silica sand and unreacted minerals have a little effect on quality of plaster, but have abrasive action on the treatment equipment of PG. Organic matters have a great effect on the product, making it gray, slowing down the setting rate and reducing the strength of product. Cocrystallized phosphate can also slow down the setting rate of gypsum. Therefore, PG must be purified first and then treated by dehydration when PG is used to produce building plaster.

The chemical compositions of PG from 16 Chinese phosphate fertilizer companies analyzed by Institute of Solid Wastes Pollution Control Technology, Chinese Academy of Environmental Sciences, are listed in Table 4.1.

Table 4.1: Chemical compositions of PG samples.

Production enterprise	Chemical composition (mass fraction, %)					pH
	$CaSO_4 \cdot 2H_2O$	Total P_2O_5	Soluble P_2O_5	MgO	F^-	
Xifeng coarse whiting plant of Guizhou Kailin group	50.7–86.2	0.01–13.30	0.01–0.68	0.23–0.53	0.12–0.48	2.1–4.6
Yunnan Honghe state phosphate fertilizer plant	72.9–85.1	0.01–3.60	0.001–2.10	0.20–0.26	0.15–0.46	2.3–3.6

(continued)

Table 4.1 (continued)

Production enterprise	Chemical composition (mass fraction, %)					pH
	CaSO$_4$·2H$_2$O	Total P$_2$O$_5$	Soluble P$_2$O$_5$	MgO	F$^-$	
Yunnan phosphate fertilizer plant	72.5–87.3	0.01–2.96	0.001–1.28	0.22–0.37	0.01–0.36	2.5–6.5
Yunfeng chemical industry company	61.8–78.6	0.02–4.29	0.004–1.92	0.23–0.27	0.17–0.76	2.4–4.8
Jiangxi Guixi chemical plant	69.7–86.2	0.001–0.96	0.001–0.10	0.21–0.24	0.09–0.48	3.6–5.3
Dayukou chemical plant of Jingxiang phosphorization company	68.0–83.7	1.51–3.00	0.09–0.74	0.27–0.31	0.001–0.004	2.4–2.6
Sichuan Yinshan chemical industry group	72.9–85.6	0.01–2.08	0.001–0.009	0.25–0.36	0.13–0.67	2.7–3.5
Compound fertilizer plant of Shanxi chemical fertilizer plant	76.0–81.8	0.53–3.34	0.05–0.63	0.23–0.26	0.001–0.017	2.2–2.8
Shandong Lubei enterprise group company	84.4–88.6	0.15–0.35	0.01–0.04	0.27–0.29	0.11–0.20	2.5–3.2
Ammonium phosphate plant of Feicheng City, Shandong	71.1–88.9	0.25–11.2	0.01–1.00	0.25–0.52	0.12–0.75	2.2–5.4
Shandong red sun group	75.8–88.2	2.20–9.47	0.01–1.95	0.29–0.50	0.09–0.67	2.2–2.6
Shenyang chemical fertilizer plant	79.2–85.7	0.01–0.59	0.001–0.28	0.23–0.27	0.02–0.29	2.4–5.2

Table 4.1 (continued)

Production enterprise	Chemical composition (mass fraction, %)					pH
	CaSO$_4$·2H$_2$O	Total P$_2$O$_5$	Soluble P$_2$O$_5$	MgO	F$^-$	
Phosphate fertilizer plant of Nanhua group	49.8–91.8	0.20–3.89	0.001–1.55	0.22–1.34	0.02–2.04	2.3–2.8
Jiangsu Taixing phosphate fertilizer plant	71.3	0.67	0.47	0.269	0.134	2.7
Tongling chemical industry group company	62.4–89.6	0.006–1.0	0.004–0.51	0.16–0.21	0.08–0.66	2.8–5.4
Zhanhua enterprise group company	61.7–96.4	0.006–17.1	0.004–5.7	0.15–0.22	0.1–1.51	1.9–5.1

The main impurities listed in Table 4.1 are fluoride and P$_2$O$_5$ (undecomposed phosphate and free phosphoric acid, expressed as P$_2$O$_5$) showing strong acidity. The lowest pH value is 1.9, the highest fluoride content is up to 2.04%, the total P$_2$O$_5$ content is up to a maximum of 17.1% and the content of soluble P$_2$O$_5$ is up to a maximum of 5.7%. The leaching water produced by PG storage site is obviously a major drawback for the environment.

Due to the effect of impurities on the performance of gypsum products, phosphorus gypsum should be washed with water to remove impurities and purified by neutralizing free acid treatment. The key in purification of PG is as follows: first, the obtained DH must be stable with an impurity content meeting the requirements in standards after washing; second, secondary pollution caused by washing must be resolved. At present, the main internationally used purification methods are water washing, grading, lime neutralization and so on. Water washing can remove fine water-soluble impurities in PG, such as free phosphoric acid, water-soluble phosphate and fluorine. The graded treatment can remove fine insoluble impurities from PG, such as silica sand, organic matter and fine PG crystals. These highly dispersible impurities can affect the setting time of building gypsum, while the black organic matters also affect the appearance of building gypsum products. Graded treatment is effective in removing phosphorus and fluoride. Coarse particles of quartz and unreacted impurities can be removed by wet sieving of PG. The lime neutralization method is particularly simple and effective in removing residual acid from PG.

If the raw materials used for the production of phosphoric acid is high-grade selected apatite, the produced PG has a relatively high purity. About 80–90% of soluble impurities can be removed only by washing, filtering or centrifugal dehydration. The utilization rate of PG can reach 97%.

When preparing phosphate by using phosphate with soluble impurities, insoluble matter and high content of organic matters, the produced PG is a polymerized crystal and its purification method is more complex. One method is to use a three-stage hydrodynamic separator to separate the PG slurry, the removal rate of water-soluble impurities can reach more than 95% and the utilization rate of PG is 70–90%.

When particles' size of PG is particularly small and water is not sufficient, flotation can be used to separate impurities. The extent of organics separation by flotation is very high, removal rate of water-soluble impurities is 85–90% and recovery rate of gypsum is 90–96%.

To reduce the heat consumption in the subsequent drying process, the content of free water in purified gypsum suspension can be minimized by vacuum filtration or centrifugal filtration.

(2) Fluorgypsum

Fluorgypsum is an industrial solid waste produced while preparing hydrofluoric acid by using fluorite (also called fluorspar, fluorspar powder or fluorite powder) and sulfuric acid. Fluorite (CaF_2) reacts with sulfuric acid (H_2SO_4) in a certain proportion by heating. The reaction formula is as follows:

$$CaF_2 + H_2SO_4 \rightarrow CaSO_4 + 2HF \uparrow$$

Hydrogen fluoride gas is condensed and collected, that is, hydrofluoric acid. And the industrial solid waste is fluorgypsum, in which the main chemical composition is type II anhydrous calcium sulfate ($II\text{-}CaSO_4$). The residue sulfuric acid in the solid waste can be neutralized with lime to form $CaSO_4$. Production of 1 t hydrofluoric acid produces about 3.6 t of anhydrous fluorgypsum.

Freshly produced fluorgypsum can slowly hydrate to form DH during stacking. The fluorgypsum stored naturally for more than 2 years is mainly composed of DH with impurities of CaF_2 and anhydrite. Without radioactive pollution, fluorgypsum can be directly used as raw resources with good results.

Fluorgypsum can be used to prepare decorative gypsum board and hollow slabs, or as a cement retarder. Some are used for the production of plaster and self-leveling gypsum, and anhydrous fluorgypsum can be used to produce solid bricks.

(3) Desulfurization gypsum

Desulfurization gypsum, also known as flue gas desulfurization gypsum, sulfur gypsum or FGD gypsum, is an industrial by-product produced through desulfurization purification treatment for sulfur fuel (coal, oil, etc.) combustion. Its definition is as follows: desulfurization gypsum derived from the FGD industry is a wet dihydrate calcium sulfate crystal with fine particles and high grade. Therefore, desulfurized gypsum is a by-product produced from the treatment of SO_2 waste gas discharged from burning coal or heavy oil with a large amount by desulfurization or isolated prewashing cycle method in thermal power stations, steel plants, metallurgical plants and various chemical plants. The main composition is $CaSO_4 \cdot 2H_2O$ with a purity higher than 90%.

According to statistics of the World Health Organization and the United Nations Environment Program, the amount of SO_2 mainly from combustion of sulfurous fuels emitted into the atmosphere is about 2×10^8 t annually. The most serious problem caused by SO_2 is acid rain. Since 1970s, many countries in the world have formulated and implemented policies to control SO_2 emissions, led by Japan and the United States.

Coal is the main fuel for electricity production in today's world. Coal combustion in power plants is the primary reason for SO_2 pollution and acid rain. Coal consumption in thermal power plants accounts for about one-third of the total coal production in China. Emission of SO_2 into the atmosphere caused by coal-fired power plants accounts for about 50% of total annual SO_2 emission in China, and about 75% of industrial SO_2 emission.

The most widely used and most effective controlling technology of SO_2 in coal-fired power plants is FGD, which is currently being applied commercially on a large scale worldwide. FGD technology can be divided into three categories as wet, semidry and dry according to process characteristics. Among them, the wet desulfurization process is mainly applied.

The lime/limestone-gypsum method is the most mature desulfurization process in the world. The main features of the process are as follows:
(i)   Efficiency of desulfurization is above 90%.
(ii) It is operated stably.
(iii) It has a good adaptability to coal variety.
(iv) The absorbents are rich in resources with low cost;
(v)   The by-product of desulfurization is desulfurization gypsum, which is easy to use.

The mechanism of lime/limestone-gypsum desulfurization and formation of desulfurized gypsum are as follows: lime or limestone powders form slurry through importing smoke treated by dust-removal process into the absorber, and the flue gas is washed by spraying in the absorber. The slurry reacts with SO_2 in the flue gas to form calcium sulfite ($CaSO_3 \cdot 0.5H_2O$). After introducing large quantity of air, calcium sulfite oxidizes to dihydrate calcium sulfate ($CaSO_4 \cdot 2H_2O$). The reactions in the lime–gypsum process are as follows:

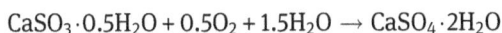

$$CaO + H_2O \rightarrow Ca(OH)_2$$

$$Ca(OH)_2 + SO_2 \rightarrow CaSO_3 \cdot 0.5H_2O + 0.5H_2O$$

$$CaSO_3 \cdot 0.5H_2O + 0.5O_2 + 1.5H_2O \rightarrow CaSO_4 \cdot 2H_2O$$

The reactions in the limestone–gypsum process are as follows:

$$CaCO_3 + SO_2 + 0.5H_2O \rightarrow CaSO_3 \cdot 0.5H_2O + CO_2 \uparrow$$

$$CaSO_3 \cdot 0.5H_2O + 0.5O_2 + 1.5H_2O \rightarrow CaSO_4 \cdot 2H_2O$$

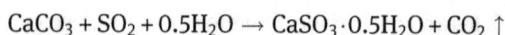

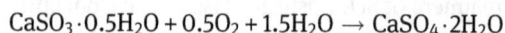

Desulfurization gypsum is the final product with fine particles, high grad and a residual water content of 5–15%, which is formed by dehydrating the gypsum suspension discharged from an absorber with concentrator and centrifuge.

Wet lime/limestone–gypsum method accounts for more than 80% of the overall desulfurization projects. By this process, 2.7 t desulfurization gypsum can be produced by each absorption of 1 t SO_2. In a coal-fired power plant with a capacity of 300,000 kW, the annual discharge of desulfurization gypsum is about 3×10^4 t, even if the sulfur content in coal is calculated at 1%. The amount of by-product desulfurization gypsum is therefore huge.

Desulfurization gypsum has same physical and chemical characteristics as that of natural gypsum, but there are differences in the original state, mechanical properties and impurity compositions.

Desulfurization gypsum is the wet, loose fine particles containing 10–20% of free water after washing and filtering treatment. The color of desulfurization gypsum is white yellow when normally desulfurized. Sometimes, the color of desulfurization gypsum is black because of the desulfurization instability, resulting in more impurities such as coal ash.

Comparison of chemical composition of desulfurization gypsum with that of natural gypsum and PG in several Chinese plants is listed in Table 4.2. The phase compositions and specific surface areas of the three types of gypsum are listed in Table 4.3. It can be seen that the purity of desulfurization gypsum is 75–90%, which is lower than the purity of gypsum in other countries.

The main impurity in gypsum is $CaCO_3$. Under a polarizing microscope, we can see that some of $CaCO_3$ are present in the form of limestone particles, because some of the particles do not react. Due to the differences in impurities and grinding characters of gypsum, the coarse particles grinded from natural gypsum are mostly impurities;

Table 4.2: Comparison of chemical composition of desulfurization gypsum, natural gypsum and PG (mass fraction, %).

Composition type	SiO$_2$	Al$_2$O$_3$	Fe$_2$O$_3$	CaO	MgO	Na$_2$O	K$_2$O	SO$_3$	Crystal water
Natural gypsum from Yingcheng	—	0.48	0.48	31.25	—	—	—	43.15	19.06
Desulfurization gypsum from Chongqing	1.82	0.39	0.2	31.24	0.64	0.05	0.13	44.25	18.56
Desulfurization gypsum from Taiyuan	3.26	1.90	0.97	31.93	—	0.09	0.15	40.09	16.64

Table 4.3: Comparison of phase composition and specific surface area of three types of gypsum.

Gypsum	Specific surface area (Blaine test) S (m²/kg)	Water content (%)	Phase composition in dry state (mass fraction, %)		
			DH	HH	CaCO₃
Natural gypsum powders	250	1.7	89.00	3	3.00
Desulfurization gypsum from Chongqing	106	9.8	91.78	0	4.41
Desulfurization gypsum from Taiyuan	150	7.5	75.87	0	16.61

on the contrary, the coarse particles grinded from desulfurization gypsum are mostly FGD gypsum, and fine particles are impurities.

At present, almost all desulfurization gypsum is used in building materials industry, for the production of plaster powders, α-gypsum powders, gypsum products, gypsum mortar and cement additives.

On the basis of physical characteristics of desulfurization gypsum, such as presenting as wet powders, high water content, unreasonable particle size distribution, lack of whiteness, desulfurization gypsum is suitable for the production of gypsum wall panels, blocks, and so on and should not be used as a decorative material.

(4) Other by-product gypsums

(i) Industrial waste gypsum

Fluorgypsum and PG are produced directly from chemical reactions. In addition, in chemical production, industrial gypsum solid waste can also be formed when adding materials containing calcium to neutralize excessive sulfuric acid. For example, after the reaction is completed, by which an oxidant – dye salt S (also known as nitrobenzene sulfonic acid) for printing and dyeing, is produced by using benzene and sulfuric acid as raw materials, gypsum is formed from the neutralization reaction between remaining sulfuric acid and hydrated lime. The reaction formula is as follows:

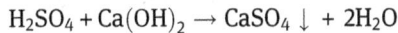

$$H_2SO_4 + Ca(OH)_2 \rightarrow CaSO_4 \downarrow + 2H_2O$$

The gypsum is filtered and separated from the product. This kind of gypsum is neutral yellow powder, so it is also known as "yellow gypsum," belonging to AH. In this product, there is also the component of DH because of the existence of a large amount of adsorbed water and free water.

Another example, when producing dye intermediates, Cliff acid [(5 or 8)-aminonaphthalene-2-sulfonic acid], by using naphthalene and sulfuric acid, dolomite powders

($CaCO_3 \cdot MgCO_3$) are used for neutralization. The excess sulfuric acid reacts with calcium carbonate contained in dolomite, and the gypsum is produced at the same time.

The reaction formula is as follows:

$$H_2SO_4 + CaCO_3 = = CaSO_4 + CO_2 \uparrow + H_2O$$

This type gypsum is a yellowish white neutral powder with more free water.

(ii) Citrate gypsum

Citrate gypsum is an industrial waste residue produced during the production of citric acid, producing about 1.5 t citrate gypsum with the production of each 1 t of citric acid. Citric acid fermentation liquid is produced by the fermentation of starch raw materials, and then neutralized with calcium carbonate before adding sulfuric acid solution. The extracted solid waste residue is citrate gypsum by the following reaction:

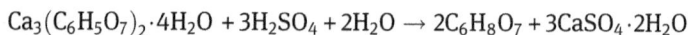

$$2C_6H_8O_7 \cdot H_2O + 3CaCO_3 \rightarrow Ca_3(C_6H_5O_7)_2 \cdot 4H_2O \downarrow + 3CO_2 \uparrow + H_2O$$

$$Ca_3(C_6H_5O_7)_2 \cdot 4H_2O + 3H_2SO_4 + 2H_2O \rightarrow 2C_6H_8O_7 + 3CaSO_4 \cdot 2H_2O$$

The original state of the industrial waste from citric acid residue is a paste with a moisture content of 40–50% and it is gray. The content of $CaSO_4 \cdot 2H_2O$ in it, $w(CaSO_4 \cdot 2H_2O)$, is more than 85%. The main impurities in citrate gypsum are unreacted calcium citrate and a small amount of untreated citric acid.

Only a small amount of citrate gypsum is used in domestic cement industry, construction gypsum products and construction gypsum mortar in China. The majority has not been used. This is because that the small amount of citric acid from citric acid gypsum and citrate is extremely strong retarders for building gypsum. These impurities can be removed by water washing and calcination at high temperatures.

(iii) Mirabilite gypsum

Mirabilite gypsum is an industrial solid residue from extraction of sodium sulfate or production of mirabilite using glauberite by separating the paragenesis of mirabilite and gypsum. In China, mirabilite gypsum is annually discharged about 3.6×10^6 t, and mainly used as cement retarder, and for production of gypsum blocks.

(iv) Salt gypsum

Salt gypsum is also known as saline-alkali sediment or saltpeter sediment, a by-product produced in salt industry or from concentration of seawater in salt fields. Here, the content of $CaSO_4 \cdot 2H_2O$, $w(CaSO_4 \cdot 2H_2O)$, is higher than 95%. Salt gypsum is annually collected about one million tons in China. It was used to produce gypsum fiberboard at the Institute of Building Materials Science, Fujian Province, gypsum hollow slabs at the Institute of Building Materials Science, Guangdong Province and lime sand bricks at Zigong Salt Plant, Sichuan Province. These products are all qualified. Because of the lack of gypsum resources, many cement plants in Hainan Province use salt gypsum as a cement retarder.

4.2 Production of gypsum cementitious materials

Gypsum is an air-hardening inorganic cementitious material. In its preparation process, DH is transformed into different dehydrated phases under certain temperatures and pressures. The production process is simple with low-energy consumption. Gypsum has excellent characteristics such as light weight, fast setting time, low radioactivity, sound insulation, heat insulation and good fire resistance.

4.2.1 Dehydration transformation of gypsum

4.2.1.1 Gypsum and dehydrated phases

Thermodynamically, a phase is a state of aggregation of matter that has an internal uniform structure. A phase is characterized by its homogeneous physical and chemical properties. There is a physical interface between two phases, and the macrophysical and -chemical properties are abrupt on both sides of the interface. The dehydration transition temperature of gypsum is also called as the transition point or phase transformation point that is closely related to dehydration conditions.

From a thermodynamic point of view, gypsum and its dehydrated products are all in one phase in the $CaSO_4-H_2O$ system. They are in the equilibrium system of $CaSO_4-H_2O$ under specific conditions. Gypsum is currently accepted to have five phases and seven variants, namely,

Dihydrate gypsum ($CaSO_4 \cdot 2H_2O$);

α-Hemihydrate gypsum (α-$CaSO_4 \cdot 0.5H_2O$), β-hemihydrate gypsum (β-$CaSO_4 \cdot 0.5H_2O$);

III-α-Anhydrite (III-α-$CaSO_4$),III-β-anhydrite (III-β-$CaSO_4$);

II -Anhydrite (II -$CaSO_4$);

I -Anhydrite (I -$CaSO_4$).

Transformation temperatures were proposed by H. Lehmann, as shown:

$$CaSO_4 \cdot 2H_2O \xrightarrow[115°C(\alpha)]{107°C(\beta)} CaSO_4 \cdot 0.5H_2O \xrightarrow[110°C]{200°C} III-CaSO_4 \xrightarrow{250°C} II-CaSO_4$$

$$\xrightarrow{1193°C} I-CaSO_4 \xrightarrow{1450°C} CaSO_4$$

(1) Dihydrate gypsum

DH, also called gypsum or raw gypsum, has a stable phase in nature. Most industrial by-product gypsums are DHs. DH is not only the raw material for dehydrated products, but also the final product of rehydration of dehydrated products. The final hydration product is also known as the recycled gypsum.

(2) Hemihydrate gypsum (HH)

HH has two variants as alpha and beta. When DH is heated under pressurized steam or in acid and salt solutions, α-HH (α-$CaSO_4 \cdot 0.5H_2O$) can be formed. When DH dehydrates in dry condition, dehydration occurs when heated at 65–75 °C, and a part of crystal water is removed when heated at 107–170 °C to obtain β-HH (β-$CaSO4 \cdot 0.5H_2O$), namely, building gypsum.

(3) III-Anhydrite

III-Anhydrite, called soluble anhydrite, also has α and β variants, formed by heating α- and β-HH to dehydrate, respectively. If steam pressure is too low during dehydration of DH, anhydrite can be formed from DH directly without transformation of HH.

(4) II-Anhydrite

II-Anhydrite is insoluble or hard to dissolve. It is the stable final product at normal temperature formed by dehydration of DH, HH and III-anhydrite at high temperatures. Natural stable anhydrite belongs to this type.

Time required to complete a phase transformation is long in laboratory. In industry, it is desired to complete phase change with minimum energy consumption and within a short time. The calcination temperature in industrial production is much higher than that in laboratory to promote phase transformation. The following are common dehydration transformation temperatures in industry.

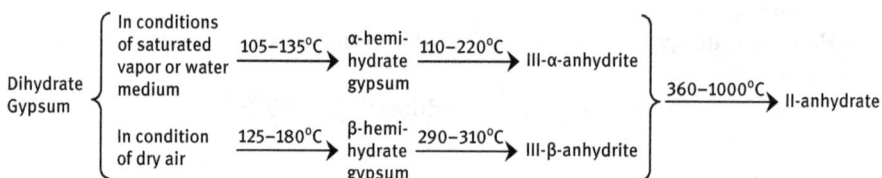

(5) I-Anhydrite

I-Anhydrite, also called α-anhydrite, is a phase that can only exist at temperatures above 1,180 °C. I-Anhydrite will change to II-anhydrite below 1,180 °C. I-Anhydrite does not exist at room temperature.

4.2.1.2 Structure and characteristics of gypsum

According to nuclear magnetic resonance and infrared absorption spectroscopy analysis, the crystal water in DH is composed of at least two types of water, structural water and zeolitic water. It is generally believed that structural water is the water removed from DH during transformation of HH while zeolitic water is retained and can be removed only when HH changes into anhydrite III. Thus, the amount of HH is related to the ratio of these two types of water. Nuclear magnetic resonance and infrared analysis confirm that proportions of the two types of water in different DHs from different sources are different. There are differences in the properties of gypsum from different places and different crystalline states.

When DH changes into HH, two changes occur in its structure.

(1) Water with 3/4 molecules in the water layer between two ion layers is lost.

(2) Ions of Ca^{2+} and SO_4^{2-} dislocate with each other to form a staggered layer of calcium and sulfur.

The crystal structure section of DH is shown in Figure 4.2, and the crystal structure section of HH is shown in Figure 4.3.

In DH, Ca^{2+}–SO_4^2 spacing is 0.31 nm, Ca–O spacing is 0.257–0.259 nm and Ca–Ca spacing is 0.628 nm. It can be deduced from the staggered Ca^{2+}–SO_4^2 layer that there is a channel with a diameter of about 0.3 nm in HH as a channel for water molecules. This is the reason why HH is easy to hydrate.

III-Anhydrite is divided into α-type and β-type, formed by dehydration of α- and β-HHs, respectively. It is considered that anhydrite belongs to the hexagonal system. The

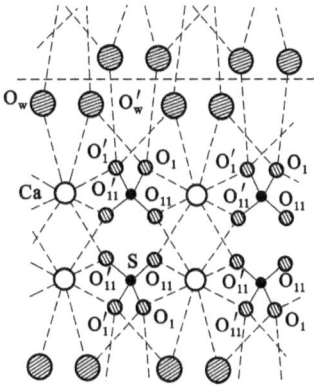

Figure 4.2: Crystal structure section of dihydrate gypsum.

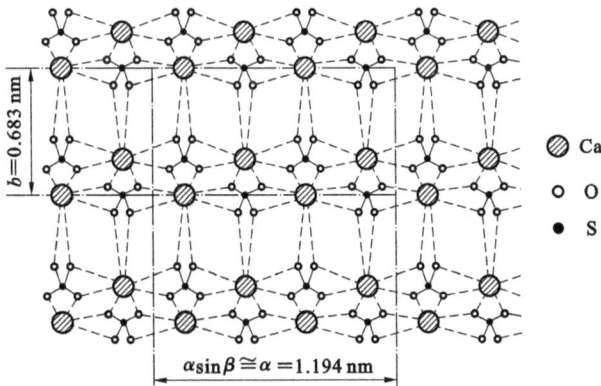

Figure 4.3: Crystal structural section of HH ($\perp$ c axis).

main difference between it and HH is that a half-molecular water in layers is removed with the transformation of structure from a triangular system to a hexagonal system.

The transformation temperatures for the two stages of DH dehydration are, respectively, as follows.

In the first dehydration stage, it is 128 °C that is the transition point where DH changes into HH.

In the second dehydration stage, it is about 163 °C that is the transition point where HH transforms into anhydrite.

4.2.1.3 Formation mechanism of dehydrated gypsum phases

(1) Formation mechanism of α- HH
Dissolution–crystallization mechanism is currently accepted according to the fact that α-HH must be formed in a saturated water vapor medium or an aqueous solution. When DH is heated in a saturated water vapor medium or an aqueous solution, dehydration occurs. In suitable conditions, 1.5 molecules of crystal water can be removed from the crystal lattice of DH, resulting in the formation of HH crystallites. The crystallites quickly dissolve in liquid water surrounded by water. When the concentration of HH in liquid phase reaches saturation, dissolved HH crystallizes rapidly to form coarse and dense crystals of α-HH.

(2) Formation mechanism of β-HH
Formation mechanism of β-HH can be divided into once-formation mechanism and twice-formation mechanism based on the formation process. In the once-formation mechanism, β-HH is formed directly from dehydration of DH:

$$CaSO_4 \cdot 2H_2O \rightarrow \beta\text{-}CaSO_4 \cdot 0.5H_2O + 1.5H_2O$$

In the twice-formation mechanism, DH dehydrates to form III-anhydrite directly, then III-anhydrite immediately absorbs the expelled water to become HH:

$$CaSO_4 \cdot 2H_2O \rightarrow \beta\text{-}CaSO_4III \xrightarrow{+H_2O} \beta\text{-}CaSO_4 \cdot 0.5H_2O$$

The formation mechanisms of β-HH are in dispute.

(3) Formation mechanism of II -anhydrite
Under the condition of low temperature (about 100 °C) and saturated water vapor, DH or III-anhydrite is transformed into II -anhydrite following dissolution–crystallization mechanism.

The transformation mechanism of II-anhydrite into I-anhydrite is unclear. It was found that there was liquid water at about 1,000 °C when natural gypsum was calcined to change into II-anhydrite by Lelong's experiment. In the condition with a certain water vapor pressure and a proper temperature range, DH can be directly transformed into II-anhydrite. Lelong also confirmed that dehydration was completed in 9 days at a low temperature of 100 °C.

4.2.2 Preparation of HH

Different gypsum products with different properties can be produced by heating natural DH (or chemical by-products mainly composed of DH) at different temperatures using different heating methods.

The preparation methods of α-HH are usually pressurized water vapor method and aqueous solution method.

(1) Pressurized water vapor method

The formation mechanism of α-HH is different from that of β-HH, which is formed by dissolution–crystallization process under pressurized water vapor.

Under normal circumstances, α-HH can be obtained at 120–130 °C by heating 5–8 h with a raw materials particle size of 5–8 cm. It can also be obtained at 150–160 °C by heating for 1.5–3 h. In general, α-HH slowly precipitated at low temperatures has a higher hardening strength than α-HH rapidly precipitated at high temperatures. The greater the density of bulk raw gypsum, the smaller the water demand of product, and the higher the strength of the hardened body.

(2) Pressurized aqueous solution method

The patent of pressurized aqueous solution was originally acquired by Haddon. Unlike pressurized water vapor method, powdered DH is used as a raw material in the pressurized aqueous solution method.

Several types of crystal conversion agents must be used in this process, including
(i)   Water-soluble protein that dissolves with water;
(ii) succinic acid, maleic acid, citric acid and carboxylic acids having two or more carbon atoms and salts;
(iii) palmitic acid, linoleic acid and water-soluble alkali metal salts with a fatty acid having more than 15 carbon atoms.

By adding DH in an aqueous solution of 0.01–0.2% (mass fraction) crystal conversion agent, the temperature in a sealed container is increased to 120–140 °C and is kept for a period under the saturated water vapor pressure; DH transforms into α-HH like a short column or plate-shaped. Stable α-HH can be obtained by washing and drying after separating solid phase and liquid phase in the hot state.

(3) Property differences between α-HH and β-HH

There are differences in properties between α- and β-HHs due to different conditions of formation. The comparison of properties between the two types of HH is listed in Table 4.4.

Table 4.4: Comparison in properties between α- and β-HHs.

Property		α-HH	β-HH
Density ρ (g/cm³)		2.74–2.76	2.60–2.64
Refractive index		$N_g{}^c$: 1.584	N_g: 1.556
		$N_p{}^c$: 1.559	N_p: 1.550
Specific surface area S (m²/g¹)	Air permeability method[a]	0.3490	0.5790
	BET method[b]	1.0	8.2
Average size $\bar{d}$ (nm)		0.940	0.388
Setting expansion rate ε		0.30	0.26
Hydration heat q at 25 °C (J/mol)		17,200 ± 85	19,300 ± 85
Water consumption for standard consistency (%)		30–45	65–85
Setting time (min)	Initial setting	7–18	5–8
	Final setting	<30	<30
Dry flexural strength R_f (MPa)		7–12	4–6
Dry compressive strength R_c (MPa)		25–100	7–20

Note:
[a]   The area tested by air permeability method is the external surface area.
[b]   The area tested by BET method is the total surface area.
[c] N_g and N_p represent the refractive indices for optical principal axis N_g and axis N_p.

Scanning electron microscope shows that α-HH is composed of dense, coarse and complete crystals with crystal shapes of rod-like, columnar, granular, needle-like and fibrous. While β-HH is composed of loose, fine, irregular grain crystals, and grains morphology is mostly scaly, a small amount is in thin plate shape.

β-HH dehydrates rapidly in superheated unsaturated steam, resulting in loose and fine crystals. However, α-HH dehydrates in liquid water and is formed through recrystallization resulting in dense, coarse crystals. Therefore, the two variants vary greatly in the specific surface area (β-HH is about 2–8 times larger than α-HH). Consequently, density, refractive index, expansion rate, hydration heat, water consumption for standard consistency and strength for both are significantly different.

It can be considered that a certain number of defects and distortions in a crystal lattice of β-HH due to fine grains make inside water similar to the combination of zeolite water. In α-HH, water molecules are strongly combined due to coarse grains and complete crystal lattice to make inside water similar to the combination of structural water. Therefore, the differences in morphology, dispersion and inner surface area of α-HH and β-HH crystals are the reasons of property differences.

4.3 Building gypsum

Recently, research and development in building gypsum and its products has been increased rapidly because of requirements on comfort (heat and sound insulation, adjustment humidity) and safety (fireproof). Building gypsum and its products are being developed to light, high-strength compound, which is multifunctional, and environmental friendly, so that we can gradually replace traditional wall and decoration materials. The gypsum-based composite wall materials with energy-saving and environmental characteristics have become the leading products of wall materials in China.

4.3.1 Production of building gypsum

Natural bulk DH must be treated by breaking, homogenizing, milling and calcination when natural DH is used as the raw material to produce building gypsum. In industries, the order in the process is different due to different calcination processes.

For example, in the process of direct fire concurrent flow rotary kilns, the processing orders of gypsum treatment is as follows:

$$(\text{pre})\text{homogenizing} \rightarrow \text{breaking} \rightarrow \text{calcination} \rightarrow \text{grinding}$$

When gypsum is calcined in a frying wok, the treatment order is as follows:

$$\text{breaking} \rightarrow (\text{pre})\text{homogenizing} \rightarrow \text{milling} \rightarrow \text{calcination}$$

The dry calcination process is applied to produce architectural gypsum powders with β-HH as the main component. DH is heated to dehydrate, resulting in the following reaction:

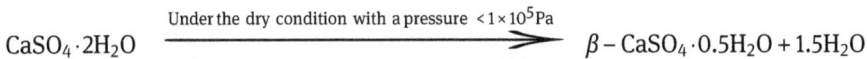

$$CaSO_4 \cdot 2H_2O \xrightarrow{\text{Under the dry condition with a pressure } <1\times10^5 Pa} \beta - CaSO_4 \cdot 0.5H_2O + 1.5H_2O$$

In twentieth century (50s and 60s), batch frying pans and external burning rotary kilns were used worldwide. Continuous frying pans were developed in 70s and cone-shaped frying pans were developed in 80s. The calcination mode of rotary kiln has also been developed from external calcination to direct-fire calcination. Air-flowing calcination was developed with the development of industrial techniques. To improve thermal efficiency, an integrated equipment for grinding and sintering was developed in Germany to integrate grinding, drying and calcination of gypsum. In China, calcination equipment are mostly external burning rotary kilns, direct fire burning rotary kilns, batch frying woks, continuous frying woks and ebullition furnaces in architectural gypsum plants.

(1) Batch frying wok
Operation of a batch frying wok is executed in mass production by indirectly heating. Heat is transmitted to materials in a pan from outer wall of the wok, bottom of the pan and fire tube while the pan is held still. Mechanical stirring action and steam process in the wok make materials boil to achieve dehydration and transformation of DH. Gypsum powders smaller than 0.2 mm are usually ground first and then calcined in a batch frying pan.

When initially used, a batch frying wok requires to be dried off at a low temperature to protect structural strength of refractories in system. In production, a small amount of materials are put into the preheated wok and kept stirring while heating. When the temperature reaches 110–120 °C, all materials are fed at one time and maintained at 110–120 °C by continuously heating. This is a decalescence process and needs 1.5–2 h to make DH dehydrate to transform into HH. When the materials present a boiling state, the temperature is increased to 140–160 °C and materials are discharged. A whole working cycle is completed. Generally a cycle is about 2 h depending on the time of dehydration process, purity of DH, structure of wok and thermal system. Feeding and discharging time lies on the quality of the equipment. It is generally short, about 3–5 min. Figure 4.4 shows a typical periodic curve of batch frying wok calcination (the labeled temperature in the figure is only a set of examples).

The quality of plaster produced by batch frying woks in China meets the quality requirements of construction gypsum powders.

(2) Continuous frying wok
A continuous frying wok works continuously; the calcination curve is shown in Figure 4.5. In production using continuous frying wok, materials are slowly put into the empty wok that is preheated and kept heating and stirring to make materials temperature reach 110–120 °C. Then the feed speed is controlled until materials are filled and the feed is stopped. After materials' temperature rises to 140–160 °C and materials present a complete boiling state, the feeding continues. When the level of materials in the wok reaches the height of overflow line, the materials begin to overflow and discharge from the bottom pipe. A balance between feeding and discharging is achieved and the level of materials is kept constant, resulting in continuous production.

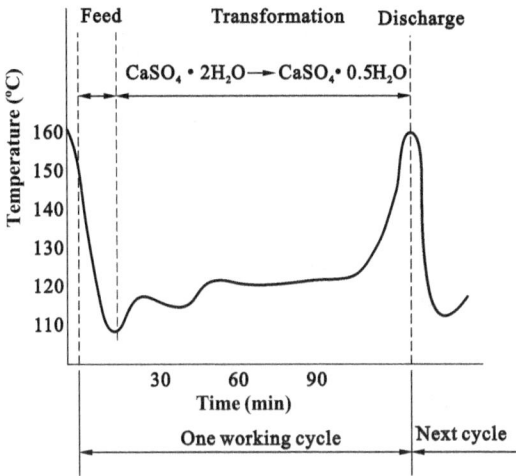

Figure 4.4: The calcination temperature curve of intermittent wok.

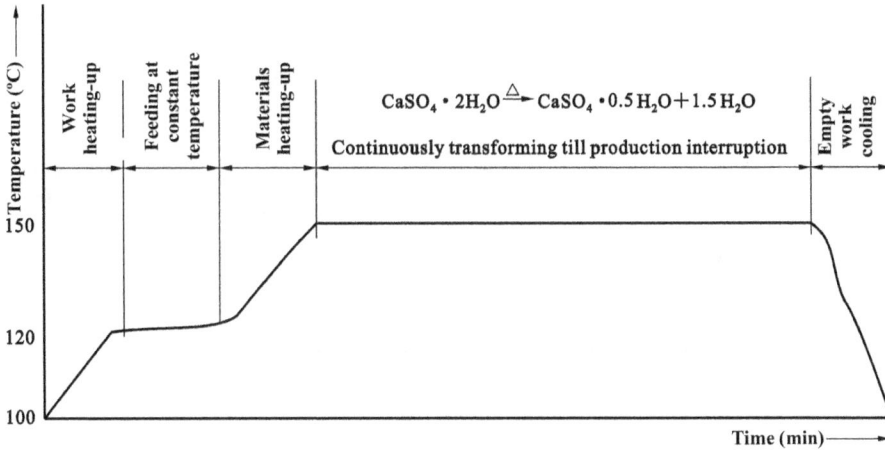

Figure 4.5: The calcination temperature curve of continuous wok.

Materials present a fluidized state because they are not only stirred by multilayer mechanical mixing, but also agitated by steam and thermal cycling gas generated by dehydration. Raw material particles into a pan immediately absorb heat and quickly dehydrate as the wok and materials are maintained at a constant temperature.

The average residence time for materials in a continuous frying wok is 1.5 h. Compared with a batch frying wok, the former has characteristics of faster chemical reaction, higher thermal efficiency and greater output, and the system is easily controlled automatically.

(3) Embedded frying wok

Heat is indirectly transferred from bottom of the wok and through fire pipe and pot wall to materials for both a batch frying wok and a continuous frying wok. Effect of heat transfer is affected by transfer carrier. If production capacity is improved by increasing the temperature, the bottom of a wok will be destroyed once the temperature exceeds the maximum temperature that the wok can withstand.

The efficiency of heat transfer in the boiling layer is very high due to agitation by mechanical stirring and steam produced in dehydration to make materials be fluidized – in a boiling state. This effect is accelerated by embedded burner. The dehydration temperature of DH is quickly achieved and calcination is sped up. Using an embedded burner, the capacity of dehumidification and dust removal must be improved to achieve a relative balance due to an increase in the airflow inside the wok.

By adding an embedded burner in a continuous frying wok, the calcined gypsum has a good homogeneity with similar phase compositions except slightly increased insoluble II-anhydrite.

The composition and properties of typical examples of calcined gypsum produced by various calcination methods are summarized in Table 4.5.

Table 4.5: The composition and properties of typical examples of calcined gypsum with various calcination methods.

Calcination method	Residual DH (%)	Soluble III-anhydrite (%)	Insoluble II-anhydrite (%)	Water requirement (%)	Setting time (min)
Batch frying wok	3	11	2	64	25
Continuous frying wok	5	7	2	66	18
Embedded frying wok	4	7	2–4	66	15

(4) Conical frying wok

A conical frying wok is composed of a cone-shaped bottom and round wall, and it is insulated with 100 mm thick rock wool without bricks. The pan contains a burner (or hot gas), which can use gas, oil or coal as the fuel. The burning hot gas is spouted directly to materials, so that the materials can be heated rapidly and dehydrated into HH. This wok is fed less at one time with a rapid heating rate, high reaction speed and high output.

4.3.2 Performance and influencing factors of building gypsum

The main component in building gypsum is β-HH, also known as plaster, which is white powder with a density of 2,600–2,750 kg/m^3 and a bulk density of 800–1,000 kg/m^3.

4.3.2.1 Performance of building gypsum

(1) Rapid setting and hardening
Building gypsum begins to set after it is mixed with water in 6 min, and the final setting time is less than 30 min. It can completely harden in about 1 week's time at room temperature in natural dry environment. In order to meet construction requirements, certain amounts of retarder are often added to building gypsum, such as pulp waste, bone glue, borax and so on.

(2) High porosity and low apparent density
On the one hand, it has good performance of insulation and sound-absorbing. When building gypsum hydrates, the theoretical water demand only accounts for 18% of the mass of building gypsum. For gypsum slurry to have necessary plasticity, 60–80% of water is usually added to. The excess water evaporates after hardening, leaving a porosity of hardened product up to 50–60%. On the other hand, a high porosity results in a low strength, for instance, hardening strength is only 3–6 MPa.

(3) Poor water resistance and frost resistance
Hardened building gypsum has a strong hygroscopicity. In wet environment, the adhesion between crystals weakens and the intensity significantly reduces. Crystals are destroyed resulted from dissolution in contacting with water. Frozen after absorbing water, it will crack because water in pores freezes to expand.

(4) Temperature and humidity control
Due to porous structure, a gypsum product has a large heat capacity and strong hygroscopicity. Ambient temperature and humidity can be adjusted with the change of temperature and humidity due to the "breathing" effect of the product.

(5) Good fireproof
When DH encounters fire, crystal water evaporates absorbing the heat. When it is used for high-temperature environment, a steam curtain will be formed on the product's surface, thus preventing spread of fire and playing fireproof effect. However, building gypsum should not be used in the environment above 65 °C for a long term; the strength of DH will be reduced because of slow dehydration decomposition.

(6) Volume microdilatancy at solidification
Building gypsum has microdilatancy during setting and hardening. The volume expansivity is about 0.05–0.15%. This feature makes the molded gypsum product have a smooth surface, clear outline, full line, angle, pattern, accurate size and no shrinkage cracks when dried.

(7) Good machinability

Building gypsum products can be sawed, nailed, drilled and pasted; the construction is flexible and convenient.

(8) Decorativeness

Gypsum is white and can be used to decorate indoor walls or ceilings in dry environment. Being exposed to moisture, it changes color to yellow and loses decorativeness.

According to the Chinese standard *Building gypsum* (GB/T 9776−2008), the technical requirements of building gypsum mainly include fineness, setting time and strength. Based on the difference in strength and fineness, building gypsum is divided into three grades: superior grade, first grade and qualified product.

4.3.2.2 Main factors affecting the performance of building gypsum

There are many factors affecting the performance of building gypsum. According to production process, the main factors are from the following aspects.

(1) Purity and impurities in raw materials

Because of different formation conditions, there are differences in defects and distortion of DH ore crystal structure, size and shape of crystals, amount and binding of crystal water, type and amount of impurities. These differences affect not only the purity and density of ore, but also the dehydration temperature and speed.

The evaluation of ore quality is often characterized by purity or grade, that is, by the amount of $CaSO_4 \cdot 2H_2O$ in raw materials. The theoretically chemical compositions of pure DH are known, namely, $w(CaO)$ is 32.56%, $w(SO_3)$ is 46.51% and $w(H_2O)$ is 20.93%. On the basis of this, any chemical composition can be converted into a content of 100% of $CaSO_4 \cdot 2H_2O$.

$$w(CaSO_4 \cdot 2H_2O) = 4.78w(H_2O) = 21.5w(SO_3) = 3.07w(CaO) = 100\% \qquad (4.1)$$

According to mineral composition specified by Chinese national standard *Natural gypsum* (GB/T 5483−2008), natural gypsum can be divided into three categories, gypsum (code as G), anhydrite (code as A) and mixed gypsum (code as M). Different natural gypsum can be divided into five grades: superfine, Special Grade, Grade 1, Grade 2, Grade 3 and Grade 4. The requirements on level and grade of natural gypsum are listed in Table 4.6. Gypsum ores containing more than 75% of $CaSO_4 \cdot 2H_2O$ can be used as the raw material for the production of building gypsum.

When determining the grade of natural gypsum, the chemical composition should be analyzed first. According to the percentage of CaO, SO_3 and crystal water, the content of $CaSO_4 \cdot 2H_2O$ could be calculated. The minimum in the three data is taken as the basis for grading. The grade formulas are as follows:

Table 4.6: Requirements on level and grade of natural gypsum.

Level	Grade (mass fraction, %)		
	Gypsum (G)	Anhydrite (A)	Mixed gypsum (M)
Special level	≥ 95	–	≥ 95
Level 1		≥ 85	
Level 2		≥ 75	
Level 3		≥ 65	
Level 4		≥ 55	

$$G_1 = 4.7785w \, (H_2O) \tag{4.2}$$

$$G_2 = 1.7005w \, (SO_3) + w \, (H_2O) \tag{4.3}$$

$$w \, (CaSO_4) = 1.7005w \, (SO_3) - 4.7785w \, (H_2O) \tag{4.4}$$

where G_1 is the grade of G product, %; G_2 is the grade of A and M product, %; $w \, (CaSO_4)$ is the percentage of $CaSO_4$, %; $w \, (SO_3)$ is the percentage of SO_3, % and $w \, (H_2O)$ is the percentage of crystal water, %.

(2) Grinding and calcination

The quality of calcination is closely related to grinding of raw materials. As gypsum is a poor heating conductor, ununiformed of fineness of ground gypsum powders will cause different degrees of calcination. In general, when the calcination temperature is low, gypsum dehydration is not complete, known as underburnt. There is more DH left in the product, causing quick setting. When calcination temperature is high, it is known as overburnt. At calcination temperatures above 200 °C, some HH will dehydrate and transform into soluble AH. Unlike properties of HH, anhydrite gypsum has the characters of lager water demand, fast setting time, lower strength of hardened products and lager expansively. If the calcination temperature is too high, anhydrite gypsum II is likely to exist, causing significantly low hydration activity.

In order to improve the quality of the product, the process of "twice grinding and one burning" is used in many milling plants, that is, there are two times of grinding before and after calcination, respectively. Especially, the performance of calcined gypsum can be improved because grinding after calcination works as a homogenization process. During the second grinding, undehydrated DH in the core of residual particles is exposed and can further dehydrate in the impact of heat generated from rolling and hammer frictions. And soluble AH will react with water vapor. In both processes, HH is transformed, thus improving the quality of product.

Fineness is one of the important indicators of building gypsum product quality. Within a certain range of fineness, product strength is improved with an increase of fineness. Beyond the range of fineness, strength is decreased. The thinner the particles, the more soluble they will be, and so will be the greater degree of super-saturation. With the increase of fineness, the supersaturation will rise to more than a certain value; a high crystallization stress occurs in the hardened gypsum, causing destruction of hardened structure and a resultant decreasing structural strength.

(3) Aging effect

Freshly fried gypsum containing a certain amount of soluble AH and a small amount of unstable DH has unstable phase compositions and is characterized by high containing energy, large dispersion and high adsorption activity. This leads to a high water demand for standard consistency, low strength and unstable condensation time. An aging stage is needed in this case. Aging is to store or to damp-heat treat the freshly fried or calcined plaster for a period to improve the physical properties. Therefore, aging is one of the measures to improve the quality of plaster products.

The concept of aging is not the same as the general storage. Aging refers to the homogenization of gypsum; it is a storage process to improve physical properties of plaster. During aging, two types of phase changes occur in the plaster.

(i) Soluble anhydrite absorbs water to transform into HH.

(ii) Residual DH continues to dehydrate into HH.

After aging, phase changes of gypsum tend to be stable, leading to homogenized materials, increased content of HH, decreased specific surface area, lowering water requirement of standard consistency, normal setting time and improving strength so that the performance of cementitious gypsum is improved and product quality is increased.

Yue Wenhai and Li Fengren divided aging into a valid period and a failure period. The valid period is the period of storage that can improve the physical properties of plaster. During this period, soluble anhydrite and residual DH are changed to HH. The failure period is the storage period to reduce the physical properties. At this time, HH begins to absorb water and transfer to DH. Experiments showed that when the adsorbed water of plaster was less than 1.5%, aging of plaster was in the valid period. During this period, aging can be sped up by stirring, rolling, opening, sealing and spraying. When the adsorbed water of plaster was higher than 1.5%, aging was in the failure period; HH began to absorb water and convert to DH. The standard consistency for plaster in the failure period was increased and strength was significantly decreased.

In order to improve the aging effect, the mandatory aging method can be used. This means that aging can be accelerated by spraying water of a suitable amount into calcined plaster with full stirring so that plaster gypsum absorbs moisture. In addition, additives like NaOH and $CaCl_2$ can be added to promote aging.

4.3.2.3 Requirements on phase composition of architectural gypsum in plaster building products

Architectural gypsum is a product obtained by calcinating DH at high temperatures and is mainly composed of β-HH. It is a new building material with wide applications. Building gypsum can be used as a bulk material to produce gypsum plaster, scraping wall putty, mortar, block, board, fiber plaster board, gypsum plank, decorative gypsum board and other gypsum construction products.

In fact, it is impossible to produce 100% of β-HH in the industry. Commonly used building gypsum, named calcined gypsum in industry, is a mixture mainly composed of β-HH together with overburnt gypsum and incompletely decomposed DH. Therefore, from the phase composition, building gypsum is a multiphase mixture.

Various building materials products of gypsum have different requirements on phase compositions of building gypsum. Plaster is specially produced to meet a specific purpose. To meet particular needs, the properties of plaster must be first made clear. The phase composition is then adjusted for providing the desired properties.

The following points should be made clear based on the property requirements of gypsum building products:
(1) the most suitable natural DH ores;
(2) the elementary components and phase composition in building gypsum;
(3) the most suitable grinding and calcination process for the production of building gypsum;
(4) the type and amount of suitable admixtures;
(5) the method of mixing and stirring and adjustment of temperature and humidity.

For example, in the production of ceramic molds and art sculpture with high-strength cast gypsum, it is required to use highly pure DH as raw materials and the produced gypsum powders do not contain iron or hard impurities and without residual DH. On the contrary, if used for spraying and plastering, the purity of DH can be lower, up to 80%.

Calcined gypsum used for plastering generally is not mixed with other blending materials; it is only a mixture of β-HH and overburnt gypsum. This calcined gypsum is also known as base-calcined gypsum. Due to the fast hydration of β-HH, the substrate plaster can achieve high strength in a short term. During the later stage of hydration, overburnt gypsum hydrates slowly, then the substrate gypsum completely hydrates and crystallizes to enhance cohesion of crystals, to compensate for contraction caused by drying and to avoid cracks. In the production of substrate gypsum, the optimum proportion of the β-HH and overburnt gypsum can be determined by purity of DH, reactivity of components and requirements on properties of building gypsum products.

Using building gypsum to produce building products members, it is best to apply calcined gypsum produced by quick burning method (with β-HH as the main component) because the applied calcined gypsum has a high reactivity at the initial setting stage and hardens rapidly. In addition, appropriate grinding equipment (mill, screen, separator, etc.) and grinding process should be chosen according to grind ability of calcined gypsum.

4.4 Hydration and hardening process of HH

4.4.1 Hydration of gypsum cementitious materials

HH, III-anhydrite and II-anhydrite dehydrated from DH will hydrate, set and harden after contacting with water; this is the valuable nature of gypsum cementitious materials. Hydration refers to the chemical reaction between gypsum cementitious materials and water in, that is, dehydrated phases of gypsum hydrate with water to become DH. Setting and hardening is the process that the hydration products condense, crystallize and develop mechanical strength. Hydration is the prerequisite for setting and hardening. M. A. Budnikov referred hydration as the first reaction process, and setting and hardening as the second reaction process.

The newly formed HH has a strong hydration activity. It usually hydrates to crystallize in 5–6 min, and dihydrate is formed in 30 min. Within 2 h, all HH are transformed into DH forming crystalline network hardenite. III-Anhydrite has the strongest hydration activity and will be immediately converted into HH when contacting with water before it is finally converted into DH. The entire hydration process is longer than the hydration process of HH. The hydration activity of II-anhydrite is low, and the ability to hydrate depends on calcination temperature. According to the solubility curve of gypsum phases, theoretically, II-anhydrite can no longer hydrate into DH when the reaction temperature is higher than 42 °C. Therefore, hydration of II-anhydrite should be carried out below 42 °C because II-anhydrite is easier to dissolve than DH, leading to a supersaturated solution of DH resulting in crystallization.

Natural anhydrite in nature, although its crystal structure is the same as that of II-anhydrite, has significantly different hydration properties, which are greatly related with lattice defects, microporous structure and surface state left by the formation of II-anhydrite. The natural anhydrite gypsum, which is not activated, has a very slow process of hydration and hardening.

The hydration process of HH is an exothermic process. Therefore, the exothermic process can be measured by a microcalorimeter to indicate hydration process of HH. The hydration process curve of β-HH is shown in Figure 4.6.

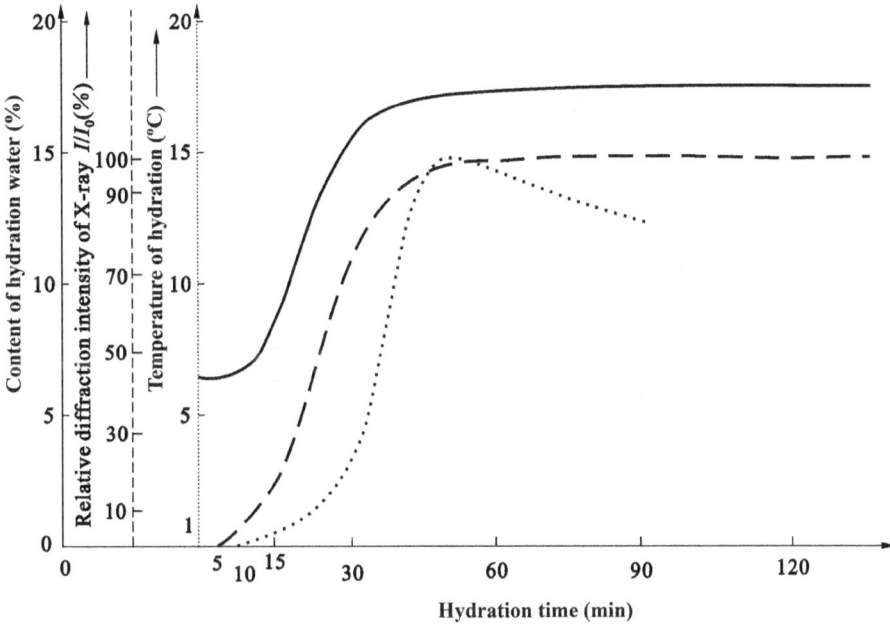

Figure 4.6: The hydration process of β-HH.

The graph shows the relationship between the bound water content, the diffraction intensity of DH, hydration temperature and hydration time. The speed and extent of hydration of HH can be quantitatively determined by measuring the bound water content at regular intervals, or by measuring the intensity of the diffraction curve with an X-ray diffractometer, or by measuring the heat change in the exothermic process with a microcalorimeter. From the exothermic curve, it can be clearly found that there is a very short induction period during the early stage of HH hydration, that is, a stage with very slow hydration rate and very low heat release. It is generally believed that nuclei of DH begin to be generated from the supersaturated solution during the induction period and the energy is continuously absorbed to form stable nuclei. When crystals grow into DH crystals releasing hydration heat, the induction period ends and an accelerated period begins. Some scholars believe that the formation of a gel during the induction period prevents chemical reactions, and the induction period ends once crystallization happens in the gel.

It is known from crystallography that the formation and growth of nuclei are closely related to the supersaturation of the liquid phase. The greater the degree of supersaturation, the faster the formation of nuclei, the slower the crystal growth, the more contacting points in crystal network and it is easier to form the initial structure of framework. For the system of HH–water, the formation of supersaturation is due to dissolution of HH, resulting in a supersaturated solution for DH. Thus, the measure of the supersaturation of gypsum slurry can be expressed as the ratio of the solubility of HH to the equilibrium solubility of DH in the same condition.

The relationship between the solubility c of gypsum and temperature is shown in Table 4.7. The equilibrium solubility of DH, the maximum solubility of HH, and the corresponding supersaturation all vary with temperature, i.e., the supersaturation decreases with increasing temperature.

Table 4.7: Relationship between solubility of gypsum (c) and temperature.

Absolute temperature	Equilibrium solubility of DH	Maximum saturated solubility of α-HH	Supersaturation of α-HH	Maximum saturated solubility of β-HH	Supersaturation of β-HH
(K)	c_∞ (g/L)	c_{max} (g/L)	c_{max}/c_∞	c_{max} (g/L)	c_{max}/c_∞
288	1.99	7.91	3.98	8.74	4.39
293	2.05	7.06	3.44	8.16	3.98
303	2.09	6.10	2.91	7.07	3.38
313	2.11	5.70	2.71	6.12	2.97
333	2.07	5.17	2.50	4.68	2.26

Formation of the crystal structure of DH during hardening process is divided into two stages.

In the first stage, with generation of new crystals, contacting by interweaving between crystals occurs, and crystals gradually grow, resulting in framework formation of crystal structure.

In the second stage, there are no newly formed contacting points of crystals, only the framework of the crystal structure grows. The growth of constituent grains leads to improve the structural strength. Because of directionally growing of crystals, the internal tensile stress can be generated, leading to a reduction in the structural strength.

According to this argument, the strength of the hardened structure not only depends on the degree of supersaturation, but also on the rate of supersaturation formation, that is, the solubility and dissolution rate of HH. With fast dissolution, supersaturation forms quickly, benefiting the formation of initial structure. If the dissolution rate is slow and the supersaturation duration is long, after the initial structure is formed, hydrated crystals are still increased and the structure will be densified. However, after a certain limit, an increase of hydrated crystals will cause increasing internal stress, resulting in the reduction in final strength. Therefore, to obtain a higher structural strength, it is necessary to create good hydration conditions, such as the appropriate temperature, fineness of material and the ratio of water to ensure appropriate growth of crystals with suitable amount and size during the formation and development of crystal structure to make the structure dense with a sufficient amount of crystals and large enough contacting areas, without generating internal stress that could damage the structure.

In addition, the strength is also related to the contact area between gypsum crystals. The greater the contact area, the higher is the strength. However, since crystalline contact points are thermodynamically unstable, they dissolve and recrystallize in a humid environment to weaken the structural strength. The greater the amount of contact points, the smaller the contact point size, the more severe the lattice at contact points deformation and the greater the possibility of structural strength reduction. Therefore, a proper control of supersaturation is necessary.

4.4.2 Hydration and mechanism of setting and hardening of dehydrated phases

For hydration and hardening process of dehydrated phases, there are two main theories: one is the dissolution–crystallization theory and the other one is the colloidal theory.

(1) Dissolution–crystallization theory (crystallization theory)
The dissolution–crystallization theory was first proposed by Le Chatelier in 1887. According to this theory, a metastable saturated solution is formed by dissolution of HH after being mixed with water (the solubility of HH is 8 g/L at 20 °C). It is highly supersaturated for DH (the solubility of DH is 2 g/L at 20 °C). Thus, DH can precipitate from the supersaturated solution. Due to the precipitation of DH, the original equilibrium state of HH dissolution is broken to lead to further dissolution of HH to compensate the content of calcium sulfate reduced by precipitation of DH in the solutions. Dissolving of HH and precipitation of DH are constantly carried out until HH hydrates completely. The chemical reaction is as follows:

$$CaSO_4 \cdot 0.5\,H_2O + 1.5\,H_2O \rightarrow CaSO_4 \cdot 2H_2O$$

Crossing growth and linking together of needlelike crystals of DH during hydration result in setting and hardening of gypsum slurry.

Le Chatelier divided the aforementioned process into three stages: (i) a chemical process of hydration, (ii) a physical process of crystallization and (iii) a mechanical process of hardening.

On the basis of the crystallization theory, a high supersaturation degree maintained sufficient time is a necessary condition for setting and hardening of HH.

(2) Colloid theory (topochemical reaction theory)

The colloid theory was first proposed by Michaelis in 1909, which aimed at explaining hydration and hardening of cement and then was applied in setting of plaster. According to this theory, hydration of plaster is carried out through an intermediate stage of colloid. Mixed with water, water directly enters into solid phase of HH and reacts to form a gel, and then needlelike crystals of DH are generated from the gel.

Afterward Cavazzi added alcohol in the filtrate of mixed HH and water to prepare gelatinous gypsum, and a gel gypsum precipitated it at the same time. Under a microscope, the gel could be seen become needlelike crystals. He speculated that gypsum was initially formed into a gel, and then fine needlelike crystals precipitated.

Triollier et al. used scanning electron microscope to observe DH obtained from hydration of β-HH in steam. He found that there was no difference in shape and size between HH before hydration and the DH generated in steam. Thus, they believed that hydrating of HH was based on the topochemical reaction theory. This hydration mechanism is divided into three stages: adsorption of water molecules on the surface of gypsum, dissolution of the adsorbed water molecules and formation of new phase.

Production of building gypsum is a process of dehydration of DH by heating to become HH. The application of building gypsum is the process where HH hydrates to form DH and hardens to provide strength by mixing with water to become a paste and formed into the required shape before hardening.

4.4.3 Factors influencing the hydration of HH

There are many factors that affect the hydration rate of HH, including calcined temperature, grinding fineness, crystal morphology, impurity and hydration conditions of gypsum. In the same conditions, the hydration rate of β-HH is greater than that of α-HH. At the room temperature, the complete hydration time of β-HH is 7–12 min and that of α-HH is 17–20 min.

The faster the hydration rate of HH, the faster the setting time of slurry. The hydration rate and setting time can be adjusted through adding admixtures in the production. If the hydration and setting of gypsum slurry are too fast, a retarder can be added.

Retarders can be divided into three categories according to the function.

(1) Large molecular substances, as the colloidal protective agent, can reduce dissolution of HH and prevent the development of nuclei, such as bone glue, protein glue, starch residue, molasses residue, hydrolyzate of animal products, amino acids and methanol compounds, tannic acid and so on.

(2) The substances can reduce the solubility of gypsum, such as glycerol, ethanol, sugar, citric acid and its salts, boric acid, lactic acid and its salts.

(3) The substances can change the crystal structure of gypsum, such as calcium acetate, sodium carbonate, phosphate and so on.

Different retarders have different retarding effects. Although the use of retarders alone can extend the setting time, sometimes it also cause strength reduction at different levels. Experiments show that the use of retarders and water-reducing simultaneously can achieve better results.

4.4.4 Hardening of gypsum paste

(1) Structure development of hardened gypsum

Practice has proved that a gypsum cementitious material only form hydrated products during hydration and the slurry does not necessarily form an artificial stone with a strength. An artificial stone with a strength can only be formed when hydrated crystals grow to link together, resulting in a network of crystal structure. Therefore, hardening of gypsum slurry is a process of the formation of crystal structure network, which is accompanied by the development of strength. For convenience, the structure strength of slurry is characterized by plastic strength P_m, which is the ultimate shear stress of slurry structure. Its physical concept is that when shear stress produced under an external force is greater than or equal to the ultimate shear stress, the perceived flow produced in slurry can be observed, and its deformation varies with time. However, the slurry shows a solid performance when the shear stress is lower than the ultimate shear stress. The greater the ultimate shear stress, the greater is the plastic strength P_m, the better the development of slurry structure and the greater the strength.

Figure 4.7 shows the development of structural strength of β-HH slurry with time. It can be seen that the plastic strength P_m is very low and strength growth is slow at

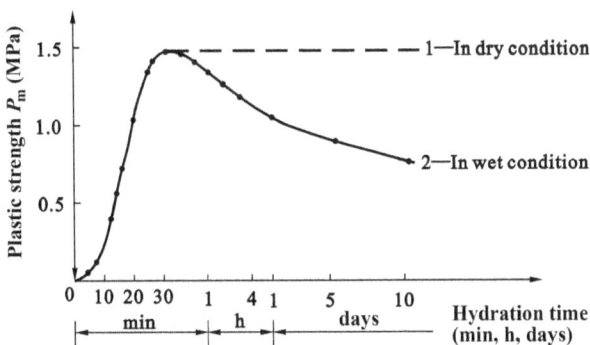

Figure 4.7: The development of structural strength of β-HH with hydration time.

the first stage, before 5 min. In the second stage, the strength increases rapidly and reaches the maximum during 5–30 min. In the third stage, the strength remains stable if the hardened slurry is kept in a dry environment as indicated by dotted line 1 in the figure; the strength decreases gradually if the hardened slurry is kept in a humid environment, as shown by curve 2.

The first stage is the stage of agglomeration structure of the gypsum slurry. There are water films between particles in the slurry. The particles interact with van der Waals molecular forces through the water films, so the strength is low. This structure has thixotropic recovery, that is, the structure can automatically recover by an external damage.

The second stage is the formation and development stage of the crystal structure network. A large number of crystal nuclei are formed, and crystals grow up, contact and join together. A crystalline structure network is formed in the whole gypsum slurry with a high strength.

The third stage reflects the characteristics of the crystal contact point in the gypsum crystal structure network. In a dry environment, the crystalline contact points remain stable and the strength is constant. If the structure is in a humid environment, the strength will decrease. This is due to the thermodynamic instability of the crystalline contact points. Compared with the regular crystals, the lattice at the crystal contact points has a high degree of solubility due to distortion and deformation. Therefore, after damp, the contact points are dissolved, larger crystals recrystallize, the hardened body is destroyed and the strength is lowered.

(2) Factors influencing structure development of hardened gypsum

The formation and destruction of the crystalline structure of the gypsum slurry are controlled by almost the same factors, supersaturation of liquid phase. It affects the nucleation rate, the amount and crystal growth and continuous conditions, and also determines crystal stress.

When the supersaturation is high, there are many nuclei formed in the liquid phase and the grains are small with many crystal contact points, and the crystal structure network is easy to form. On the contrary, less nuclei are formed in the liquid phase and grains are coarse with less crystal contact points. More hydrates are needed to form the same structure network under the same conditions. When the gypsum slurry hydrates, the supersaturation at 20 °C is greater than that at 60 °C, the amount of hydrate consumed at the initial structure at 20 °C is less, the former is about 10% and the latter is 25%. After the initial structure is formed, hydrate continues to be generated, playing an enhance role in the structure network to further dense. However, when a certain limit is reached, the hydrate continues to increase, creating an internal stress (called crystalline stress) in the formed structural network. After the structure is dense, more crystals are formed, and higher is the crystallization stress. When the crystallization stress is greater than the limit structure it can withstand, it will lead to structural damage, and so the final plastic strength will decrease. If the supersaturation is too high, the structural strength is weakened due to the crystallization stress, and at the same time, the relationship between development of the structural strength and the hydration reaction will show that the time to reach the highest intensity is earlier than the end of the hydration. Provided that the formation of the hydrate is not sufficient to produce crystallization stress after the formation of the crystalline structure, or the resulting crystalline stress is not sufficient to cause structural damage, the time to reach the highest intensity is the same as the end time of hydration.

Figure 4.8 shows the kinetic curves of structure formation for α-HH slurry suspension at 20 and 60 °C, respectively.

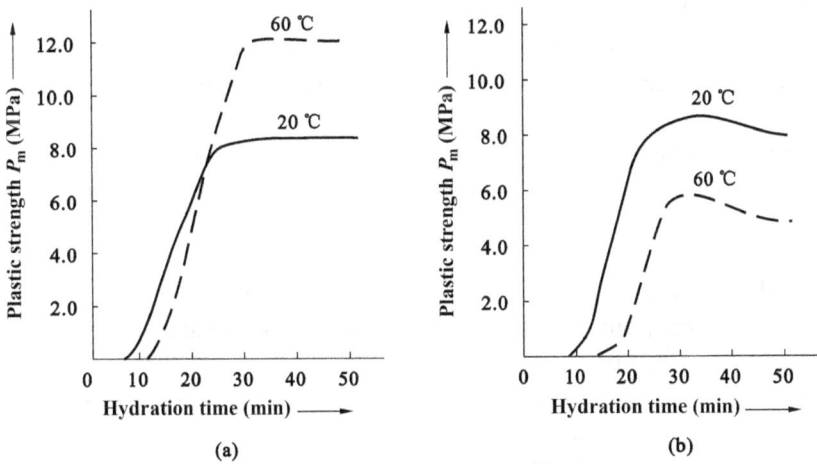

Figure 4.8: Kinetics curves of structure formation of α-HH paste: (a) $w/s = 0.4$; (b) $w/s = 0.8$.

It can be seen from Figure 4.8(a) that when the water–solid ratio is relatively small, the plastic strength at 60 °C is higher than that at 20 °C. Because increasing temperature could reduce the saturation and the crystallization stress, the structure could protect from damage and have a higher strength. When the water–solid ratio is relatively high, the situation is just the opposite, as shown in Figure 4.8(b). This is because when the water–solid ratio is relatively high, not only the porosity in the hardened structure is increased, but also the amount of hydrates required to form the crystal structure network is greatly increased due to the large slurry filling space. Thus, the possibility of crystallization stress within the crystal structure can be decreased. In order to maintain the normal development of hardened structure, it is necessary to increase the degree of supersaturation and its duration to promote a sufficient amount of hydration product in the slurry.

4.5 Activation of anhydrite and its applications

4.5.1 Activation methods of anhydrite

Because of the slow hydration and hardening of anhydrite, it does indeed cannot be set for a long time and cannot produce strength. When using anhydrite to produce cementing materials and products, it must be activated to improve its hydration and ability of setting and hardening to make it a building material with a cementitious ability and strength. Activation methods of anhydrite are divided into three categories, namely, physical activation, chemical activation and physical–chemical activation.

(1) Physical activation
Heat treatment can be applied to change the crystal structure of anhydrite, destroy lattice, increase lattice distortion and defects and improve hydration activity.

The grinding method can reduce particles size, increase the specific surface area, increase irregularity of the surface structure and improve the surface performance and hydration activity.

(2) Chemical activation
Hydration of chemically pure anhydrous calcium sulfate (II-AH) alone is very slow. The hydration speed can be greatly accelerated when adding 1% pure alum as an activator. In addition, the ability to hydrate and harden can be also improved through adding different acids, alkali, salts and other chemical substances as anhydrite activators to stimulate the activity. Commonly used activators are divided into two categories according to their chemical properties.

(i) Sulfate activators (acid activators)
They are Na_2SO_4, K_2SO_4, $NaHSO_4$, $KHSO_4$, $CuSO_4$, $FeSO_4$, $Al_2(SO_4)_3$, $ZnSO_4$, $KAl(SO_4)_2 \cdot 12H_2O$ and so on.

(ii) Alkaline activators
Alkaline activators are materials such as lime, cement, caustic soda, alkaline blast furnace slag, calcined dolomite, flammable shale, other alkaline ash and so on.

In addition, certain nitrates, chromates and dichromates can also be used as activators.

Acidic activators can accelerate the early hydration rate of anhydrite, shorten the setting time and improve the hydration and hardening ability of anhydrite. Alkaline activators have a little effect on the early hydration rate, but have a significantly enhanced effect on later hydration hardening. The use of alkaline activators can improve the water resistance of anhydrite, resulting in a cementitious material with the ability of air-hardening and hydraulicity.

The fineness of anhydrite and the effect of HH on the early hydration rate are listed in Table 4.8.

Table 4.8: Fineness of anhydrite and effect of HH on the early hydration rate.

| No. | Natural anhydrite | | Activator | | Shares of HH | Hydration rate (%) | |
	Specific surface area S (cm^2/g^1)	Shares	Type	Shares		1 day	3 days
1	3,780	100	$K_2Cr_2O_7$	1	–	20.5	35.4
2	3,780	100	K_2SO_4	1	–	14.8	29.5
3	7,800	100	$K_2Cr_2O_7$	1	–	42.3	75.0
4	7,800	100	K_2SO_4	1	–	37.6	69.5
5	7,800	100	$K_2Cr_2O_7$	1	10	58.0	90.2
6	7,800	100	$K_2Cr_2O_7$	2	5	60.9	93.7

4.5.2 Anhydrite cementing materials

Anhydrite cementing materials, also known as anhydrite cements, are made of grinding of anhydrite with activators. At present, the research of anhydrite cement is still in the development stage and has not yet been applied in practice. Because of the advantages of the simple process without calcination, energy saving and use of waste residues, anhydrite cement occupies an important position in construction industry worldwide.

4.6 Structure and properties of hardened gypsum paste

4.6.1 Structure of hardened gypsum paste

The engineering properties of gypsum products are mainly determined by the internal structure. The hardened HH is mainly composed of hydrated (DH) crystals that link together to form a porous network structure. The properties of the hardened body mainly depends on several structural features, namely,
(1) the characteristics of interaction between the newly hydrated grains;
(2) the amount and properties of crystal contacting points between newly hydrated grains;
(3) the amount and size distribution of pores in the hardened paste.

The structure network in the hardened gypsum paste can be divided into two categories according to the properties of forces between particles. The first one is the agglomerates formed between particles by van der Waals forces. The second is crystalline structure formed between particles through contact points by chemical bonds.

The former has a very low structural strength, but the latter has a high structural strength. During the initial period of hardening of the gypsum slurry (within 5 min), it is the formation stage of agglomeration structure, and the surface of hydrated particles is encapsulated by a water film, so that particles are interacted by van der Waals force and the strength is low. At the structure formation stage of gypsum slurry, if the crystal structure network is damaged by an external force or internal stress, HH further hydrates without a high enough supersaturation, and a new crystal structure network cannot be established to link particles with enough crystal contact points so that hydrated particles can only interact by van der Waals resulting in a low strength of the product.

After the formation of crystal structure network in hardened gypsum paste, many of its properties depend on the nature and amount of contacting points between crystals. On the one hand, the strength of hardened gypsum paste depends on the strength of single contact point and the number of contact points in unit volume. On the other hand, due to thermodynamical instability of crystal contact points, they will dissolve and recrystallize in the humid environment, resulting in reducing the strength.

In addition, the gypsum slurry is a porous body; the porosity and size distribution of pores are also very important structural factors.

4.6.2 Strength of hardened gypsum paste

Strength is an important indicator in engineering nature of materials. The theoretical strength of a solid material is usually several hundred times or even several thousand times higher than the actual strength. For example, the theoretical tensile strengths (R_m) of rock salt and quartz are close to 2,000 and 10,000 MPa, respectively, while the actual tensile strengths are lower than 5 and 100 MPa. The is because there are surface cracks and internal microcracks in the crystal. When the material is damaged by an external force, stress concentration occurs near the crack, which leads to the destruction of the material.

In general, the probability of defects in coarse crystals is too high, and so the strength of coarse crystals is lower than that of fine crystals, which is the size effect of reduction in strength.

The relationship between strength and diameter of a single crystal of gypsum has been studied by several scholars. The size of the gypsum crystal studied varies within the following ranges: 5–20 µm in length and 5–00 µm in diameter. The results show that the strength is significantly decreased when the grain size is increased. If the size of the single crystal of gypsum is 10 μm, the ultimate tensile strength can reach to 175.0 MPa in the (001) direction. When the gypsum slurry is observed with a microscope, the size of common DH is generally 25–50 μm, and the ultimate tensile strength of the gypsum single crystal is 30–20 MPa corresponding to this size range. And now, the tensile strength of hardening gypsum slurry is only about 2.0 MPa when prepared in laboratories. This fact, on the one hand, shows that the potential to improve strength of gypsum products is very large, and, on the other hand, it also shows that the structural factors that determine the strength of hardened gypsum slurry are not only related to the single crystal of hydrated products (DH) but also related to the nature and quantity of crystal contact points. The nature and quantity of crystal contact points in the gypsum slurry are mainly related to the supersaturation. Higher strength can be obtained by controlling the appropriate supersaturation. At the same time, the strength of hardened gypsum paste is also related to porosity and pore size distribution.

4.6.3 Water resistance of hardened gypsum paste

At present, gypsum products in construction have a common weakness, which is poor water resistance. For example, in dry conditions, the compressive strength R_c of gypsum products is 6.0–10.0 MPa, while strength loss is up to 70% or even greater when it is in the saturated water state. The is because crystal contact points between crystal particles dissolve when the product is in a humid environment for a long time, resulting in gradual dissolution and structural collapse of the hydrated product crystals.

There are several reasons for poor water resistance of gypsum products.

(1) Gypsum has a high solubility
After damping, the crystal structure has been destroyed due to dissolution and strength decreases.

(2) The porosity in gypsum products is high
Crystal dissolution and structural damage are caused by water absorption in pores.

(3) Microcracks exist in gypsum products
Fine cracks on surface resulted from forming and processing of products make water enter into the cracks, leading to the wedging effect to separate tiny units in the crystal structure and the structure is damaged.

After absorbing water, the strength of a gypsum product decreases. In addition, the product will expand and break apart in frozen conditions because water in pores freezes to expand. Therefore, measures must be taken to improve water resistance of gypsum products, such as improving the density of products, brushing or dipping waterproof layer on the surface of products and so on to reduce the penetration of water into the products.

4.7 Applications of gypsum cementitious materials

4.7.1 Gypsum cementitious materials

(1) Gypsum binder
Gypsum binder is based on plaster and prepared by mixing with a small amount of organic additives. For brick walls, concrete and aerated concrete walls, it is extremely easy to paste with binder. Gypsum binder is widely used because of strong bonding, construction convenience and low cost.

(2) Gypsum plaster
Gypsum plaster is a kind of plaster materials of interior walls and roof surface, which is made of gypsum cementitious materials. Using gypsum mortar composed of building gypsum, water, sand and retarder for plastering, the surface of walls is smooth, delicate, white and artistic. The initial setting time of gypsum plaster is longer than 1 h, and the final setting time is less than 8 h.

(3) Gypsum-based leveling mortar
Gypsum-based leveling mortar can flow automatically and form a smooth surface relied on its own gravity on the concrete floor cushion. It is the base material of laying carpets, wood flooring and so on. After construction, the size of ground is accurate, can be wiped after pouring after 24 h and carried out on it after 48 h.

(4) Gypsum putty
Gypsum scraping wall putty takes the building gypsum powder and talcum powder as the main raw materials and mixes with a small amount of gypsum modifier into powder. It is mixed well with water and levelled by scraping method in use. Scraping wall putty made of gypsum has the characteristic of quick setting, high bond strength, white and delicate and could not collapse after absorbing water.

4.7.2 Gypsum products

Gypsum products have the properties of light weight, adiabatic, sound absorption, noncombustibility and sawable nails. In addition, there are a wide range of raw materials for producing gypsum products, with low fuel consumption, short production cycle and simple processing equipment. Gypsum products belong to green building materials, for example, the fuel required for the production of 1 t gypsum cementitious materials is a quarter of that required for the production of 1 t Portland cement; the construction investment of gypsum products is half of that required in cement industry and the steel used for equipment is one-third of the cement plant.

(1) Gypsum board

Gypsum board, which is the one of important new lightweight sheets, has a broad development prospects in China. Gypsum board thickness is 9–18 mm, and is mainly used for flat and interior wall decoration. It can be directly attached to walls or nailed to wooden keel on both sides of the wall, also used for the roof and so on.

A gypsum board, using a building gypsum as the main raw material, is a kind of light construction board. The slurry obtained by mixing the building gypsum with a small amount of additives and water is continuously poured on the surface of a bottom paper, and a top paper is used to cover it after forming. Then the gypsum board is produced through edge banding, flattening, solidification, cutting and drying. A gypsum board is divided into three types: ordinary, water-resistant and fire-resistant. An ordinary gypsum board uses fiber-reinforcing materials like glass fiber and pulp. Covering papers should have a certain strength and can be firmly bonded with the core of gypsum board. A water-resistant gypsum board is made by adding waterproof, moisture-proof admixtures in the core of ordinary gypsum board and using water-resistant paper. A fire-resistant gypsum board is made of a fire-resistant core by adding inorganic refractory reinforcing fibers in ingredients of ordinary gypsum board. An ordinary gypsum board is used as interior walls, partitions and ceilings; a water-resistant gypsum board is used for kitchen, bathroom and other ceiling and partition in humid environments; a fire-resistant gypsum board is mainly used for building with high fire requirements.

A fiber-reinforced gypsum board is made of building gypsum, wood shavings or paper fibers and a small amount of chemical additives. The board made by adding wood shavings into building gypsum is called the gypsum particleboard (or wood fiber gypsum board); the board using paper fiber is called the paper fiber gypsum board. A gypsum particleboard has the advantages of gypsum board, and also has the advantages of ordinary particleboard. It has many advantages such as less water, high strength, good workability and not easy to break during construction. Gypsum particleboard is a kind of green board because it releases little formaldehyde not like ordinary particle board.

A gypsum hollow slab, using building gypsum as a raw material, is bar board with the size specification of (2,400–3,000) mm × 600 mm×(60–120) mm, having 7 or 9 holes. Adding appropriate light porous filling material (such as expanded perlite, expanded vermiculite, sawdust, etc.) or a small amount of fiber materials, a gypsum hollow slab is produced by mixing with water, pouring, vibration molding, loose core, demolding and drying. A gypsum hollow slab can replace traditional bricks used as interior wall materials. It has the advantages such as light weight leading to light weight buildings and low basic bearing. Besides, the length of a gypsum hollow slab can be customized by the floor height, resulting in high construction efficiency.

Currently produced gypsum boards in China include decorative gypsum board, gypsum insulation board, gypsum block, gypsum bagasse board and so on.

(2) New wall materials

Gypsum is mainly used for the production of gypsum blocks, gypsum plates and other new wall materials with unique functions. Due to the characteristics of light weight, high strength, water resistance and heat resistance, gypsum wall composites are gradually replacing the traditional wall materials, such as light steel keel paper gypsum board rock wool composite wall, fiber gypsum board or gypsum particle-board and the keel composite wall, inflated (or foam) gypsum insulation board or block composite wall, gypsum and polystyrene foam board and straw and other composite large plate.

(3) Decorative materials

Gypsum decoration materials continue to be developed, such as various high strength, moisture-proof, fireproofing, environmental gypsum decorative plate, relief art plaster angle, gypsum lines, fireplaces, lamp panels, lamp holders, Roman columns, doors and windows and other decorative products. They are widely applied due to sound absorption, antiradiation and fireproofing. These products are made of high-quality building gypsum as the base material, mixed together with reinforcing fibers, cementing agents, and water and then formed through injection molding, hardening and drying.

A decorative gypsum board or gypsum board as the substrate can also be processed into perforated gypsum board with a sound-absorbing effect by adjusting the thickness of plate, barrel hole size, hole distance, air layer thickness, and so on. It can adapt to different frequencies and be used for the studio, lecture hall or theater.

(4) High-strength gypsum

The water demand of a high-strength gypsum (α-HH) is about 35–45%, only about half of that for a building gypsum. Therefore, after hardening, α-HH has a high density and strength. The 3 h compressive strength of α-HH can reach 9–24 MPa and the 7 h compressive strength can reach 15–40 MPa; the initial setting time is not earlier than 3 min and the final setting time is 5–30 min.

The high-strength gypsum is applied in plastering engineering with high-strength requirements, decorative products and gypsum boards. Mixed with waterproofing agents, it can be used in a high-humidity environment; mixed with organic materials, such as polyvinyl alcohol aqueous solution and polyvinyl acetate emulsion, it can be dubbed into a binder, which is characterized by zero shrinkage.

4.8 Acceptance check, storage and transportation of gypsum

The building gypsum can easily absorb moisture and has a fast setting and hardening rate. During transportation and storage, attention should be paid to avoid damp. During long-term storage, the strength of gypsum will be reduced. The strength generally decreases about 30% after 90 days storage. If the storage period is longer than 90 days, the grade of gypsum should be re-examined and determined.

Building gypsum is usually packed with paper bags or other composite bags, which have the effect of moisture-proof and cannot be damaged. The product mark, manufacturer name, production lot number and date, quality grade, trademark and moisture mark must be clearly marked on the packaging labels. During transportation and storage, it should not be damp and mixed with debris and different grades of gypsum should be separated. The storage period is 3 months since the date of production.

Problems

4.1 What are the basic properties of building gypsum? Why its products are only suitable for indoor use, and should not be used for outdoor?

4.2 How is the phase transformation process carried out when gypsum is heated?

4.3 Please describe the hardening mechanism of gypsum briefly.

4.4 Write the phase transformation temperatures of dehydrated gypsum phases. What are the dehydrated phases of gypsum at room temperature?

4.5 What are the differences in physical performance between α-HH and β-HH? What are the differences in hydration performance between them?

4.6 Briefly introduce preparation processes for α-HH and β-HH.

4.7 What are the characteristics of gypsum products?

4.8 How to improve the hydrating activity of anhydrite as a cementitious material?

4.9 What are the factors affecting hydration of gypsum?

4.10 Why can HH hydrate? Please describe the hydration process.

4.11 What are the main applications of gypsum cementitious materials?

5 Lime

Lime is one of the earliest mineral cementing materials used in constructions. Ancient Greeks used it for architecture in the eighth century BC, and Chinese began to use lime in the seventh century BC. Lime can be divided into quicklime and slaked lime, or calcareous lime [w (MgO) ≤ 5%] and magnesia lime [w (MgO) > 5%] according to the content of magnesium oxide. Lime is still widely used as a building material due to the wide range of resources, simple manufacturing process, low cost and good cementing performance.

5.1 Raw materials for production of lime

Natural rocks consisting of calcium carbonate as the main component, such as limestone, chalk and dolomitic limestone, can be used to produce lime. In addition, chemical industry by-products can also be used to produce lime. These raw materials mainly contain $CaCO_3$ and a small amount of impurities such as $MgCO_3$, SiO_2 and Al_2O_3.

Calcium carbonate is a compound composed of $CaCO_3$. It is a common substance on earth, found in rocks, carapace and snail shells. It exists in the mineral form of calcite and aragonite in nature. Calcite belongs to the triangular system with a hexagonal crystal. Pure calcite is colorless and transparent, usually white. It consists of 56% CaO and 44% CO_2, with a density of 2.715 g/m³ and the Mohs hardness of 3, and it is brittle in nature. Aragonite belongs to the orthorhombic system with a rhombic crystal. It is gray or white, with a density of 2.94 g/cm³ and the Mohs hardness of 3.5–4, and it is dense in nature. The crystal size of calcite is very important to the physical properties of limestone. Dense limestone with low porosity has a structure of fine grains and a high strength. The densities of limestone, dolomite limestone and dolomite are about 2.65–2.80, 2.70–2.90 and 2.85–2.95 g/cm³, respectively. Its bulk density depends on porosity.

The commonly used natural raw materials for lime are dense limestone, marble, oolitic limestone, shell limestone and chalk, and so on.

(1) Dense limestone
It is more commonly used, containing higher than 90% CaO with a dense structure, hard texture and higher sintering temperature.

(2) Marble
Marble contains the highest content of $CaCO_3$, and can be used as a decorative material directly. Chemically pure CaO can be obtained when broken pieces and marble unsuitable for decoration are used to produce lime by calcining.

https://doi.org/10.1515/9783110572100-005

(3) Oolitic limestone

It is composed of bonded spherical limestone and has lower mechanical strength.

(4) Shell limestone

Shell limestone comes from sea and is bonded with different sizes of shells. It is soft in texture and low in hardness.

(5) Chalk

Chalk is soft and is mainly composed of aphanitic or amorphous, loose, fine $CaCO_3$ with a content of 80–90%.

Lime can also be produced as chemical industry by-products. For example, natural raw materials can be replaced with acetylene sludge (the main component is $Ca(OH)_2$) obtained from the production of acetylene by calcium carbide, and residue (the main component is $CaCO_3$) generated during manufacturing pure alkali by ammonia alkali process.

5.2 Production of lime

5.2.1 Calcination of calcium carbonate

Lime is a product mainly composed of CaO. It is produced by heating and incinerating raw materials, in which $CaCO_3$ is the main component, and CO_2 is removed. During calcination at high temperatures, $CaCO_3$ should be decomposed and CO_2 should be exhausted as much as possible. The chemical reaction is as follows:

$$CaCO_3 \xrightarrow{\quad 900-1,000\,^\circ C \quad} CaO + CO_2 \uparrow -176.68\,kJ$$

The decomposition process of $CaCO_3$ is reversible. To make the reaction to the positive direction, the calcination temperature must be increased and CO_2 must be exhausted in time.

The decomposition of limestone takes place slowly from surface to interior. Calcination of large particles of limestone is more difficult than that of small particles and requires much longer time. Natural raw materials often contain impurities such as clay. When the content of clay exceeds 8%, hydraulic minerals, such as β-C_2S generated from solid phase reaction, the nature of lime will be changed from air hardening into hydraulicity. Therefore, the content of clay impurities must be controlled in lime production.

After decomposition of limestone, 44% CO_2 of the original mass is lost, while the apparent volume of calcined lime only decreases by 10–15% than that of limestone. The resultant lime product has a porous structure.

At ordinary pressure, the theoretical decomposition temperature of calcium carbonate is 898 °C. In practice, the calcination temperature is influenced by many factors, such as type of raw materials, structure, density, particle size, impurity content and heat loss of kiln. The practical calcination temperature is significantly higher than the theoretical value, and is generally controlled at 1,000–1,200 °C or higher. To obtain high-quality quicklime, it is necessary to control calcination temperature and time appropriately.

The quality of lime greatly depends on the composition of raw materials, calcination temperature and calcination equipment, especially calcination condition. $CaCO_3$ cannot be completely decomposed, and more underburnt lime will be produced at low calcination temperatures with insufficient time, or with oversized particles. Low CaO content in underburnt lime degrades quality, reduces utilization rate of lime and lowers cementing performance. Overburnt lime will be produced at higher calcination temperatures with longer time. Overburnt lime has a dense structure and smaller internal specific surface area with coarse grains. The produced CaO is fused with impurities in the raw material such as SiO_2 and Al_2O_3 to form a cover of glassy glaze on the surface of product so that hydration process is very slow. In applications, after normally calcined lime is hardened, fine overburnt lime particles start to absorb moisture and hydrate slowly. Heat is released causing cubic expansion and resulting in local swelling or cracking and splintering in hardened lime paste, lowering quality of engineering projects.

The decomposition temperature of magnesite (the main component is $MgCO_3$), as an impurity in raw materials for lime production, is much lower than that of $CaCO_3$. MgO is therefore overburnt during calcination, and affects the quality of lime. When the content of magnesite in raw materials is high, the calcination temperature should be reduced as low as possible in a condition of guaranteeing complete decomposition of $CaCO_3$. For silicate products, the content of magnesite in raw materials must be limited to avoid causing volumetric instability.

The decomposition of $CaCO_3$ is carried out in kilns. There are many categories of lime kilns. Based on the state of fuels, there are mixed fuel and raw materials kilns (i.e., burning solid fuels such as coke, coke powder, coal, etc.) and gaskilns (including burning blast furnace gas, coke oven gas, calcium carbide exhaust gas, coal gas, natural gas, etc.). According to kiln shapes, there are shaft kilns, rotary kilns and sleeve kilns. Meanwhile, the operation of kilns can be exerted under positive pressures or negative pressures. Shaft kilns and rotary kilns are commonly used for calcination of limestone. Shaft kilns are widely used in the production of lime because of its reliability, low fuel consumption and high quality of product. Rotary kilns are often used in modern production of lime due to high degree of automation and stable yield and quality, shown as in Figure 5.1.

Rotary kilns are used for calcining loose, soft or small pieces of limestone with low strength, such as chalk and shell rock. Lime can be produced by rotary kilns with high activity, uniform quality, low rate of overburnt or underburnt and

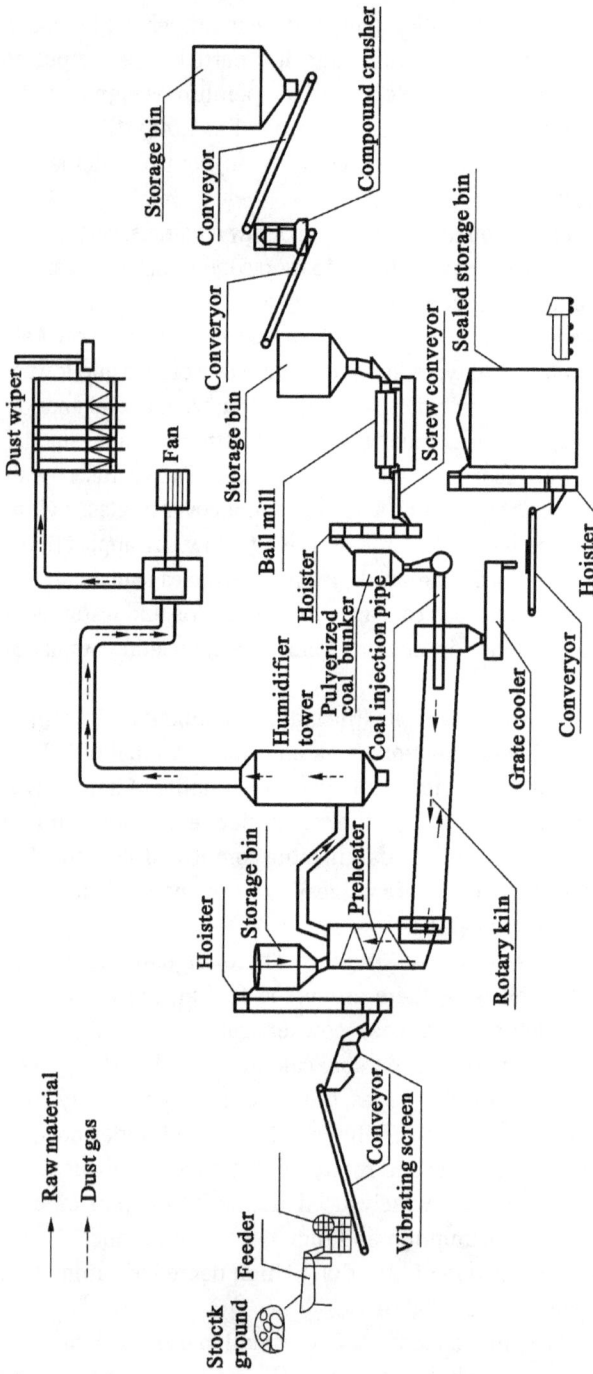

Figure 5.1: Rotary kiln system for lime production.

high yield (up to 1,100 t/day). Similar to the rotary kiln used in cement production, the lime kiln consists of a preheater, a rotary kiln body and a lime cooler. In addition, it is also equipped with a fuel combustion system and a hydraulic system (Figure 5.1).

There are four ways of heat transfer during the calcination process of limestone: heat transfer by conduction through the wall surface of a high-temperature furnace, heat transfer by radiation of high-temperature gas to raw materials, heat transfer by convection around high-temperature gas and raw materials and heat transfer by conduction from the surface of raw materials to inside. Here, heat transfer by radiation is the main way.

5.2.2 Types of lime

Lime has a great variety. According to different characteristics, there are different classification methods.

(1) Classification according to performance

(i) Air hardening lime
It is produced by calcining limestone composed of high content of $CaCO_3$ and clay impurities less than 8%. Most of the currently used lime in China belongs to this category.

(ii) Hydraulic lime
Hydraulic lime is produced by calcining limestone containing clay impurities more than 8% (such as argillo calcareous minerals). In addition to the main component of CaO in the product, it also contains a certain amount of calcium silicate ($2CaO \cdot SiO_2$) and calcium aluminate ($12CaO \cdot 7Al_2O_3$) resulting in hydraulicity, and it is known as hydraulic lime.

(2) Classification according to production process

(i) Lump lime
It is obtained by calcination of raw materials mainly consisting of CaO.

(ii) Quicklime powders
The fine powders are obtained by finely grinding lump lime. The fineness is required that the residue from 0.125 mm square mesh screen must be 7–18%. The primary component is CaO.

(iii) Slaked lime powders
These are obtained by hydrating lime using an appropriate amount of water, known as slaked lime. The main component is $Ca(OH)_2$.

(iv) Lime slurry

It is a thick paste obtained by hydrating quicklime using an appropriate amount of water (about 3–4 times the volume of lime), also known as lime paste. The main ingredients are $Ca(OH)_2$ and water. A white suspension can be obtained with more water, known as lime milk. At 15 °C, a transparent liquid with dissolved 0.3% $Ca(OH)_2$ is known as lime water.

(3) Classification according to hydration time and temperature

The hydration time (Δt) of lime means the time required to reach the highest temperature for the lime–water solution by timing from mixing lime and water in standard conditions. The highest temperature of the solution achieved during hydration is the hydration temperature (T_{max}). The heating rate of lime (i.e., the apparent slaking rate, $\Delta T/\Delta t$) refers to the ratio of the total increased temperature (ΔT) of the solution to the hydration time (Δt). It is an important indicator to characterize the hydration rate of lime.

According to the hydration time (Δt) of lime, lime can be divided into the following three categories.

(i)   high-speed slaking lime: $\Delta t < 10$ min;
(ii) mid-speed slaking lime: 10 min $\leq \Delta t \leq 30$ min;
(iii) low-speed slaking lime: $\Delta t > 30$ min.

According to the hydration temperature (T_{max}), lime can be divided into the following two categories.

(i) low-heat lime: $T_{max} \leq 70$ °C;
(ii) high-heat lime: $T_{max} > 70$ °C.

(4) Classification according to the content of MgO in quicklime

Raw materials of lime often contain $MgCO_3$. During calcination of lime, $MgCO_3$ decomposes to form MgO. According to the content of MgO, quicklime can be divided into calcareous lime and magnesia lime, listed in Table 5.1.

Table 5.1: The contents of MgO in different lime products.

Type	Quicklime	Quicklime powders	Slaked lime powders	Dolomite slaked lime powders
Calcareous lime	w (MgO) $\leq 5\%$	w (MgO) $\leq 5\%$	w (MgO) $\leq 4\%$	—
Magnesia lime	w (MgO) $> 5\%$	w (MgO) $> 5\%$	$4\% \leq w$ (MgO) $\leq 24\%$	$24\% \leq w$ (MgO) $\leq 30\%$

Hydration of magnesia lime is slow. Its strength is slightly higher after hardening. Most lime used in construction is calcareous lime.

5.2.3 Characteristics of architectural lime

Compared with other cementitious materials, the characteristics of lime are as follows.

(1) Good water-retaining property and plasticity

When lime hydrates to form lime slurry, the fine and colloid dispersed grains of $Ca(OH)_2$ covered with a thick water film can be formed automatically. Lime mortar made of lime has good water retention and plasticity. The plasticity of cement mortar can be significantly improved by adding lime slurry into cement mortar.

(2) Slow hardening and low strength

During hardening, it is hard for CO_2 in air to enter inside of the hardening paste. Carbonization is slow, leading to a low hardening strength. For example, the 28 days compressive strength of lime mortar (lime/sand mass ratio = 1:3), R_c, is only 0.2–0.5 MPa.

(3) High dry shrinkage

The evaporation of large quantities of free water can cause a significant contraction during hardening. Except brushing with a thin layer by lime milk, lime cannot be used alone. Shrinkage is often reduced by mixing with sand and paper tendons.

(4) Poor water resistance

Because the main component in hardened lime slurry is $Ca(OH)_2$ and it is slightly soluble in water, hardened lime dissolves due to absorbing moisture resulting in lower strengths and scattering in water.

(5) High hygroscopicity

Quicklime absorbs moisture in air easily and hydrates into slaked lime. For long-term storage, lime must be stored in an airtight, moistureproof and waterproof condition.

5.2.4 Technical specifications for architectural lime

The important indexes to evaluate quicklime quality include contents of effective CaO and MgO, the amount and fineness of slurry produced by hydrating. According to the Chinese standard of *Architectural lime* (JC/T 479—2013), calcareous lime and magnesia lime can be divided into different grades based on the chemical content of CaO and MgO, and the categories, names and codes are listed in Table 5.2. According to the criterion, chemical composition of architectural quicklime must agree with the data listed in Table 5.3 and physical properties must correspond to those listed in Table 5.4.

Table 5.2: Categories, names and codes of architectural quicklime.

Category	Name	Code
Calcareous lime	Calcareous lime 90	CL 90
	Calcareous lime 85	CL 85
	Calcareous lime 75	CL 75
Magnesia lime	Magnesia lime 85	ML 85
	Magnesia lime 80	ML 80

Table 5.3: Chemical composition of architectural quicklime (mass fraction %).

Code	CaO + MgO	MgO	CO_2	SO_3
CL 90	≥ 90	≤ 5	≤ 4	≤ 2
CL 85	≥ 85		≤ 7	
CL 75	≥ 75		≤ 12	
ML 85	≥ 85	> 5	≤ 7	
ML 80	≥ 80			

Table 5.4: Physical properties of architectural quicklime.

Code		Slurry yield (dm^3/kg)	Fineness	
			Residue (%): 0.2 mm screen	Residue (%): 90 μm screen
CL 90	Lump lime	≥ 2.6	—	—
	Powder lime	—	≤ 2	≤ 7
CL 85	Lump lime	≥ 2.6	—	—
	Powder lime	—	≤ 2	≤ 7
CL 75	Lump lime	≥ 2.6	—	—
	Powder lime	—	≤ 2	≤ 7
ML 85	Lump lime	—	—	—
	Powder lime		≤ 2	≤ 7
ML 80	Lump lime	—	—	—
	Powder lime		≤ 7	≤ 2

According to the Chinese standard of *Architectural slaked lime powders* (JC/T 481–2013), slaked lime powders are classified according to the content of CaO and MgO after deducting free water and bound water. The categories of architectural slaked lime powders are listed in Table 5.5. The chemical composition and physical properties of slaked lime are listed in Tables 5.6 and 5.7, respectively.

Table 5.5: Categories of architectural slaked lime.

Category	Name	Code
Calcareous slaked lime	Calcareous slaked lime 90	HCL 90
	Calcareous slaked lime 85	HCL 85
	Calcareous slaked lime 75	HCL 75
Magnesia slaked lime	Magnesia slaked lime 85	HML 85
	Magnesia slaked lime 80	HML 80

Table 5.6: Chemical composition of architectural slaked lime (mass fraction, %).

Code	CaO + MgO	MgO	SO_3
HCL 90	≥ 90	≤ 5	≤ 2
HCL 85	≥ 85		
HCL 75	≥ 75		
HML 85	≥ 85	> 5	
HML 80	≥ 80		

Table 5.7: Physical properties of architectural slaked lime.

Code	Free water (%)	Fineness		Stability
		Residue (%): 0.2 mm screen	Residue (%): 90 μm screen	
HCL 90	≤ 2	≤ 2	≤ 7	Qualified
HCL 85				
HCL 75				
HML 85				
HML 80				

5.2.5 Activity of lime

Normal calcined lime has a porous, loose internal structure and the grain size is 0.3–1 μm. The activity of lime mainly depends on the internal specific surface area and size of grains. The structure and quality of lime are greatly influenced by calcination temperature and time. As calcination temperature increases, or calcination time lengthens, lime is sintered gradually and its density increases constantly. Figure 5.2 shows the change in internal specific surface areas of lime by sintering $CaCO_3$ at different temperatures with different times. Obviously, the internal surface area of lime decreases gradually with the increasing temperature and time.

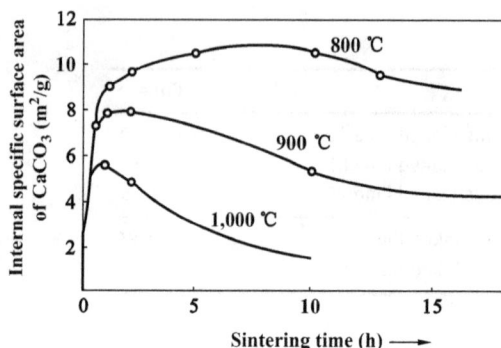

Figure 5.2: The relationship of internal specific surface area of CaCO₃ with sintering time.

The apparent density of normally calcined lump lime is 800–1,000 kg/m³, and the density is 3,340 kg/m³ when it is sintered completely. Table 5.8 lists the grain size of lime at different calcination temperatures. It can be seen that the higher the temperature, the larger the grain size and the lower the activity.

Table 5.8: Grain size of lime at different calcination temperatures.

Calcination temperature (°C)	Grain size (μm)
900	0.5–0.6
1,000	1–2
1,200	6–13

In addition, impurities of clay in limestone with lower melting points will melt at higher calcination temperatures and flow into pores in CaO particles or wrap on the surface to form enamel. The direct contact between lime and water is blocked by the enamel, leading to a great reduction of lime activity.

Overburnt lime produced at high temperatures has a dark color and a dense structure with coarse grains, low specific surface area and low activity. Its surface is often covered with glass glaze formed by melting of clay impurities, causing very slow hydration. This causes cracking or uplift of the hardened body leading to destroyed. Therefore, in order to avoid the risk of overburnt lime, it can be put into ash storage pool with water and kept for more than 2 weeks so that there is sufficient time for overburnt lime to react with water and to avoid destructiveness. This method is not commonly used because of the longer process, specially built ash pool and potential danger. At present, lump lime is usually ground into fine powders. Large contact area with water due to very fine particles benefits rapid hydration between lime and water. The damage caused by overburnt lime can be avoided. Ground lime easily absorbs moisture and CO_2 in air to form $CaCO_3$ losing cementing ability. After grinding, it is better to use lime powders as soon as possible. The sintering temperature must be controlled, that is, not to be too high during calcination so as to minimize the amount of overburnt lime in products.

If calcination temperature is too low, more underburnt lime will be produced, which is the undecomposed limestone. It has no hydration activity, cannot react with water and will reduce the active ingredient in lime slurry. Therefore, proper calcination temperature and time are very important to lime quality.

The quality of architectural lime primarily depends on the contents of active components of CaO and MgO and impurities. Active components in lime are the ones reacting with water to hydrate, and the content reflects cementious ability of lime. The more the content of active components, higher is the slurry yield. Underburnt lime and various impurities have no cementitious ability, while overburnt lime is harmful to volume stability. The smaller the size of lime powder, the better the workability, the faster the hardening and the better the quality. With increasing the content of fine particles in lump lime, lime quality is getting worse. There are restrictions on the content of free water and a requirement of volume stability for architectural slaked lime powder.

5.3 Hydration of lime

5.3.1 Hydration of lime

Lime reacts with water to produce $Ca(OH)_2$ rapidly. This process is termed hydration of lime, also known as slaking of lime. The product $Ca(OH)_2$ is known as slaked lime or hydrated lime. The reaction is as follows:

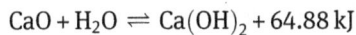

$$CaO + H_2O \rightleftharpoons Ca(OH)_2 + 64.88\,kJ$$

Lime hydrates to release a lot of heat. The heat released at the first 1 h is 10 times of that of hemihydrate gypsum or nine times of that of ordinary Portland cement. During hydration of lime, its volume expands to 1–2.5 times. The aforementioned reaction is reversible, and the reaction direction depends on temperature and vapor pressure in the environment. The right reaction takes place at room temperature, and it turns to the left reaction when the temperature reaches 547 °C, by which $Ca(OH)_2$ decomposes into CaO and H_2O. When the vapor decomposition pressure reaches one atmospheric pressure, $Ca(OH)_2$ can also partially decomposes at lower temperatures. To ensure the reaction to the right direction, the hydrating temperature and vapor pressure must be controlled.

The hydrating ability of lime can be influenced by calcination temperature, hydrating temperature and admixtures. Hydrating rate significantly increases with increasing hydrating temperature. Within 0–100 °C, the reaction is accelerated as temperature increases. The hydrating rate is double when temperature increases by 10 °C. If the temperature is raised from 20 to 100 °C, the hydrating rate will increase by $2^8 = 256$ times. The heat released from reaction greatly elevates the temperature; hence, the reaction is accelerated. At temperatures higher than 100 °C, the reaction rate decreases showing reversibility. $Ca(OH)_2$ absorbs heat and decomposes into CaO and H_2O to affect the quality of hydration.

The hydrating ability of lime can also be affected by calcination conditions. After mixing the lime having 15% underburnt lime with water, it takes 5 min to reach the highest temperature. While for the lime having 15% overburnt lime, it takes 27 min to reach the highest temperature after it is mixed with water. Moreover, the highest temperature of the latter is much lower than that of the former.

The hydrating ability for lime containing a certain amount of underburnt lime is much higher than that of lime containing a certain amount of overburnt lime, not because the underburnt part of limestone has an ability to hydrate. In practical production, the core temperature of limestone is not necessarily to reach the decomposition temperature, while the surface reaches it due to large particles of limestone to be incinerated. If the core temperature of limestone reaches decomposition temperature, the surface portion may be overburnt. The hydrating ability of lime containing some underburnt lime calcined at lower temperatures is high due to large internal specific surface area and small CaO grains. On the contrary, the hydrating ability of lime containing some overburnt lime calcined at higher temperatures is low because of its smaller internal specific surface area and coarse CaO grains.

Based on the data from J. Wooller, the hydrating rate of CaO crystals with a particle size of 0.3 µm is about 120 times higher than that of 10 µm CaO at room temperature, and the hydrating rate of 1 µm CaO is 30 times faster than that of 10 µm CaO.

Admixtures also affect the hydrating rate of lime. For example, chlorides salt (NaCl and $CaCl_2$) can accelerate lime hydration, and phosphates, oxalates, sulfates and carbonates can retard lime hydration.

The expansion pressure caused by cubic expansion during lime hydration can be higher than 14 MPa. When producing lime products and silicate products by using lime, slaked lime can be used in constructions to avoid the destructive volumetric deformation caused by this detrimental expansion. Theoretically, the required water for lime hydration is 32% of its mass. To ensure CaO hydrates completely, the practical water consumption is significantly higher than the theoretical value because of higher temperature and more water evaporation during hydration. There are two types of hydration products, slaked lime powder and slaked lime paste, by using different amounts of water.

Slaked lime powder can be obtained by adding proper amount of water (about 60–80% of the mass of lime). The principle for the specific amount of water is to ensure lime to hydrate completely without excessive moisture as conglobation. Hydration is executed in an airtight container to minimize heat loss and moisture evaporation and to prevent carbonization. Slaked lime powder is produced by water-spray method in layers on construction sites. Pieces of lime are spread on a flat floor having no water absorption in layers, and each layer with a thickness of 20 cm is once sprayed by water till 5–7 layers. Finally, it is to be covered with sand or soil to maintain temperature and prevent water evaporation so that lime can fully hydrate without carbonation. It can be taken out for use after it has rested for over 14 days at this condition. Lime soil mixed by slaked lime powder and clay, and trinity mixture fill mixed by slaked lime powder, clay and broken bricks or gravels, slag and other aggregates are utilized in road base, foundation of constructions and ground projects.

Slaked lime paste is produced by adding lots of water into lime. Lime is mixed with water in ash pool or slaking tank to slake into thin milky lime slurry flowing into ash storage pool by filtering to remove unhydrated particles or impurities. The surface of lime slurry should be covered with a layer of water to isolate air and prevent carbonation. The sediment rested over 14 days in this condition, named as lime paste, is taken out for construction after removing the upper water. Lime mortar or mixed cement lime mortar made of lime paste can be used in masonry engineering and plaster works of industrial and civil constructions.

The aim to rest the aforementioned two kinds of slaked lime for over 14 days during hydration is to eliminate the damage of overburnt lime and obtain soft and good plastic slaked lime. In this period, air must be isolated to avoid carbonation.

Slaked lime is used in constructions to avoid destruction caused by heat release and cubic expansion of lime hydration. But hardening of slaked lime is slow and its strength is low. A powder product obtained by grinding lump lime with a ball mill is called finely ground lime. The finely ground lime hydrates with uniform heat liberation and without obvious cubic expansion, so it can be used directly without slaking by mixing with water (about 100–150% of lime mass). Hydrating and hardening become a continuous process. Because of very fine particles, cubic expansion caused by overburnt lime can be effectively inhibited.

5.3.2 Characteristics of lime hydration

Compared with cementing materials such as gypsum or cement, hydration of lime is of the characteristics of high heat liberation, rapid rate of heat release, large water demand and cubic expansion. Structure formation of $CaO–H_2O$ slurry and properties of hardened product will be restricted by these characteristics.

(1) High heat liberation of hydration

Lime has a strong hydration capacity, representing high heat liberation when mixed with water. The exothermic times and heat of several cementing materials during hydration are listed in Table 5.9. It can be seen that the heat released by lime hydration is 10 times that of hemihydrate gypsum, and in the first 1 h, the released

Table 5.9: Amount and time of released heat of several cementing materials during hydration.

Cementing material	Time of heat release	Quantity of heat q (J/g)
Lime	1 h	1,156
Hemihydrate gypsum	1 h	113
Ordinary Portland cement	1 day	126
Ordinary Portland cement	3 days	360
Ordinary Portland cement	28 days	402

heat is almost nine times of that of ordinary Portland cement in one day and three times of that of ordinary Portland cement in 28 days. The hydration heat is obviously significant. For example, the heat released by 1 kg pure CaO is enough to heat 2.8 kg water from 0 °C to boiling point even ignoring heat loss.

One of the reasons why lime cannot set and harden like other cementing materials is that the strong exothermic reaction of lime paste makes water become into steam to destroy structure formation of setting and crystallizing and to produce loose slaked lime without binding ability.

(2) High amount of required water

The theoretically required amount of water for CaO reacting with H_2O to generate Ca $(OH)_2$ is 32%. In practical reactions, water evaporates due to large amounts of released heat, and the evaporated water accounts for about 37%. To produce dry and loose slaked lime powder, about 70% water is required to make entire CaO become $Ca(OH)_2$.

(3) Cubic expansion

In addition to the above exothermic reaction, lime hydration is accompanied by a significant increase in volume.

(i) Volume change

The expansion rate of specimens made at a water–lime ratio of 0.33 was 44% tested without applied load in laboratory, and most expansion occurred within 30 min after mixing lime with water. A pressure over 14 MPa, called expansion pressure, must be applied to eliminate the expansion completely. The characteristic of water absorption and expansion pressure of lime is used in lime piles, which are often used to reinforce water-bearing weak foundations.

When lime and water react chemically, the absolute volume of solid phase is increased by 97.92% after the reaction. Concerning the lime–water system, the total volume after reaction does not increase but decreases by 4.54%, as shown in Table 5.10. The chemical reaction formula is as follows:

$$CaO + H_2O = Ca(OH)_2$$

Table 5.10: Volume change before and after reaction in lime–water system.

Reactant and product	Relative molecular mass	Specific density	Absolute volume of system /cm³		Absolute volume of solid phase /cm³		Absolute volume changerate (%)		Required relative amount of water
			Before	After	Before	After	System	Solid phase	
CaO	56.08	3.34	34.8	33.2	16.7	33.2	−4.54	+97.92	0.321
H₂O	18.02	1.00							
Ca(OH)₂	74.10	2.23							

Therefore, a chemical shrinkage occurs along with the reaction of lime and water. In fact, chemical shrinkage is inevitably generated during hydration of cementing materials including cement and gypsum.

(ii) Reasons of volume increasing during lime hydration

First, material transfer occurs during lime hydration. Two types of material transfer immediately happen after mixing lime with water. One is that water molecules or OH^- come into lime particles to react with them and produce hydration products. Another is that hydration products shift to the original space occupied by water. If the two speeds are compatible, the volume of lime–water system will not expand. However, since lime is loose in structure with a large specific internal surface area, its hydration rate is much higher than the shifting rate of hydration products. As products surrounding lime particles have not transferred, new products are generated inside in quantities to break through the reaction layer, resulting in a mechanical jump of particles and specimens expanding to crack. The lime slurry splinters into powders at a low water–lime ratio.

Second, voids volume brings an increment. During hydration, the increase of solid phase volume leads to an increase in voids volume, resulting in cubic expansion. The increase in solid phase volume includes two aspects. The volume of solid $Ca(OH)_2$ produced by the reaction of CaO and H_2O is 97.92% larger than that of CaO. Next, lime particles are dispersed by hydration. The property of water molecules adsorbed on the surface of dispersed particles is similar to that of solids and so the adsorbed water molecules are regarded as an increase of solid phase volume. Assuming that solid particles are stacked together in an ideal compact hexagon pattern during hydration, the solid volume accounts for 74% of the total volume, and the voids volume is 26%. If the volume of solid particles increases by 1%, the voids volume will increase $(26/74) \times 1\% = 0.35\%$. Therefore, void sizes between solid particles will increase after solid particles enlarge.

(iii) Factors affecting volume change

The main methods to control volume change are to change lime fineness, water–lime ratio, hydrating temperature and add admixtures (gypsum).

The fineness of lime: the finer the size of lime, the smaller is the volume change caused by hydration, as shown in Figure 5.3. Materials transfer is more uniform in the system with a larger dispersion of lime. At the same time, the greater the lime dispersion, the smaller is the increment of voids volume.

The water–lime ratio: as water–lime ratio increases, the space filled with water in slurry is large enough to cushion hydration expansion. The expansion value during hydration is reduced, as shown in Figure 5.4. However, a large water–lime ratio will lower the hardened strength of lime slurry. In production of lime products, the optimum water–lime ratio must be determined by experiments in order to avoid expansion and to obtain the desired strength.

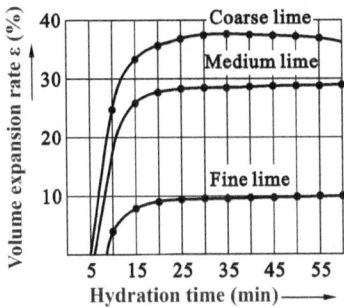

Figure 5.3: The influence of lime fineness on volume change during hydration: (water–lime ratio w/c is 0.7); (Coarse lime: 0.212 mm≤d≤0.300 mm; medium lime: 0.09 mm≤d≤0.212 mm; fine lime: d<0.09 mm).

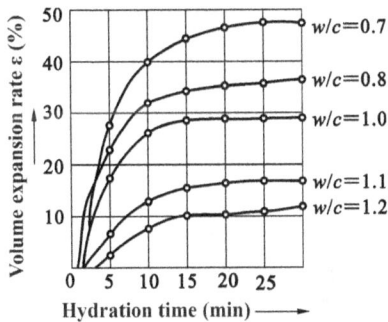

Figure 5.4: The influence of water–lime ratio (w/c) on volume change during hydration.

The hydrating temperature: the hydrating temperature and the volume of lime slurry increase with hydration. Meanwhile, because hydration is accelerated, the maximum of expansion limit will be reached in an earlier time. This is because of the fact that at temperatures below 100 °C, hydration can be accelerated by increasing temperature.

Admixture of gypsum: gypsum can inhibit cubic expansion of lime slurry. On the one hand, gypsum is an inhibitor for lime hydration that retards lime hydration. As a result, the hydrating rate is compatible with the rate of products transferring. On the other hand, the thickness of adsorbed water layer is compressed due to gypsum and the increment of voids volume is reduced.

5.4 Structure formation of lime paste

5.4.1 Dissolving and dispersing of lime by water action

Mixed with water, hydration on the surface of finely ground lime particles occurs immediately to form $Ca(OH)_2$. The reaction products dissolve in water to enter into the solution immediately and dissociate into positively charged Ca^{2+} and negatively charged OH^-.

$$Ca(OH)_2 \rightarrow Ca^{2+} + 2OH^-$$

When solution of $Ca(OH)_2$ is saturated, the action of water on unhydrated lime does not cease. Water molecules and OH^- penetrate into microcracks in lime particles and form an adsorbed layer on both walls of cracks. The surface tension inside lime particles is reduced by the adsorbed layer. The splitting of lime particles into finer particles along these cracks is accelerated under the action of thermal motion, named as adsorbed dispersion. At the same time, water molecules or OH^- react directly with CaO to form $Ca(OH)_2$ crystals. During this time, $Ca(OH)_2$ is formed by rearrangement of OH^- and Ca^{2+}, O^{2-} in lime particles but not by dissolution. When CaO is converted to $Ca(OH)_2$, the increase of solid phase volume makes lime particles disperse, known as chemical dispersion.

5.4.2 Setting structure formation of lime paste

Lime–water dispersed paste system is formed by adsorbed dispersion and chemical dispersion of lime particles. When distances between colloidal particles with a water diffusion layer are large, particles gradually precipitate and approach each other by gravity. If the distances kept getting closer after they are closed to a certain degree (about 10^{-6} mm), energy is needed to eliminate water molecules in the diffusion layer. In this case, the water diffusion layer can be regarded as a wedge between individual particles to keep two adjacent particles apart. The force to prevent particles approaching is called a wedging force. The wedging force increases rapidly when particle distances are less than 10^{-6} mm.

When the surface of lime particles is covered with a water diffusion layer, CaO inside particles must absorb the weakest linked water in diffusion layer to form new colloidal substances and to keep particles getting closer. And the thickness of water interlayer between particles is reduced and the colloidal diffusion layer is compressed. At this time, the collision of particles under the action of thermal motion is increasing. When a collision occurs in the most active sites, such as end, edge, corner or defects, the molecular attraction may exceed the wedging force so that particles bind together and a setting structure network is gradually formed. Adsorbed water and free water are distributed in the structure network. The structure network caused by setting will be strengthened with further compression of diffusion and further closing of colloidal system.

This setting structure has a characteristic of thixotropism. The structure of system is destroyed under external forces and suspension is back into flowing.

After removing applied forces, particles collide with each other due to Brownian motion and bond together again, and the setting structure of system is recovered.

Setting structure of lime can be strengthened by adding electrolytes such as gypsum. With an appropriate content of gypsum, the hydrating rate of lime can be retarded and strengths of setting structure can also be increased. Lime particles disperse under water action to form colloidal particles as shown in Figure 5.5. A layer of Ca^{2+} and OH^- absorbed around solid nucleus constitutes the adsorption layer, and a layer of OH^- concentrated in the outermost layer forms the back ions diffusion layer. Ions in this layer are hydrated and keep a large portion of water.

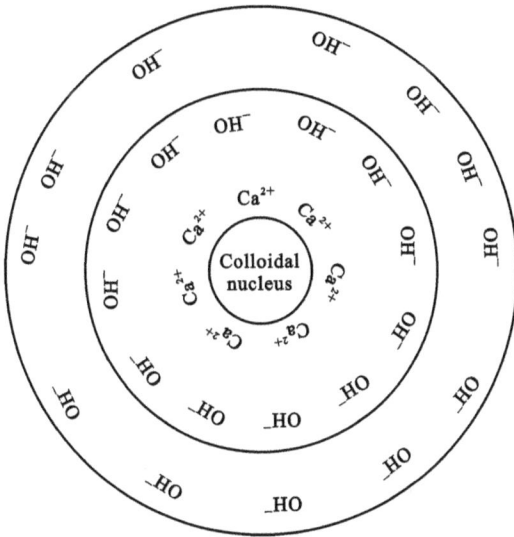

Figure 5.5: Diagrammatic sketch of lime micelle.

In micelles of lime paste, the interaction force between monovalent OH^- in the back ions diffusion layer and the surface of colloidal particles is weak, resulting in a thick diffusion layer. Some of the added gypsum in lime slurry gets into the solution. The diffusion layer is compressed by replacing monovalent OH^- with divalent SO_4^{2-}, which is more easily attracted by Ca^{2+}. An electrostatic shrinkage occurs in the diffusion layer of lime colloidal particles to cause a reduction in the thickness of bonding layer. It also destroys the balance between Van der Waals' force and wedging force. Van der Waals' force is increased to make tighter bonding of lime particles and the setting structure of lime paste is strengthened.

5.4.3 Crystalline structure formation of lime paste

It is a spontaneous process for lime paste to change its setting structure into a crystalline structure. The crystalline structure is formed by dissolution of finely dispersed $Ca(OH)_2$ grains, growth and binding coarse $Ca(OH)_2$ grains together.

Figure 5.6 shows structure formation kinetics and hydration kinetics of lime paste (solid phase: 25% CaO+75% $CaCO_3$, specific surface area: 0.2 m²/g) at different water–solid ratios (w/s). The plastic strength of paste P_m gradually increases with hydration. Approaching the end of hydration, the strength of hardened structure decreases along with time, and the greater the water–solid ratio, the more is significant reduction in strength.

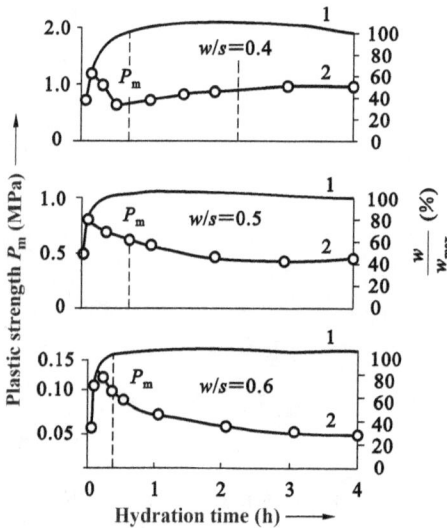

Figure 5.6: Kinetics of lime paste structure formation (curve 2) and hydration kinetics (curve 1).

The structure of paste is often destroyed during changing setting structure into crystalline structure because lime hydration is rapid, accompanied by intense heat releasing and significant cubic expansion. After the crystalline structure is formed, the strength of hardened paste can also be decreased due to factors causing structural failure such as dissolution of crystals contact points and internal stresses of crystals. Therefore, hydration of lime must be controlled by using a proper water–lime ratio, discharging the liberated heat during hydration, inhibiting cubic expansion and avoiding stirring after set to effectively utilize lime.

5.5 Hardening of lime paste

Lime paste hardens in air with strength due to gradual crystallization and carbonation. The hardening consists of three interlace processes, drying, crystallization and carbonization.

5.5.1 Crystallization and attached strength by drying

During drying of lime paste, excess water evaporates or is absorbed by masonry to form a network of pores. The remaining free water in pores has a concave meniscus at the narrowest point due to its surface tension, resulting in a capillary pressure to make lime particles contact firmly and generate an attached strength. The strength is similar to clay strength obtained with loss of water. It is low and the strength will lose when encountering water.

The degree of supersaturation of solution is increased due to evaporation of water in lime paste to promote $Ca(OH)_2$ crystallization. There is no significant increase in strength caused by crystallization because of less $Ca(OH)_2$ crystals.

A large amount of free water in lime paste is evaporated during drying and hardening. The capillary pressure generated by loss of capillary water reduces pore size, resulting in a shrinkage of hardened paste. Large deformation caused by hardening shrinkage of lime paste leads to cracking and local failure in the hardened paste. To reduce shrinkage deformation, lime is usually used by mixing with sand, paper tendons, straw and hemp but not used alone.

5.5.2 Carbonization

$Ca(OH)_2$ reacts with CO_2 present in air under water condition to produce insoluble $CaCO_3$, as follows:

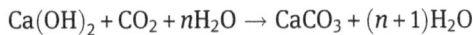

$$Ca(OH)_2 + CO_2 + nH_2O \rightarrow CaCO_3 + (n+1)H_2O$$

Newly formed $CaCO_3$ crystals are interconnected or coexist with $Ca(OH)_2$, forming a tight crystals network to further improve the strength of hardened paste. When carbonization occurs, the solid volume of $CaCO_3$ is slightly larger than that of $Ca(OH)_2$, making hardened paste more tight and firm.

Obviously, carbonation is very beneficial for enhancement and stability of strength. CO_2 content in air is low and the formation of a carbonized layer on the surface does not make it easy for CO_2 to penetrate into deep interior and hinders evaporation of internal water. Therefore, natural carbonization is very slow. It must be noted that carbonization cannot be carried out in absolute drying in the absence of moisture, nor can the paste be completely covered with a certain thickness of water because minimal CO_2 can dissolve in a certain depth of water. The rate of carbonation of lime depends not only on CO_2 concentration in environment but also on the concentration of dissolved lime and contacting area between lime solution and air. Carbonization of lime paste can be accelerated by an artificial approach. Carbonation can be accelerated by treating lime products using the gas discharged from lime kiln containing about 30% CO_2 (1,000 times higher than CO_2 content in air), and high strength can be achieved as soon as possible.

One transform cycle of lime is completed from hydration to setting, hardening and carbonization, resulting in the original product, $CaCO_3$.

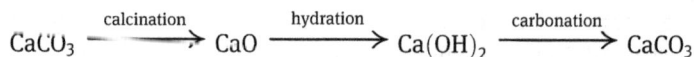

$$CaCO_3 \xrightarrow{\text{calcination}} CaO \xrightarrow{\text{hydration}} Ca(OH)_2 \xrightarrow{\text{carbonation}} CaCO_3$$

$CaCO_3$ is stable in nature. The final strength of lime paste obtained by carbonization is called carbonation strength. The floor of trinity mixture made by using air hardening lime can resist water because of a layer of carbonization film on the surface of floor. Some ancient lime masonry constructions in China still have high strengths not because the quality of ancient lime is particularly high, just as the carbonation by long years.

The main hydration products of hardened lime paste are $Ca(OH)_2$ and a small amount of $CaCO_3$ on the surface. Due to low strength of $Ca(OH)_2$, the strength of hardened paste is also low. For example, the 28 days compressive strength of lime mortar at a lime to sand ratio of 1:3 is only 0.2–0.5 MPa. Low strength and poor water resistance are the common nature of inorganic air hardening cementing materials. The poor water resistance is mainly due to high solubility of hydrated products, thermodynamic instability and higher solubility of crystals contact points caused by lattice deformation and twist. In humid environment, internal dissolution and recrystallization occur in hardened paste, leading to a significant irreversible reduction in the strength of hardened paste. In water, hardened lime paste having lower strength will break and collapse. Lime must not be used in a humid environment, and lime mortar cannot be used for foundations and outer wall plastering projects in masonry buildings.

5.6 Applications of lime

As a traditional building material, lime is still used as an important building material and widely used in various constructions and production of building materials.

5.6.1 Manufacture of architectural quicklime powders

Architectural quicklime powders are produced by finely grinding quicklime and used in constructions. The residue on 80 μm square opening sieve is less than 30%. Consumption of mixing water during hydration increases due to large specific areas of lime powders. Hydration rate can be accelerated by 30–50 times. With uniform cubic expansion during hydration, it can be used directly without prehydration or resting for 14 days to improve efficiency. Moreover, weaknesses of slow hardening can be improved by fast hydration of powder lime and hardening is accelerated by released heat. Higher strength can be reached after hardening. At the same time, hydration of overburnt lime particles can be accelerated due to grinding, and the workability and uniformity of slurry are also changed due to finely grinding of underburnt lime particles. It should be noted that finely ground lime powder can easily absorb H_2O and CO_2 in air and produce $CaCO_3$ to lose part of cementing ability. Therefore, lime powder should be used immediately after grinding.

5.6.2 Production of slaked lime powder or slaked lime slurry

Lime milk can be used for painting indoor walls and ceilings in constructions because it is cheap, white and its construction is convenient. Slaked lime milk is composed of slaked lime powder or slaked lime slurry mixed with a large amount of water. A small amount of ultramarine pigment can be added to eliminate faint yellow caused by impurities containing iron and to make it be pure white. Slaked lime powder or slurry can be prepared with lime hydration by artificial or mechanical processes.

(1) Production of slaked lime powder

(i) Artificial slaking process
There are two ways of the artificial process: spraying and soaking.

In spraying process, lump lime is spread in layers over not water sucking base with a layer thickness of 20 cm. Then water is sprayed accounting for 60–80% of lime weight. This process is repeated until lump lime is made into powders. The total thickness of lime layers is from 1 to 1.5 m. As excess water in the upper layers flows down, the lower layers of lime should be sprinkled with less water. The produced slaked lime powder should be sieved and stored for use.

In soaking process, lump lime is put into a basket made of metal wire and lump lime together with basket is immersed in water. When bubbles stop coming up and lime milk begins to appear, the basket is taken out of water. After water drains away, lime is poured out and piled 1–1.5 m high.

(ii) Mechanical slaking process
In mechanical process, lump lime is put into a slaking tank after sprayed with hot water. A large amount of hot steam liberated by hydration makes materials be fluidized. Overflowing materials are collected and screened as final products. The commonly used continuous process is shown in Figure 5.7.

Lump lime is broken into particle sizes less than 15 mm by a hammer crusher. The crushed lime is then fed into storage bin by hoister before being supplied to a screw

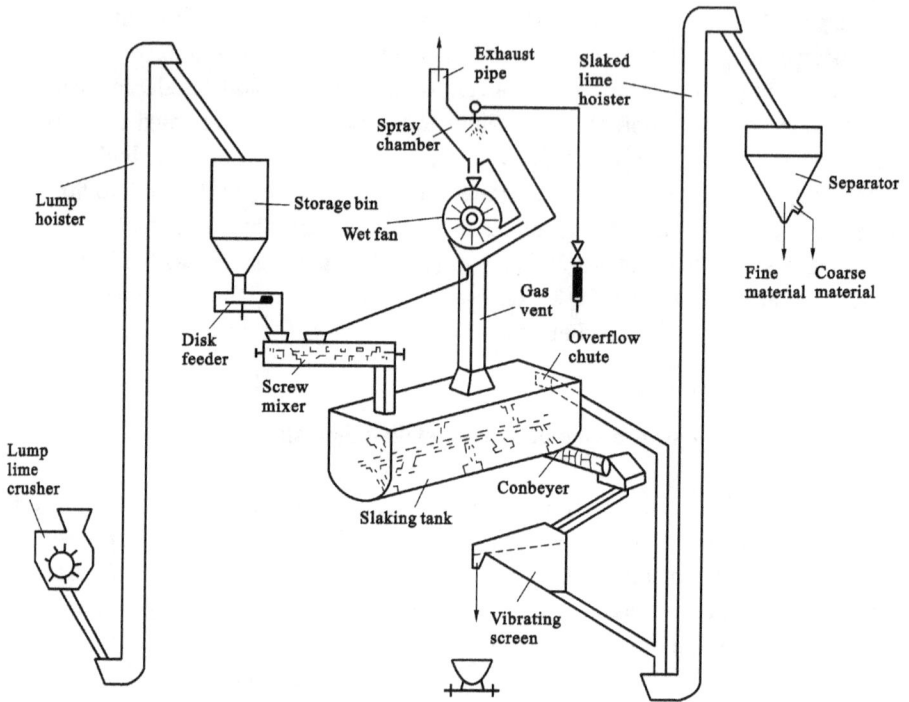

Figure 5.7: Flow chart of lime hydration process.

mixer by quantitative feeding. Hot water from a dust wiper is added into the mixer to make all lime particles be in contact with water. Wetted materials coming out of mixer fall into a slaking tank and hydrate there. Because of a large amount of steam produced during slaking, hydrated materials are fluidized. The hydrated particles become particularly light and are discharged through overflow chute, while unhydrated particles remain longer in slaking tank. Coarse residues are discharged with a screw conveyor and sent to a vibrating screen. Sieved fine particles are merged into the slaked lime from overflow chute. For those that cannot pass through the sieve, they are residues that cannot hydrate.

The amount of water added is very important in the preparation of slaked lime powder. By using less water, materials temperature will be very high, resulting in partial decomposition of formed $Ca(OH)_2$ and formation of coacervate, preventing moisture into inside particles. Lime cannot hydrate completely with less water. Slaked lime powder will absorb moisture with more water and its quality will be affected.

(2) Production of slaked lime slurry
The process by which lump lime directly hydrates into lime slurry is mostly carried out at sites.

In the artificial process, lump lime is put into ash pool and mixed with a large amount of water to hydrate into a lime aqueous solution flowing into storage pit by sieving. More excess water in the solution overflows from the pit walls, resulting in the formation of lime slurry before final formation of lime paste. Lime slurry is kept in a pit for at least 2 weeks before use. The longer the storage time, the better is the quality of lime slurry.

In the mechanical process, 5 mm sized lump lime is put into a hydrating device with a mixer by adding 40–50 °C warm water and it hydrates into a lime aqueous solution. This solution flows into a clarifying tank to be concentrated into lime slurry.

5.6.3 Preparation of lime mortar and grout

Lime mortar and lime cement mixed mortar prepared by using lime paste as a raw material with good workability are widely used in masonry and plastering works of industrial and civil constructions. In order to avoid rapid water loss from lime mortar resulting in dry powders without cementing ability, bases with high water absorption (such as ordinary clay bricks) must be wetted before lime mortar is applied.

Various grouts, such as lime hemp mortar and lime paper tendons mortar, are usually produced by uniformly mixing lime paste or slaked lime powder with water and other materials and are used for plastering in constructions.

Lime milk paint can be prepared by adding a large amount of water with lime paste or slaked lime powder. The properties of paint can be modified by admixtures. Various colors can be obtained by adding basic color. Water resistance can be improved by adding a small quantity of cement, granulated blast furnace slag or coal fly ash. Powdering of paint-coat can be modified by adding casein, calcium chloride or alum. Lime milk paint can be used for painting indoor walls and ceilings, for which decorative requirements are not high.

5.6.4 Preparation of lime soil and trinity mixture fill

Slaked lime powder is uniformly mixed with fine clay to produce lime soil, with a mass ratio from 1:2 to 1:4. Trinity mixture fill is prepared by mixing slaked lime powder, clay, sand and gravels or slag at a mass ratio of 1:2:3. After adding an appropriate amount of water, high density in structures of lime soil or trinity mixture fill can be achieved by striking, ramming or rolling in layers. Such structure layers having certain hydraulic and cementing properties can be used as lime soil walls or squares, cushion or simple surface of roads, and they have higher strengths and better water resistance than that of lime or clay.

5.6.5 Production of silicate products

Silicate products are dense or porous products produced by batching, mixing and molding before steam curing or autoclaved curing by using raw materials of finely ground lime or slaked lime powder and siliceous materials, such as quartz sand, coal gangue, fly ash and granulated blast furnace slag. When calcareous materials and siliceous materials are hydrothermally synthesized, hydrated calcium silicate salts are the main cementing materials, and so they are collectively referred to as silicate products. Commonly used silicate products are various fly ash bricks and blocks, slag bricks and blocks, autoclaved lime-sand bricks and blocks, autoclaved lime-sand concrete hollow slabs and air-entrained concrete.

5.6.6 Manufacture of carbonized products

Green bricks are produced by mixing ground lime powder, sand, ove tails or mountain flour and a small amount of gypsum and blending with water before compression molding. Green hollow slabs are produced by adding 30–40% of short glass fibers in ground lime, mixing with water and vibration molding. Treated by artificial carbonation using CO_2 waste gas generated from lime kilns, the two green bodies become light carbonized bricks and carbonized slabs. Strengths of lime products will be greatly increased by carbonization. For example, the strength of lime-sand products can be increased by 4–5 times after carbonization. For carbonized lime hollow slabs, the apparent density is 700–800 kg/m^3 (porosity is 34–39%), flexural strength is 3–5 MPa, compressive strength is 5–15 MPa and thermal conductivity is less than 0.2 W/(mK). These products can be sawed, planed and nailed, and they are suitable for use as ceilings and clapboards of nonload-bearing indoor walls.

5.7 Acceptance check, storage, safekeeping and transportation of lime

Lime will absorb moisture in air to hydrate into $Ca(OH)_2$ automatically when kept for too long, resulting in losing the cementing ability. During storage of lime, it must be moistureproof and storage time must be shortened as much as possible. When stored, it can be made into lime paste sealed or covered with sand to isolate from air, thus preventing hardening. The quality guarantee period of lime is 30 days. If it is not used up, the storage period should be changed into incubation period, in which lime paste must be covered by water with a certain thickness.

Because a large amount of heat will be released and cubic expansion will occur when lime becomes damp, safety measures ensuring waterproof must be taken during storage and transportation of lime.

Architectural quicklime powder and architectural slaked lime powder are generally packed in bags, meeting requirements in standards and marked with plant name, product name, trademark, net weight and batch number, such as kraft paper bags, composite paper bags or plastic woven bags. They must be stored in a dry storehouse by types and grades. Lime reacts with water to release a lot of heat. Therefore, lime must not be stored and transported with flammable and combustible goods, so as not to lead a fire.

Problems

5.1 What are the factors that affect lime hydrating activity?

5.2 Please list harmfulness caused by overburnt lime, and analyze the reasons and propose possible measures to avoid harmfulness.

5.3 What are the characteristics of lime hydration? And how to reduce deleterious affects on its use?

5.4 What are the main components in quicklime and slaked lime, respectively? What are their applications?

5.5 Please explain the reasons that lime paste has good plasticity, slow hardening, low hardened strength and poor water resistance.

5.6 What is the influence of calcination temperature and time on lime quality?

5.7 Please give a brief introduction of factors affecting lime hydration.

5.8 Please explain the hardening mechanism of lime.

5.9 In some walls made of lime mortar, swelling or radial cracking around particles sometimes occurs, or even walls sometimes break apart locally. What are the reasons?

5.10 In demolition of an ancient temple constructed with bricks and lime mortar masonry, it was found that the strength of lime mortar was particularly high. It is asserted that the quality of ancient lime is better than that of modern lime. Is the assertion true? Why?

5.11 Lime itself is not water resistant, but some of well-constructed lime soils have a certain degree of water resistance. What are the reasons?

6 Magnesia cementitious materials

Magnesia cementitious materials are air-hardening binding materials with MgO as the main component, and the natural magnesite is commonly used as the raw material. Reserves of magnesite in China are equal to 3.4×10^9 t and reserves account for world 30% of total reserves. Magnesia cementitious materials are powders produced by fine grinding calcined magnesite at high temperatures, with a color of white or pale yellow, and the density of 3.1–3.4 g/cm^3. The density of magnesia cementitious materials is an important index to determine whether calcination is normal. If the temperature of calcination is too high, MgO will produce contraction and grains become compact after sintering, and the density of the product is high, the rate of hardening is slow and the activity is low. On the contrary, if the temperature of calcination is too low, decomposition of $MgCO_3$ will not be sufficient, and the density of the product is low, cementing property is bad and the strength is low.

Magnesia cementitious materials can set and harden above 10 °C after being mixed with $MgCl_2$ aqueous solution. $MgCl_2$ is a by-product of mother liquor of salt bittern after potassium chloride and bromine are extracted. It is distributed in coastal salt mines and inland salt areas.

Regulated by Chinese construction material industry standard *Glass-magnesia flue pipes* (JC/T 646—2006), for magnesia cementitious materials, the compressive strength R_c must be higher than 90 MPa, and the bending strength R_f must not be less than 65 MPa. Regulated by Chinese construction standardization association, *Technical specification for glass fiber reinforced magnesia cement ventilation pipes* (CECS 95—1997), the water absorption rate should not be less than 13%. Regulated by Chinese building materials industry standard of *Glass-magnesia plates* (JC 688—2006), the drying shrinkage rate ε should be less than 0.6%, and they should not be flammable. Magnesia cementitious material products are therefore widely used in construction, environmental protection, packaging, agriculture and light industries.

6.1 Definition of magnesia cementitious materials

The material obtained by mixing magnesia cementitious materials with $MgCl_2$ aqueous solution is called magnesium oxychloride cement, magnesia cement for short. It is also known as Sorel cement since it was created by S. Sorel in 1867. It is composed of MgO obtained by firing magnesite (mainly $MgCO_3$) at 750-850oC and grinding, and $MgCl_2$ aqueous solution. To improve its water resistance and durability such as efflorescence resistance, improve workability, and adjust setting time, modifiers are usually added to magnesia cementitious materials. Therefore, it is a quaternary system consisting of light burned MgO, $MgCl_2$, water and modifier.

https://doi.org/10.1515/9783110572100-006

6.2 Raw materials for the production of magnesia cementitious materials

The main component in magnesia cementitious materials is MgO, and raw materials are natural magnesite (the main component is $MgCO_3$), natural dolomite (mainly composed of double salt of $MgCO_3$ and $CaCO_3$), serpentine (the main component is water magnesium silicate $3MgO \cdot 2SiO_2 \cdot 2H_2O$) and other minerals. In addition, slag of smelting light magnesium alloy and sea water can also be used as raw material.

In China, dolomite is a more important resource than magnesite for the development of magnesia cementitious materials due to the richer reserves and wider distribution. But in natural deposits, some transition often exists between dolomite and limestone; it is generally called dolomite only when the $MgCO_3$ content is greater than 25%.

(1) Magnesite
The main component of magnesite is $MgCO_3$, where MgO accounts for 47.47% and CO_2 accounts for 52.19% in its molecular composition. The relative density is 2.9–3.3 g/cm^3, the Mohs hardness is 3.5–4.1 and it belongs to the trigonal system.

(2) Dolomite
The crystal structure of dolomite also belongs to the triangular system. And it is mainly composed of a double-salt of $MgCO_3$ and $CaCO_3$. Theoretically, the content of MgO is 21.27%, ignition loss is 45.73%, the Mohs hardness is 3.5–4.1 and the density is 2.8–2.9 g/cm^3. In order to avoid the influence of CaO produced from $CaCO_3$ on the use of MgO, it is necessary to prevent $CaCO_3$ decomposition when dolomite is calcining. Therefore, the control of the calcination condition is crucial, and it is required to keep the temperature in a range of 730–780 °C for 20–30 min.

(3) Bischofite
Resources of kalium are abundant in the salt lake of Qinghai Tibet Plateau in China, especially in the area of Geermu in Qinghai Province. After extraction of potassium, bischofite ($MgCl_2.6H_2O$), the by-product of a large amount of carnallite ($KCl \cdot MgCl_2 \cdot 6H_2O$), can produce MgO by pyrolysis. The equation is given as follows:

$$MgCl_2 \cdot 6H_2O \rightarrow MgO + 2HCl + 5H_2O$$

To reduce the cost of this method to produce MgO and promote the development and utilization of magnesium oxychloride cement in Northwest China, the pyrolysis process can be carried out by using industrial waste heat.

6.3 Calcination of magnesia cementitious materials

MgO is the pyrogenic product by $MgCO_3$ obtained during the calcination process. Whether it is magnesite, dolomite or basic magnesium carbonate [$Mg(OH)_2 \cdot 4MgCO_3 \cdot 5H_2O$] as raw materials, the decomposition reaction is an endothermic reaction. MgO has a porous structure for CO_2 and H_2O produced during the pyrolysis process. Because decomposition of raw materials and crystallization of MgO are dependent on calcination temperature and time, the calcination should be under control to make raw materials decompose completely without producing overburnt MgO (this means that the internal porosity in the product achieves the highest), resulting in a product having the highest activity.

Generally, $MgCO_3$ begins to decompose at 400 °C; the decomposition reaction is carried out at 600–650 °C. In the actual production, the csalcination temperature is often set at 800–850 °C. The heat required to decompose 1 kg of $MgCO_3$ is about 14.4×10^5 J. The decomposition reaction is as follows:

$$MgCO_3 \rightarrow MgO + CO_2 \uparrow$$

Generally, to fully decompose $MgCO_3$ in dolomite and to avoid decomposition of $CaCO_3$, the calcination temperature is often set at 650–750 °C in the production of caustic dolomite (also known as light burning dolomite). The obtained magnesia cementitious material is mainly active MgO and inert $CaCO_3$. In this temperature range, the decomposition of dolomite is carried out based on the following two steps: first is the decomposition of double salt, and the second is the decomposition of $MgCO_3$. The reactions are as follows:

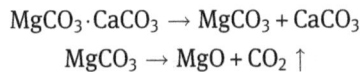

$$MgCO_3 \cdot CaCO_3 \rightarrow MgCO_3 + CaCO_3$$
$$MgCO_3 \rightarrow MgO + CO_2 \uparrow$$

The structure of MgO mainly depends on the calcination temperature and the calcination time of the raw material. The activity of calcined MgO depends on the raw materials, the calcination temperature and the calcination time; the most influential of which is the calcination temperature. On the premise to ensure raw materials to be completely decomposed, the lattice of products is larger, and there is a larger gap between grains at lower calcination temperatures. Therefore, the internal specific surface is relatively large, that is, the area reacting with water is large, and the reaction is fast. If the calcination temperature is increased or the calcination time is prolonged, the size of the lattice is reduced, the density of crystal particles is increased, and hydration reaction is delayed. Therefore, the larger the specific surface area of MgO, the faster is the hydration rate, and the higher is the activity of MgO.

The relationship between the specific surface area of MgO and calcination temperature and calcination time is listed in Table 6.1.

Table 6.1 shows the dispersion of MgO obtained at different calcination temperatures and times. With the increase of calcination temperature, the specific surface

Table 6.1: Relationship between specific surface area of MgO and calcination.

Number	Calcination temperature (°C)	Calcination time (h)	Specific surface area (m²/g)
1	450	5	126
2	680	4	32
3	1,000	2	15
4	1,300	3	3

area of MgO decreases significantly. When the temperature is higher than 1000 °C, the rate of recrystallization increases and the dispersion decreases rapidly.

The relationship between the hydration degree of MgO and calcination temperature with different hydration times is listed in Table 6.2.

Table 6.2: Relationship between hydration degree of MgO and calcination temperature with different hydration times.

Hydration time (day)	Calcination temperature (°C)		
	800	1,200	1,400
1	75.40%	6.49%	4.72%
3	100.00%	23.40%	9.27%
30	–	94.76%	32.80%
360	–	97.60%	–

Table 6.2 shows the hydration rate of MgO paste with different internal specific surface areas (expressed by hydration degree (%)). The larger the specific surface area of MgO, the faster is the hydration rate. This indicates that the activity of MgO is higher.

The mass of the calcined magnesia can be estimated by its density. The average density of the underburnt magnesia is $3.00 \ g/cm^3$, and the overburnt is $3.70 \ g/cm^3$. To obtain the fired MgO with activity, the density should be controlled in the range of $3.10–3.40 \ g/cm^3$. The general light calcined magnesia (also known as light calcined powders) refers to the irregular, highly active MgO with uncompleted crystallization possessing lattice defects obtained by heating $MgCO_3$ under 1,000 °C when using gas.

The magnesia (also known as bitter powder) is different from light calcined magnesium oxide. Magnesia is prepared by grinding the mixture of the coal ash and the semifinished product or finished product obtained from direct contacting of magnesite with the anthracite coal. Thus, the content of MgO in the product is not high, with more impurities and a poor activity.

The differences between light burned magnesia and magnesia are listed in Table 6.3.

Table 6.3: Differences between light burned magnesia and magnesia.

Product	MgCO$_3$ content in raw material (mass fraction, %)	Price of raw material (Yuan/t)	Combustion way	Calcination temperature (°C)	MgO content (mass fraction, %)	Production cost (Yuan/t)	Price of product (Yuan/t)
Light burned magnesia	≥45	32–36	Gas burning	750–850	80–90	270–290	310–340
Magnesia	35–40%	22–25	Coal burning	700–750	70 ± 5	160	220–260

The reverberatory kiln for gas-fired MgO is shown in Figure 6.1. The investment in the construction of reverberatory kiln is small, and the price of light burned MgO is low. But the particle sizes of magnesite calcined in the kiln are required to be large (about 17–25 cm), leading to a long calcination time and high calcination temperatures, and the quality of the product is unstable. This situation can be improved by a dynamic calcination process. First, the raw materials are broken into 45–125 μm, and the dry powders of raw material are sprayed into the kiln at the top of the kiln to achieve an instant quick heat transfer. After flowing in several tens of seconds, the material calcination is finished, and the product will be discharged at the bottom of the kiln. The dynamic calcination process can achieve a continuous production and the product can be produced with high activity.

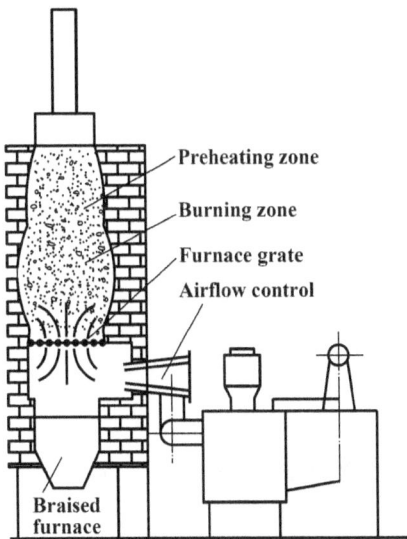

Figure 6.1: The reverberatory kiln for gas-fired MgO.

The dynamic calciner for firing magnesia is shown in Figure 6.2.

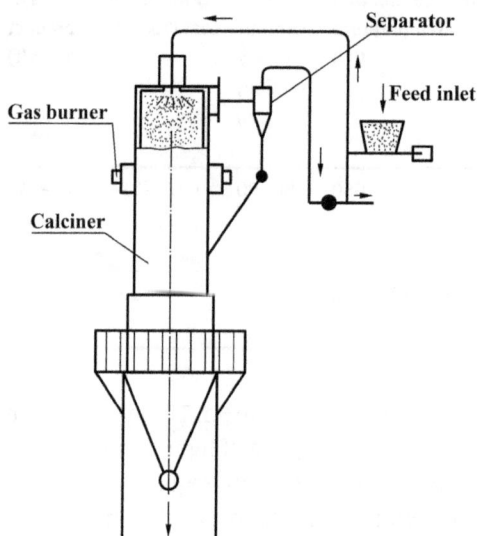

Figure 6.2: Schematic diagram of a dynamic calciner for firing magnesia.

6.4 Hydration of magnesia cementitious materials

6.4.1 Mixing liquid for hydration

(1) Water

The following reaction occurs after mixing MgO with water:

$$MgO + H_2O == Mg(OH)_2$$

The surface of MgO is encapsulated with a colloidal film formed by produced $Mg(OH)_2$ that is in a colloidal state, preventing water molecules from continuing to infiltrate into MgO particles. Therefore, the hydration process is delayed, the condensation is very slow and the strength after hardening is very low. During MgO hydration, a large amount of hydration heat will be generated to turn water into steam, resulting in cracks in structure. When MgO hydrates at room temperature, its maximum concentration is 0.8–1.0 g/L. The equilibrium solubility of hydration product $Mg(OH)_2$ at room temperature is about 0.01 g/L. As a result, its relative supersaturation is 80–100, which is much higher than that of other cementitious materials.

This causes the following two problems.

(i) The hydration process will take a long time because the solubility of MgO is relatively small. If the calcination temperature is increased to reduce its specific surface area, its dissolution rate and solubility are lower, and the hydration process is very slow. And longer hardening period will results in an increase in production costs.

(ii) The internal surface area can be increased by further finely grinding MgO, the dissolution rate and solubility of MgO will corresponding increase and the hydration process will be accelerated. However, this causes a higher supersaturation, resulting in a greater crystallization stress to destroy the crystalline structure network and to decrease strength significantly. MgO is not suitable for mixing with water alone.

In order to effectively use magnesia cementitious materials, it is necessary to accelerate the dissolution of MgO or reduce the supersaturation of the system. The effective way to reduce the supersaturation is to improve the solubility of the hydrated product or to rapidly form double salts.

(2) Magnesium chloride aqueous solution

In the actual use of magnesia cementitious materials, it is usually prepared with aqueous solutions of $MgCl_2 \cdot 6H_2O$, $MgSO_4 \cdot 7H_2O$, $FeCl_3 \cdot 6H_2O$ or $FeSO_4 \cdot 7H_2O$. It is common to use $MgCl_2$ solution by mixing with MgO to make a slurry, and the main hydration products are magnesium oxychloride and magnesium hydrate. Chemical equations are as follows:

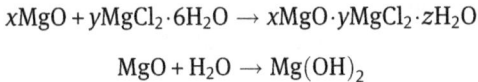

$$x MgO + y MgCl_2 \cdot 6H_2O \rightarrow x MgO \cdot y MgCl_2 \cdot z H_2O$$

$$MgO + H_2O \rightarrow Mg(OH)_2$$

The solubility of magnesium oxychloride in water is higher than that of $Mg(OH)_2$, which reduces the supersaturation of the solution and promotes the continuous hydration reaction. When the reaction proceeds continuously and the produced magnesium oxychloride is saturated, the hydration product will no longer dissolve, but precipitate directly as a colloid state to form a gel. Afterwards it gradually grows into fine grains by recrystallization to make slurry set and harden producing strength. The concentration and density of $MgCl_2$ solution have great influence on the strength and hygroscopicity of magnesia cementitious materials. The greater the concentration, the slower is the setting and hardening process, and the higher is the strength of the final product.

The hydration of MgO in $MgCl_2$ solution can be described by the change in releasing rate of hydration heat q with hydration time t. The common curve of the releasing of hydration heat of magnesia cement is shown in Figure 6.3.

The kinetics of hydration of magnesia cement is basically similar to that of ordinary Portland cement. After mixing, magnesia (MgO) and $MgCl_2$ solution react immediately to release heat. The first exothermic peak q_1 appears as shown in Figure 6.3, but the time t_1 during period I is short (only 5–10 min). Then, the reaction is in period II with lower reaction rates, and t_2 generally continues for a

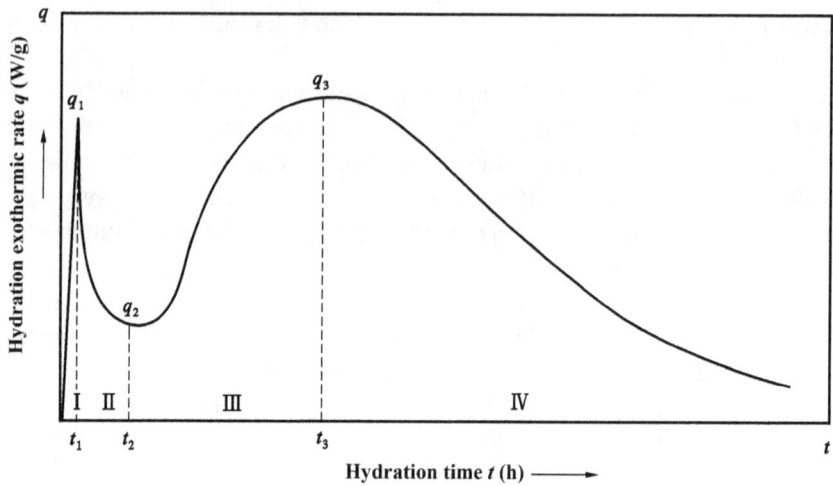

Figure 6.3: General exothermic curve of hydration of magnesia cement.

couple of hours. Later, the reaction is accelerated again, moving to period III, and the second exothermic peak q_3 appears. Finally, the reaction rate decreases with time and is gradually stabilized in period IV. Therefore, the hydration process of magnesia cement can be divided into four stages: preinduction period (I), induction period (II), acceleration period (III) and deceleration and stabilization period (IV).

The exothermic rates during hydration for magnesia cement pastes prepared using the same raw materials with different proportions are summarized in Table 6.4. It can be seen that the difference in proportion has little effect on the first exothermic peak time t_1, but the end time of induction period t_2 and the time of the second exothermic peak t_3 change with proportion.

Table 6.4: Hydration heat of magnesia cement pastes prepared with different proportions.

Number	Proportion $n[Mg(OH)_2]: n(MgCl_2): n(H_2O)$	t_2 (h)	t_3 (h)	Hydration heat (J/g)		
				8 h	10 h	12 h
1	2:1:8	4.5	12.5	147	253	382
2	3:1:8	4.0	12.5	139	231	339
3	4:1:8	3.0	8.0	225	317	395
4	5:1:8	3.0	6.5	275	378	468
5	6:1:8	1.5	5.0	333	432	515

In the case if $n[Mg(OH)_2]/n(MgCl_2)$ remains unchanged and only if $n(MgCl_2)/n(H_2O)$ changes, the concentration of $MgCl_2$ solution also changes, and the hydration heat is therefore influenced, as shown in Table 6.5.

Table 6.5: Hydration heat at different ratios of $n(MgCl_2)/n(H_2O)$.

Number	Proportion $n[Mg(OH)_2]$: $n(MgCl_2)$: $n(H_2O)$	t_2 (h)	t_3 (h)	Hydration heat (J/g)		
				8 h	10 h	12 h
1	3:1:7	4.0	12.5	104	184	290
2	3:1:8	3.5	12.3	139	231	349
3	3:1:11	1.5	6.0	321	409	485
4	5:1:5	4.0	11.5	170	273	202
5	5:1:8	3.0	6.5	276	373	468
6	5:1:11	1.5	5.5	341	433	500

It can be seen that with the decrease in the concentration of $MgCl_2$ solution, the induction period in the hydration process is shortened, the acceleration period is cut short and the released hydration heat is increased. This is because when the $MgCl_2$ solution is diluted, the content of MgO is relatively increased to make new phases form easily. As a result, the hydration time is shortened and the released heat is increased. But at this time, the pores in the hardened body are correspondingly increased, and this will reduce the strength of the product, thus the concentration of $MgCl_2$ solution should not be too low.

6.4.2 Hydrated products of magnesia cement

The hydrated phases from system of $MgO-MgCl_2-H_2O$ include $3Mg(OH)_2 \cdot MgCl_2 \cdot 8H_2O$ (3·1·8 phase or phase 3 for short), $5Mg(OH)_2 \cdot MgCl_2 \cdot 8H_2O$ (5·1·8 phase or phase 5 for short) and $Mg(OH)_2$. Among them, phase 3 and phase 5 are two major crystal phases in the system. Magnesium chlorobenzoate [$2MgCO_3 \cdot Mg(OH)_2 \cdot MgCl_2 \cdot 6H_2O$, 2·1·1·6 for short] will be formed after the hardened body of magnesium cement slurry is placed in the air. $MgCl_2$ may be leached and converted into hydromagnesite [$4MgCO_3 \cdot Mg(OH)_2 \cdot 4H_2O$, 4·1·4 for short] after magnesium chlorocarbonate reacts with water for a long time.

It can be seen from the above case that hydration products of magnesia cementitious material cannot be stable in the air. The primary reason for the poor water resistance and durability of magnesia cement products is that phase 5 and phase 3 are unstable. Therefore, the key for improving the water resistance and durability of magnesia cement is to improve the stability of phase 5 and phase 3.

In the process of hydration of magnesia cement, the relationship between the activity and dispersion of MgO and the ratio of MgO to $MgCl_2$ and H_2O is an important factor affecting the hydration process and hydration products. This is due to the fact that when $MgCl_2$ aqueous solution is mixed with MgO, the hydration products change with the ratio of the two substances.

(1) When $n(MgO)/n(MgCl_2)$ is less than 4, the phase 5·1·8 is formed initially, with a small amount of $MgCl_2$. The phase 5·1·8 gradually changes to the phase 3·1·8 by losing water with time, and the change of rate is accelerated with a decrease in the ratio of the two substances of MgO to $MgCl_2$.

(2) When $n(MgO)/n(MgCl_2)$ is from 4 to 6, the formed phase 5·1·8 is stable.

(3) When $n(MgO)/n(MgCl_2)$ is higher than 6, $Mg(OH)_2$ and the phase 5·1·8 are formed at room temperature. At this time, the phase 5·1·8 is unstable and it will be converted into 3·1·8.

In addition, irrespective of the ratio of $n(MgO)/n(MgCl_2)$, the hydrate can react with CO_2 to form $MgCl_2 \cdot 2MgCO_3 \cdot 2Mg(OH)_2 \cdot 6H_2O$, and its amount increases with time to indicate that the carbonization resistance of these hydrides is not good. Researches in the conditions of $4 \leq n(MgO)/n(MgCl_2) \leq 6$ show that the carbonization process is slow and happens mainly on the surface, which has a little effect on the strength of the hardened body. The practically applied concentration of $MgCl_2$ is about 12–30 °Bé (Baume degree), and the density is 1.15–1.20 g/cm^3.

6.4.3 Strength of hardened magnesia cement

The structure of hardened magnesia cementitious materials prepared with $MgCl_2$ solution is a heterogeneous porous structure, similar to other cementitious materials. The structure characteristics depends on the type and quantity of hydrates, interaction between hydrates, porosity and pores distribution. It has the characteristics of fast hardening and high strength under the dry condition. Its setting and hardening rates are very fast, and the tensile strength of 1 day can reach 1.5 MPa.

Figure 6.4 shows the comparison of the strength of magnesia cement and Portland cement under dry conditions. It can be seen that the strength development of magnesia cement in a dry environment is faster than that of Portland cement, and it has a higher strength. But because of the transformation of hydration products, the strength decreases after 24 h. With the prolongation of hydration, the hydration phase tends to be stable, and the supersaturation in the magnesia cement system is kept longer; newly formed hydrated phases restore the structure and the strength is gradually increased.

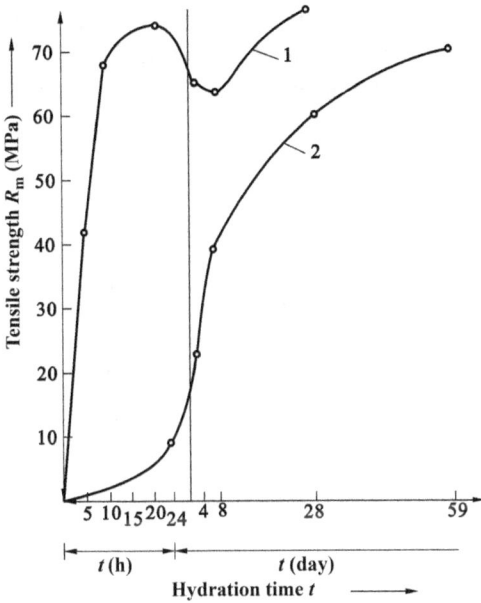

Figure 6.4: Comparison of strength development between magnesia cement and Portland cement: 1, magnesia cement: $n(MgO)/n(MgCl_2)$ = 5.68, $w(sand)/w(MgO)$ = 2.75, $w(H_2O)/w(MgO)$ = 1.3; 2, Portland cement: $w(sand)/w(cement)$ = 2.75, $w/\,c$ = 0.5.

The water resistance of hardened magnesia cement is poor, because the hydration products, especially the crystalline contact points between crystals, have strong hygroscopicity and high solubility. Under wet conditions, the strength of hardened magnesia cement is reduced rapidly, and it is easy to get damp and warping. Magnesia cement is only suitable for use in dry conditions, and it is an air-hardening cementitious material. Magnesia cementitious materials must be cautiously applied in structural engineering.

6.5 Properties of magnesia cementitious materials

6.5.1 Hydration activity

Magnesia cement is a kind of air-hardening cementing material; it hydrates and hardens very fast. Magnesia cement is characterized by high hydration heat, light weight, fast-hardening and high strength. The tensile strength R_m of the paste of MgO mixed by using a solution of $MgCl_2$ with a density ρ of 1.2 g/cm^3 can reach no less than 1.5 MPa after 24 h. Its hydration heat at 30 days is 3–4 times that of Portland cement, and the maximum exothermic temperature can reach 140 °C. This is the reason that its products can be made without using thermal means of sintering and steaming curing. According to Chinese *Standard for acceptance of construction quality of building ground engineering* (GB50209—2010), the technical requirements for MgO are summarized in Table 6.6. The strengths of magnesia cement products at different ages are listed in Table 6.7.

Table 6.6: Technical requirements for magnesia.

Property	Setting time (min)		Volume stability	Tensile strength R_m of paste at 24 h (MPa)	Content of MgO (%)	Fineness (residue from square sieve, %)	
	Initial setting	Final setting				0.08 mm	0.3 mm
Qualification	≥20	≤360	qualified	≥1.5	≥75	≤25	≤5

Table 6.7: Strengths of magnesia cement (hardening at room temperature).

Age	5 h	10 h	15 h	20 h	24 h	3 days	7 days	28 days
Compressive strength R_c (MPa)	9.2	20	23.8	40	44	46	47.8	53.3
Strength formation rate (%)	17	37.5	44.5	75	82.5	86.3	89.7	100

Products with a density ρ of 0.5–1.89 g/cm^3, which is 28–70% of Portland cement, can be produced by using magnesia cement, whose specific strength can be 2–5 times as high as that of Portland cement products. For example, the bending strength R_c of a wallboard with a density ρ of 1.8 g/cm^3 is 67 MPa, while it is about 17 MPa for the same type board of Portland cement.

6.5.2 Cohesive properties

Magnesia cement has a strong adhesive force and high cohesiveness to many kinds of materials such as plant fibers, bamboo reinforcement, wood debris, rattan materials and so on. It can be made into substitute wood products. The pH value of its hydration product that presents weak alkalinity is 8.0–9.0; it will not corrode wood and other organic materials, and so the range of applications is wide.

After using MgCl$_2$ solution to mix the magnesia cement, the salt solution has strong corrosion to the steel. Therefore, steel bar cannot be used in the product. The hydration products of the magnesia cement are close to neutral and have no corrosive action, and because of this, the performance of the magnesia cement can be improved by adding fillers or organic materials. To improve its water resistance, it can be mixed with some pumice, tuff and other pozzolanic addition. Some organic materials such as sawdust, wood shavings and other fillers can be added so that the products have the dual properties of wood and stone, that is, the product not only has the strength of stone but also has good processing property of wood.

6.5.3 Heat and frost resistance

MgO and $MgCl_2$ have their own properties of heat resistance and low temperature resistance. The melting point of MgO is 2,270 °C; it has high refractoriness and is the raw material for producing magnesia refractory bricks. The hydration product of the magnesium cement [$5Mg(OH)_2 \cdot MgCl_2 \cdot 8H_2O$] contains eight crystalline water, and the moisture content is 37%. When it is heated, the vaporization heat that contains eight crystalline water must be consumed first, and the decomposition of the hydration product at high temperature can release chlorine, which has quenching property that can quickly extinguish fire. Therefore, magnesium cement has the function of fire and heat resistant and can produce fireproof board.

$MgCl_2$ belongs to the antifreeze halogen salt, which has low temperature resistance, and is usually able to withstand the low temperature of −30 °C.

6.5.4 Wear resistance

The wear resistance of magnesia cement is three times that of ordinary Portland cement, and so it is suitable for producing floor tiles, high wear-resistant products and abrasives such as polished tiles, grinding blocks and so on.

6.5.5 Efflorescence

When air humidity is high, the surface of magnesia cement products soaks up moisture and gets wet, and then it will be covered with water droplets and even dripping water; this phenomenon is called dehalogenation. When air humidity decreases, the water on the surface of products evaporates, leaving a white spot; this is called efflorescence. It will reduce the strength of products, affect the appearance and pollute the environment.

The key reason for efflorescence is the improper preparation such as wrong ratio of raw materials, inappropriate forming and improper curing. These will produce excess or residual $MgCl_2$, and $MgCl_2$ is a kind of hygroscopic agent.

When the powder is precipitated on the surface of the magnesia cement, the efflorescence occurs. The composition of the efflorescence is as follows:

(1) Sodium chloride frost

It is from the impurities in a halogen block ($MgCl_2 \cdot 6H_2O$) and is soluble in water and can be washed away. Thus, we can purge away sodium chloride dust from the magnesia cement by reducing its content.

(2) Precipitation frost of light filling material, talc powder and light calcium carbonate

This phenomenon often occurs in magnesia slurry. Although the additives can reduce the density of the product, when the moisture is greater, the filling forms a frost on the surface of the product by chromatography. To avoid efflorescence, we can add the filling whose density is similar to the density of light burn powder to reduce moisture in the slurry.

(3) $Mg(OH)_2$ and $Ca(OH)_2$ frost

The reason of formation of $Mg(OH)_2$ and $Ca(OH)_2$ cream is always that the ratio of the amount of MgO to $MgCl_2$ is improper or the light burning powder raw material has excessive loss on ignition or the CaO in light burning powder raw material reacts with water to form $Ca(OH)_2$. To avoid efflorescence, it is important to choose qualified and stable light burning powder raw materials that contain trace amounts of CaO.

(4) $MgCl_2 \cdot 6H_2O$ frost

Its formation is related to the slurry ratio of magnesium cement and have greater harm. To avoid efflorescence, we can adopt actions such as adjusting ingredients ratio in raw material, establishing a sound maintenance system and adding additives.

6.5.6 Resistance to water

Hardened magnesia cement is hygroscopic due to the porous structure; it is easy to damp and frost. Moreover, its hydration products change easily. The strength of magnesia cement product will be weakened if the product gets wet. In order to improve its water resistance, we can add phosphoric acid, phosphate, water-soluble resin or appropriate filling material such as brick powder, talc powder, fly ash, ground granulated silica sand and so on to increase the degree of product's density. The other way to improve its water resistance is by mixing up magnesium oxide with magnesium sulfate, iron vitriol and other solutions, but the strength of product made by this way is slightly weaker than the one made by adding magnesium chloride solution.

The use of additives can improve the water resistance of magnesia cement products from two aspects.
(i) Hydration products are modified to form mixtures and crystalline phases that are difficult to dissolve in water.
(ii) Additives have cementing ability to increase the density of product by filling capillary pores in the hardened magnesia cement; the water resistance is therefore improved.

6.6 Applications of magnesia cementitious materials

Magnesium oxide is a fast hardening material with a fairly high strength. The magnesia cement has a good cohesiveness with the plant fiber. Meanwhile, compared with other cementitious materials such as Portland cement and lime, the alkalinity of magnesia cement is weaker. Therefore, it has no corrosive action on organic materials and fibers. In building engineering, the common magnesia cement products are particleboard, wood chip board, artificial marble, magnesium fiber composite products and porous products.

6.6.1 Magnesia cement particleboard and excelsior board

The wood shavings, linen, leather or other fibrous material with magnesia cement by mixing, pressing forming and hardening can be made into particleboard or fiberboard, which can be used to build walls, partitions, ceiling and so on.

The preparation process of magnesium oxychloride cement excelsior board is shown in Figure 6.5.

The sound absorption of magnesia cement excelsior board is good, and the effect of sound absorption is enhanced with an increase of board thickness. Table 6.8 lists the acoustic performance of excelsior board with a thickness of 25 mm.

6.6.2 Magnesia cement floor

Panels used for laying on the ground, which are made of magnesia cement, sawdust, pigments and other fillers, are the so-called magnesia cement floor. They can be pressed into various kinds of plates, or directly paved at the bottom to make jointless flooring by compressing and decorating. Magnesia cement floor has the characteristics such as good thermal insulation, noise-free, no ash, smooth surface, good elasticity, fireproof, and wear resistant. It is used as a floor material for civil buildings and workshops of textile mills for replacing wooden floor. The ground with a bright-colored and beautiful pattern can be made through assembling by adding different pigments.

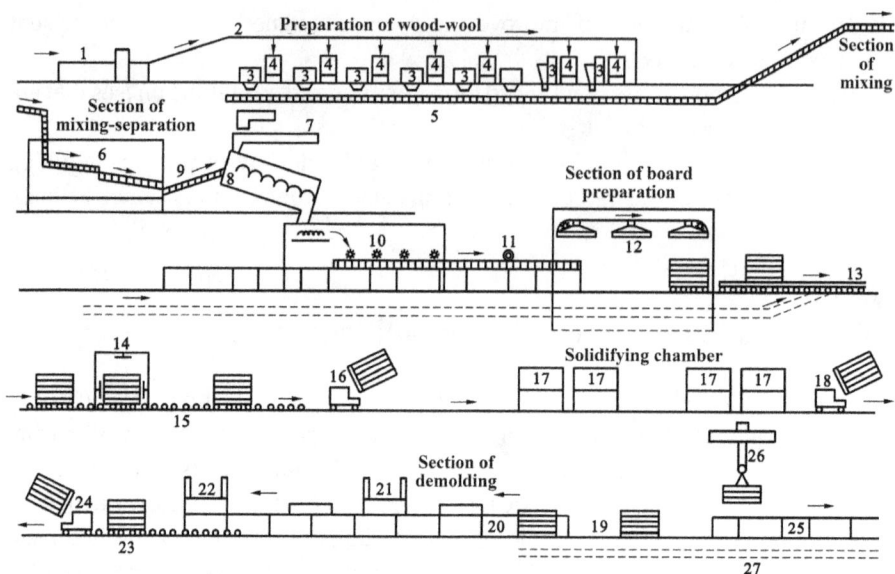

Figure 6.5: The preparation process of magnesium oxychloride cement wood-wool board: 1, multichips saw; 2, conveyor belt toward excelsior chutting machine; 3, excelsior chutting machine; 4, storehouse of wood; 5, conveyor belt to transfer wood-wool to mixing room; 6, device for wood mineralization; 7, conveyor belt for cement weighing; 8, mixing roller; 9, conveyor belt to transfer wood-wool to mixing roller; 10, combing chamber; 11, preloading roller; 12, plate collector; 13, plate stacks roller; 14, pressing machine; 15, roller for plate stacks out of pressing machine; 16, van; 17, solidifying chamber; 18, van; 19, plate stacks for demolding; 20, roller for demolded plates; 21,22, edge cutting machine; 23, roller; 24, van; 25, conveyor belt to transfer baseplate for combing chamber; 26, hoist and 27, conveyor belt to transfer model frame.

Table 6.8: Sound absorption of magnesium oxychloride cement wood-wool board with a thickness of 25 mm.

Frequency (Hz)	250	500	1,000	2,000	4,000
Coefficient of sound absorption	0.67	0.48	0.44	0.72	0.73

6.6.3 Magnesia cement concrete and magnesia cement products

The magnesia cement concrete is made by blending magnesia cement, aggregates of sand and gravel, or fiber materials for making unimportant plates. When an appropriate amount of foaming agent is added to magnesia cement, a porous light material called foam magnesia cement can be formed.

The alkalinity of magnesia cement is low. Magnesia cement has good cementation with wood and other plant fibers. However, it has poor bonding with Portland cement, leading to shedding. Various salt solutions for mixing magnesia cement have a strong corrosive effect on steel bars. Therefore, magnesia cement products cannot be reinforced with steel bars, instead, with bamboo, reed, glass fibers and so on. Magnesia cement products are not suitable for use in humid environments, and magnesia cement plates are not suitable for use in grounds that are often affected by dampness, water and acid erosion.

6.6.4 Magnesia fiber composites

Magnesia fiber composite is a material composed of magnesia cementitious material as the matrix, glass fibers or bamboo bars as the reinforcement. It is characterized by high strength, corrosion resistant, air tight and heat resistant (>300 °C). A material to resist temperatures above 900 °C can be produced by using high temperature–resistant glass fibers to make chimneys, air ducts, shaped tiles and interior partition walls.

6.6.5 Magnesia concrete solar stove

Magnesia cement can be used to make solar stove casing. The 28 days tensile strength can reach 3–3.5 MPa, and the compressive strength is up to 40–60 MPa for a product made of sawdust and magnesia cement in a ratio of 1:3, by mixing with a solution of $MgCl_2$. Based on the aforementioned ratio, a small amount of plant fibers (e.g., sisal) is added; a concrete is made by mixing with magnesium chloride solution, which can be used to make the bottom casing of a solar stove reflector. Compared to Portland cement stoves, it has the characteristics of light weight and high strength, and it weighs only 1/3–1/2 of a Portland cement stove with the same area. The dissolvable salt, $MgCl_2$, can be easily washed with water, and the concrete has a strong hygroscopicity in air with poor water resistance, so it easily deforms with inappropriate curing.

6.7 Storage and transportation of magnesia cementitious materials

Magnesia cementitious materials will lose activity under the action of moisture in air. Attention must be paid to moisture-proof during storage and transportation. Magnesia cementitious materials should not to be stored for a long time so as to prevent them from absorbing moisture in air to become $Mg(OH)_2$, and recarbonized into $MgCO_3$ to lose the cementitious ability.

Problems

6.1 Following problems frequently occur during the production of magnesia cement products: (i) hardening is too slow; (ii) hardening is too fast, and products easily absorb moisture and get damp. What do you think are the reasons? How to improve?

6.2 What are the effects of calcination temperature and time on the quality of light burned MgO?

6.3 What are the characteristics of magnesia cement? Please state its applications.

6.4 What is the hydration process of magnesia cement? What are its hydration products?

6.5 Why are magnesia cements not water resistant? How to improve?

6.6 What are the issues that must be paid attention to during application of magnesia cements? How to avoid?

6.7 Why cannot magnesia cement be mixed with water? What are the suitable mixing liquids?

6.8 There are several issues that need to be paid attention to when dolomite is used to produce magnesium oxide. What are they?

7 Other binders

7.1 Water glass

Water glass is a complex silicate material (including sodium silicate, potassium silicate, lithium silicate, etc.) that contains both soluble phase (dispersed ions and molecules) and colloidal phase (dispersed colloidal particles) in water. Because the most widely used water glass is sodium silicates, it is often referred to as sodium water glass, unless specifically clarified. Water glass as a binder has many advantages such as strong adhesive strength, relatively high strength, excellent acid resistance and heat resistance and so on.

7.1.1 Technical parameters of water glass

The main technical parameters of water glass are modulus, density, solid content and extraneous salts concentration. The commonly used chemical formula of water glass is $M_2O \cdot nSiO_2 \cdot mH_2O$, where M_2O is the alkali metal oxide (Na_2O, K_2O, Li_2O, Rb_2O, etc.), and n is the molar ratio SiO_2 to Na_2O. It is also called the modulus, and it is calculated as follows:

$$n = n(SiO_2)/n(Na_2O) \tag{7.1}$$

The ratio of SiO_2 to Na_2O by mass is also used, which is calculated as follows:

$$r = w(SiO_2)/w(Na_2O) \tag{7.2}$$

Because the molecular mass of SiO_2 is 60, while that of Na_2O is 62, the relationship between modulus and the mass ration is calculated as follows:

$$n = 1.033\,r \tag{7.3}$$

The density and mass concentration of water glass do not have strict linear relationship, but the density is usually used to reflect the concentration of water glass. The density is represented with symbol ρ and the international standard unit is kg/m^3, but it is customary to use g/cm^3 and "°Bé." The calculation formula of Baume degree and density is as follows:

$$\rho = 144.3/(144.3 - {}^\circ Bé) \quad (g/cm^3) \tag{7.4}$$

Solid content is used to indicate the total amount of solids in the water glass. Because the commercial water glass often contains 1–2% (mass fraction, %) soluble salt and less than 1 mass% insoluble substance, the solid content should include the quantity of $Na_2O \cdot nSiO_2$, soluble salt and insoluble substance.

https://doi.org/10.1515/9783110572100-007

Sodium silicate system also contains other salts (called extraneous salts, e.g., NaCl, Na_2SO_4, Na_2CO_3, etc.), having significant negative effects on density, viscosity, surface tension aging rate, hardening rate and bond strength of water glass, and so the salt concentration is one of the important parameters of sodium silicate.

7.1.2 Classifications and preparation methods of water glass

Water glass mainly includes sodium water glass, potassium water glass, lithium water glass, rubidium water glass and quaternary ammonium water glass.

(1) Sodium water glass

Sodium water glass is the most commonly used water glass in production, which is abundantly available and is of low cost. The microstructure of sodium silicate glass is not very clear. It is generally believed that sodium water glass with different modulus n and water content m presents different microstructure patterns. When the modulus of sodium glass is less than 1, it has little practical use. The sodium water glass is difficult to dissolve in water when its modulus is greater than 4. Therefore, the modulus of commonly used sodium water glass is between 1.5 and 3.5.

The sodium water glass presents good adhesion ability. The silica gel is precipitated in the capillary pores during hardening process, which prevents water from penetration. Sodium silicate is nonflammable, and silicate gel dries drastically at high temperature to build a three-dimensional silica skeleton, which increases the strength of sodium water glass. Thus, sodium water glass exhibits excellent heat resistance. Since the main product of hardened water glass is SiO_2 that can resist the attack of most inorganic acid and organic acid, sodium water glass exhibits excellent acid-resistance performance. However, its resistance to alkali and water is poor.

The sodium water glass is usually produced with two methods, that is, the dry process (solid phase reaction) and the wet process (liquid phase reaction). During the dry process, quartz powder (SiO_2) is mixed with soda powder (Na_2CO_3) and then the mixture is heated to 1,400 °C (1,673 K) in the reactor. Afterward, the molten sodium silicate is cooled and milled. Sodium water glass is obtained after dissolution, filtration and concentration. The equation of solid phase reaction is as follows:

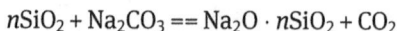

$$n SiO_2 + Na_2CO_3 == Na_2O \cdot n SiO_2 + CO_2$$

During the wet process, caustic soda solution (NaOH) and quartz (SiO_2) hydrothermally react at 160 °C (433 K) in an autoclave. After vacuum filtering and concentration, the sodium water glass is obtained. However, the liquid phase reaction only generates sodium water glass with modulus of less than 3. The equation is as follows:

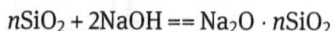

$$n SiO_2 + 2NaOH == Na_2O \cdot n SiO_2$$

Pure sodium water glass is a colorless transparent liquid. However, impurities such as Fe_2O_3, MnO_2 and Al_2O_3 are present in the commercial sodium water glass products because of impurities in the raw materials, corrosion of refractory brick or reaction container, sulfur and phosphorus in the fuel or improper filtering. Commercial sodium water glass products usually are therefore green gray, yellow or red.

(2) Potassium water glass
The potassium water glass is very similar to the sodium water glass. The potassium water glass has better antiaging properties, but stronger moisture absorption to deliquesce.

The preparation process of potassium glass is basically the same as that of sodium water glass, which is also divided into the dry or the wet process. The solid phase reaction is to melt potash and quartz sand together in a certain proportion. And the liquid phase reaction is the same as the reaction for preparing sodium water glass, but only replacing NaOH with KOH.

(3) Lithium water glass
The lithium glass is a colorless, odorless, transparent or slightly turbid liquid, which is weakly alkaline and has a pH value of around 11. Lithium water glass is different from sodium silicate glass, which can have the modulus (n) of 4–15. Lithium water glass can be infinitely diluted with water to be dispersed with multiple molecules in colloid solutions. The polymerization occurs if the concentration of lithium silicate is high. Solid lithium silicate is obtained after further evaporation and water loss. Solid lithium silicate is insoluble in water and organic solvents, but it is soluble with other alkaline aqueous solutions and reacts with acid to produce silica gel. Since the lithium silicate cannot be dispersed and dissolved, the lithium water glass exhibits excellent antihygroscopic performance.

Because of insolubility of solid lithium silicate in the water, lithium water glass cannot be prepared with the same dry process as sodium water glass. In addition, the liquid reaction between quartz and LiOH solution is slow. Thus, lithium water glass cannot be prepared with the same wet process as sodium water glass. Currently, the preparation of lithium glass is mainly using the silica sol method, silica gel method and ion exchange method.

The silica sol method is to obtain polysilicic acid (silica sol) after exchange reaction of sodium silicate through ion exchange resin, and then neutralizes it with LiOH to make lithium silicate.

The silica gel method is used to produce hydrated silicic acid and sodium sulfate after the reaction of sodium silicate and sulfuric acid, and then neutralizes the hydrated silicic acid and LiOH to produce lithium silicate.

The ion exchange method is to generate lithium silicate from lithium carbonate, lithium sulfate and other lithium solution by cation exchange resin reaction. Then the lithium resin reacts with sodium silicate solution to obtain lithium silicate.

7.1.3 The properties of water glass

The density of water glass solution increases with the volumetric concentration, with a weak linear relation. The relationship between density and concentration of water glass is shown in Figure 7.1.

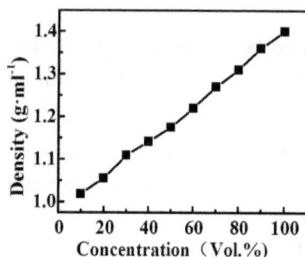

Figure 7.1: Relationship between density and concentration of water glass.

The effect of SiO_2 on density is usually less than that of Na_2O. The viscosity of the water glass is very large and varies with the modulus, concentration and salt content. At the same time, temperature has a significant effect on viscosity.

The effect of storage time on the properties of water glass is very significant, and the viscosity and tensile strength R_m decrease significantly with the increase of aging time, while the surface tension is significantly increased.

The influence of water glass storage time on its technical properties is listed in Table 7.1.

Table 7.1: Effect of storage time on the properties of water glass.

Storage time (d)	0	7	30	60	90
Viscosity (mPa/s)	10.8	9.6	9.1	8.6	7.6
Surface tension (N/m)	76	96.1	111	183	195
Tensile strength (MPa)	1.04	0.95	0.9	0.8	0.62

When the modulus is certain, the viscosity of the water glass increases with the increase of concentration, and decreases with the increase of temperature. For sodium silicate, the relationship between viscosity and concentration is shown in Figure 7.2.

Figure 7.2 shows that the viscosity increases with the increase of concentration. The increase of viscosity of high modulus water glass is much faster than

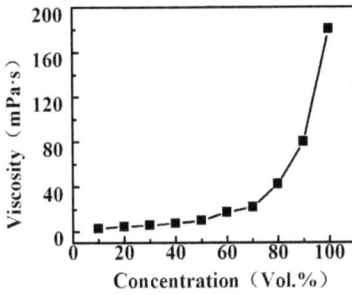

Figure 7.2: Relationship between viscosity and concentration of water glass.

that of low modulus glass. However, each modulus has a corresponding concentration threshold, but its viscosity increases linearly after the modulus exceeds its corresponding concentration threshold. The existence of soluble salts has a great influence on viscosity, especially for water glass with high modulus and concentration. With the increase of salt content, its viscosity increases rapidly.

The water glass with low modulus $n = 2.1–2.6$ presents a pH value of 11–13. Therefore, the phenolphthalein indicator is red in water glass with low modulus. However, the red color gradually fades away when the modulus is raised to more than 2.8. The high modulus water glass ($n > 2.8$) is usually called as neutral water glass. In fact, they are still alkaline, with pH of 8–10.

7.1.4 Hardening process of water glass

Liquid water glass absorbs carbon dioxide in the air, generating amorphous silicic acid, and gradually drying and hardening. The main equation is as follows:

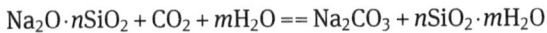

$$Na_2O \cdot nSiO_2 + CO_2 + mH_2O == Na_2CO_3 + nSiO_2 \cdot mH_2O$$

The reaction rate is low, and so the hardening process of water glass is often accelerated by heating or using sodium fluorosilicate (Na_2SiF_6) as a coagulant. And the dosage of sodium fluorosilicate is about 12–15% by mass of water glass. Lower dosage of sodium fluorosilicate results in not only long hardening time but also poor strength and water resistance. Too little hardening time would be observed by excessive amount of sodium fluorosilicate. The reaction of sodium silicate with water glass is as follows:

$$2(Na_2O \cdot nSiO_2) + Na_2SiF_6 + mH_2O == 6NaF + (2n+1)SiO_2 \cdot mH_2O$$

7.1.5 Applications of water glass

The applications of water glass have a history of more than 100 years. With the further understanding of the characteristics of water glass, its application scope is constantly expanding.

(1) Application in casting industry

Water glass is mainly used as raw material for water glass sand in casting industry. Water glass sand is one of the three molding sand (clay sand, resin sand and water glass sand), which has many advantages over clay sand and resin sand. Compared with clay sand, water glass sand exhibits good flowability, easy to compact and operate, low energy consumption, no (or less) dust pollution, nontoxic, high precision size, good module quality and less defects. Compared with resin sand, the water glass sand presents low cost, fast hardening speed and nontoxic.

The main disadvantage of water glass sand is that the sand is difficult to recycle and leads to environment damage without proper recycling methods. The water glass sand has strong tendency to absorb moisture in the air, thus resulting in poor storage stability.

(2) Application in construction engineering

The efflorescence resistance of building materials improves by water glass coating on the surface. Sodium silicate in the water glass can penetrate into cracks and pores of the materials, and precipitated silica gels terminate the pore channels, and increase the density and strength. However, water glass coating cannot be applied to gypsum products because that sodium silicate reacts with calcium sulfate to produce sodium sulfate crystals, resulting in volume expansion and damages that destroy gypsum products.

Water glass is also used as grouting material to reinforce the foundation of buildings. Water glass and calcium chloride solution are alternately irrigated in to the foundation and generate silica gels, which act as a binder wrapping the soil particles and filling in the pores. Water glass–based grouting material can improve the bearing capacity of foundation and enhance its waterproofness.

Water glass can resist the corrosion of most inorganic acid (other than hydrofluoric acid). Thus, water glass is usually used to manufacture acid-proof mortar and concrete. At the same time, water glass has a good heat resistance and can be used in the preparation of heat resistant mortar and concrete.

Water glass is used as alkali to activate granulated blast furnace slag and other potential active materials or volcanic ash materials and prepare geopolymer. Preparation of geopolymer can not only alleviate the environment impact on energy and resources crisis of cement industry but also make full use of industrial waste.

Furthermore, water glass is also widely used in chemical engineering, textile, papermaking and washing industries. Although the application scope of water glass is constantly expanding, there is a serious aging phenomenon in the use of water glass, which greatly affects its performance.

7.2 Phosphate cement

Metal oxide (Al_2O_3, MgO, etc.) and phosphoric acid or salt (such as $MgCl_2$) are the basic component of phosphate cement, which is formed by the chemical reaction. They differ from Portland cement in terms of hydration and practical applications.

Phosphate cement can be used to make a variety of heat and thermal stability materials, corrosion and electrical insulation coatings, high-performance plastic and so on. The properties of the phosphate material are similar to those of the ceramic material, and also have a number of other advantages. For example, it is not necessary to calcine at high temperature to produce phosphate cement, which is stable in many aggressive media. Phosphate cement is used in refractory bricks, mortars, road repair materials, cement pipes, sprayed foam insulation materials, heat-resistant coatings and so on.

Phosphoric acid has three forms: orthophosphoric acid (H_3PO_4), pyrophosphoric acid (H_4PO_7) and metaphosphoric acid (HPO_3). The orthophosphoric acid is the most stable phase, so it is mainly used in the production of phosphate cement. The cementation properties of phosphate cement are closely related to the cationic species in the aggregate. When the phosphoric acid reacts with the weakly alkaline oxide, it can produce stronger cementation performance. When the phosphoric acid reacts with the strong alkaline oxide, due to the fast reaction speed, the structure is porous. The acid or neutral oxide (Al_2O_3, ZrO_2, CrO_3, etc.) usually does not react at normal temperature, and it is necessary to use heating treatment or adding accelerant to increase the strength of phosphate cement. The powder and filler used in phosphate cement generally include refractory clinker, bauxite clinker, siliceous clinker, magnesia clinker and corundum, mullite, zirconium quartz, alumina, chromium slag, carbonized silicon and so on.

The phosphate cement belongs to the polymerization-coagulation hardening cementitious material whose properties are determined by the nature of the compounds in the RO–P_2O_5–H_2O (R is alkaline earth metal) system. Because of many types of phosphates and the different metal elements in the RO, the composition and properties of cement are not the same. The most widely used phosphate cements are magnesium phosphate cement and aluminum phosphate cement.

7.2.1 Magnesium phosphate cementitious materials

Magnesium phosphate cement is composed of MgO, phosphate, retarder and some mineral admixtures. Among them, MgO provides Mg^{2+}, and the acid environment needed for hydration reaction and PO_4^{3-} are provided by phosphate.

(1) Raw materials for producing magnesium phosphate cement

The raw materials of magnesium phosphate cement is commonly the magnesite. The reburned magnesite is made from magnesite ($MgCO_3$) by calcining at 1,700 °C. It is a raw material for brick making and refractory production. The finished product is used for steelmaking, electric furnace bottom and ramming lining. Using reburned magnesite in the magnesium phosphate cement is mainly due to its low reactivity to prevent excessive MgO solubility, resulting in rapid hydration and difficulties in handling.

Phosphate mainly uses ammonium dihydrogen phosphate, potassium dihydrogen phosphate and diammonium phosphate. The main role is to provide the acidic environment for the hydration and phosphate ions. The dissolution rate of phosphate and pH of the solution will directly affect the formation of hydrated minerals, strength of the material and high-temperature performance of the cement.

Magnesium phosphate cement must be able to provide sufficient construction time in applications, so it is necessary to add a retarder into cement. Common retarders include borax, boric acid, sodium tripolyphosphate and alkali metal salts. The most commonly used retarder is borax. It forms a protective film on the surface of MgO, thus preventing the reaction and effectively delaying the coagulation time of magnesium phosphate cement. Borax content is necessary to be controlled within a certain range; too much borax will cause the decline in the strength of magnesium phosphate cement.

Based on environmental requirements, fly ash is usually added as a mineral admixture to magnesium phosphate cement. The cost of fly ash is relatively low. Fly ash can improve and adjust the coagulation time, color, fluidity and workability of the phosphate cement, and can also improve the late strength. Other mineral admixtures include slag, granite, quartz sand and limestone improving the performance of magnesium phosphate cement.

The preparation of magnesium phosphate cement materials mainly includes the adjustment and optimization of the ratio of magnesium to phosphorus (the mass ratio of magnesium oxide to phosphate, w (M)/w (P)), dosage of retarder, the activity of magnesium oxide and specific surface area, water consumption and so on. Among them, the content of magnesium oxide has great influence on its gelation ability. If the content of magnesium oxide is increased, the gelation capacity can be improved. However, when the content of magnesium oxide is high, the reaction rate of cementing material is fast and the working time is short, which makes it difficult to work with.

The relationship between the compressive strength of cement paste at different hydration time and the ratio of magnesium to phosphorus is shown in Figure 7.3.

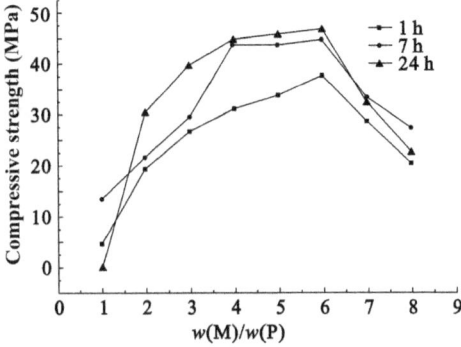

Figure 7.3: Compressive strength of cement paste with different ratio of magnesium to phosphorus.

(2) Curing of magnesium phosphate cement

A curing system has a significant effect on the strength of magnesium phosphate cement. In the presence of external water, the unreacted phosphate and hydration products of the cement dissolve and hydrolyze, resulting in increase in porosity and decrease in the strength of the material.

In the case of dry air or sufficient sealing conditions, the strength of magnesium phosphate cement was continuously increased, and the difference of the two curing conditions had little effect on the gelation capacity. If the relative humidity of the curing environment is increased, the gelation capacity of magnesium phosphate cement is significantly reduced. Therefore, magnesium phosphate cement products should be maintained in dry air, or coated to isolate water infiltration.

The relationship between compressive strength and hydration time of magnesium phosphate cement under different curing conditions is shown in Figure 7.4.

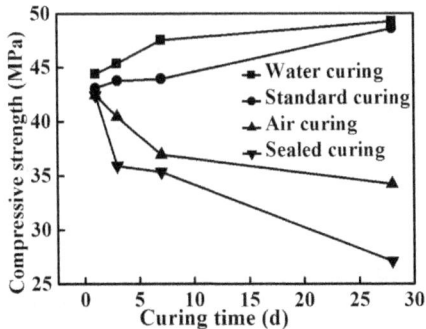

Figure 7.4: Compressive strength of magnesium phosphate cement under different curing conditions.

(3) Hardening process of magnesium phosphate cement

It is generally believed that the hydration of magnesium phosphate cementitious material proceeds in five stages (Figure 7.5).

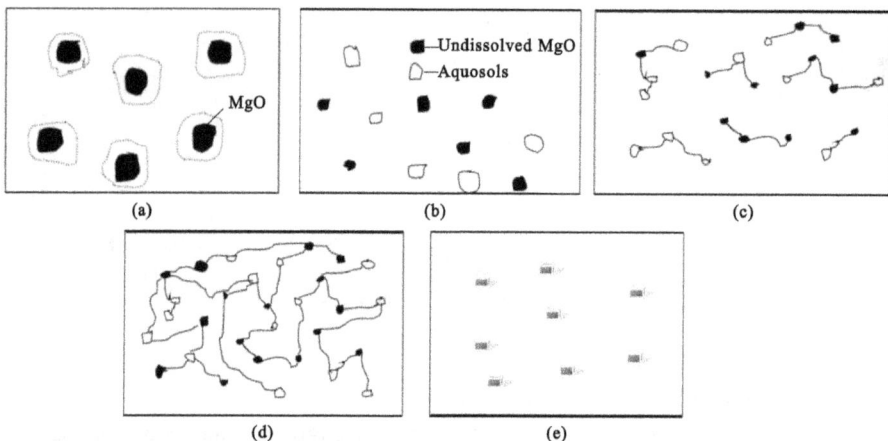

Figure 7.5: Hardening process of condensation process of magnesium phosphate cement: (a) oxide dissolution; (b) hydrated gel formation; (c) reaction and polymerization; (d) gel formation and (e) saturated crystallization.

The first stage is the reaction between MgO and the acidic solution, causing MgO to dissolve. The reaction can be expressed by the following equation:

$$MgO + 2H^+ = Mg^{2+}(aq) + H_2O$$

The second stage is the formation of sol. This process mainly involves the reaction of MgO and water to form positively charged hydroxide. The reaction rate of cement is mainly controlled in this stage. The reaction is as follows:

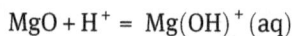

$$MgO + H^+ = Mg(OH)^+(aq)$$

The third stage is the formation of cementitious gel. The sol formed at the previous stage interacts with the phosphate ion at this stage to form the phosphates and emits heat. As the reaction proceeds, the amount of gel increases.

The fourth stage is the network of gels. Gel in a loose state is mainly linking with each other while the reaction proceeds, and it forms the gel network finally at this stage.

The fifth stage is the saturation and crystallization of the gel, and finally the formation of a ceramiclike matrix. At this stage, the slurry will be gradually sticky and the gel will be transformed into the crystal attached to the surface of the unreacted MgO particles, so that the strength of the material gradually increases.

(4) Application of magnesium phosphate cement

The application of magnesium phosphate cement is currently concentrated on the fields of rapid repairing materials, man-made panels, composite industrial waste materials, permafrost and deep well curing, spraying of building materials, immobilization of harmful and radioactive nuclear wastes.

The rapid repairing material is the broadest application of magnesium phosphate cement. It is mainly used as rapid hardening, high strength and high bonding strength material with the old concrete, mainly for road, bridge and aircraft runway fast repair.

Magnesium phosphate cement is also mixed with straw, pulp waste, paper scrap, deforestation waste, wood chips produced by wood processing and residual scraps to produce man-made boards. These boards have the properties of high resistance to mechanical shock, thermal shock and antifracture.

In the field of building materials, magnesium phosphate cement is mixed with industrial by-product to produce cement with adequate strength. Thus, magnesium phosphate cement converts solid waste into useful building materials and has a prominent advantages of environmental protection.

Magnesium phosphate cement is also used in frozen soil treatment and deep well stabilization. Its good performance at low temperature and rapid hardening properties make it superior to the conventional Portland cement in the low temperature or high geothermal environment. Magnesium phosphate cement is also sprayed onto wood board or foamed polystyrene board to form a protective layer using its high durability and high temperature performance. Magnesium phosphate cement is a potential inorganic thermal insulation material with broad application feasibilities.

Magnesium phosphate cement is effective for immobilizing hazardous wastes containing heavy metals or municipal solid waste incineration ash and other harmful wastes. The hazardous waste reacts to form insoluble substances, enter directly into the hydrated lattice, or is wrapped in the hydrated matrix, preventing it from spreading into the environment.

7.2.2 Aluminum phosphate cement

Preparation of aluminum phosphate cement mainly uses aluminum hydroxide, noncalcined and calcined alumina or synthetic alumina as raw materials. Raw materials containing aluminum hydroxide, such as refractory clay, clay, kyanite and montmorillonite, are normally used. Industrial alumina does not react with orthophosphoric acid and acts as an inert filler. Because the aluminum phosphate cement is normally used at high temperature, the change of the inert materials at high temperature must be considered as well, which will affect the performance of the hardened cement.

Typical processes for producing aluminum phosphate cement include sodium phosphate–aluminum sulfate method, phosphoric acid–sodium aluminate method, phosphoric acid–aluminum hydroxide method, solid phase reaction and gas phase reaction method.

(1) Sodium phosphate–aluminum sulfate method

Aluminum phosphate precipitation is produced by the reaction of sodium phosphate and aluminum sulfate, leaving sodium sulfate in the solution. The reaction is as follows:

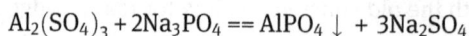

$$Al_2(SO_4)_3 + 2Na_3PO_4 == AlPO_4 \downarrow + 3Na_2SO_4$$

Sodium phosphate and aluminum sulfate are separately dissolved in hot water to form solutions, followed by removing the insoluble impurities via filtration. The two solutions are fed into the reaction kettle with a proper concentration and white colloidal aluminum phosphates precipitate. If the sodium phosphate is slightly excessive, it will be beneficial to accelerate the reaction and for precipitation of aluminum phosphate. After the reaction is completed, the solids are separated with plate-and-frame filter press from the mixture. The solid phase is left in the filter, washed with dilute hydrochloric acid and water to remove the entrained sulfate ion and dried and ground to obtain the final product. The filtered solution is sent to recover sodium sulfate.

(2) Phosphoric acid-sodium aluminate method

This method is a liquid phase reaction, which needs to be heated up to 250 °C in autoclave. The reaction is as follows:

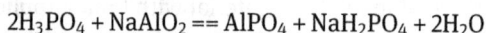

$$2H_3PO_4 + NaAlO_2 == AlPO_4 + NaH_2PO_4 + 2H_2O$$

The sodium aluminate is dissolved in hot water to form a concentrated solution and heated to 85 °C. A phosphoric acid with concentration of 85% (mass fraction) is added to produce white precipitate, and the pH of the final solution is controlled in the range of 4.2–4.5. The reaction slurry is moved into autoclave with stirring device and heated at 250 °C for several hours. Then, the solid–liquid mixture is separated by centrifuge; the solid substance is white crystal aluminum phosphate and a small amount of sodium dihydrogen phosphate. To remove the water-soluble impurities, the solid substance is washed with dilute hydrochloric acid and clear water in the washing tank, and after the filtration and drying, the finished product is obtained. The main component of the solution after the solid–liquid separation is sodium dihydrogen phosphate, which can be used as raw materials to produce other phosphate.

(3) Phosphoric acid–aluminum hydroxide method

This method is a liquid phase reaction as follows:

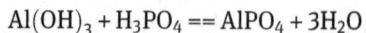

$$Al(OH)_3 + H_3PO_4 == AlPO_4 + 3H_2O$$

Phosphoric acid (60% concentration by mass) is added into the reaction kettle and heated to 85–90 °C. Aluminum hydroxide is added with consistent stirring and controlling the pH of the solution. A paste is obtained after continuous heating till the reaction is complete. Excessive water is afterward added to form the white aluminum phosphate precipitates. The slurry is pressed into a filtering machine for solid-liquid separation. The separated solid phase is washed with water, and dried, then calcined at temperatures above 800 °C, after which the hexagonal crystals of aluminum phosphate are obtained.

(4) Solid state reaction process
The preparation of aluminum phosphate with the solid phase reaction can be divided into two types of reactions, that is, (a) the reaction of ammonium biphosphate and aluminum hydroxide and (b) the reaction of phosphorus pentoxide and α-Al_2O_3.

(a) Reaction of ammonium biphosphate and aluminum hydroxide
α-Al_2O_3 or aluminum hydroxide and ammonium biphosphate are mixed with a mass ratio of 1:(1–3). The mixed material is transferred into the kneading machine, preheated for 30 min and then heated in a baking furnace of 900 °C to synthesize the aluminum phosphate. The finished product is obtained by grinding. The reaction is as follows:

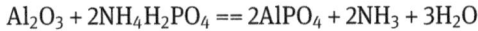

$$Al_2O_3 + 2NH_4H_2PO_4 == 2AlPO_4 + 2NH_3 + 3H_2O$$

(b) Reaction of phosphorus pentoxide and α-Al_2O_3
The α-Al_2O_3 and phosphorus pentoxide are mixed with a molar ratio of 1:1. The mixture is then heated at 500–900 °C and ground to obtain the finished product of aluminum phosphate. The reaction is as follows:

$$Al_2O_3 + P_2O_5 == 2AlPO_4$$

In the process of production, special attention is needed while mixing reactants; otherwise, the products may contain irregular aluminum phosphate.

(5) Gas phase reaction method
In a special gas phase reactor, aluminum chloride and phosphorus trichloride are combusted and gasified with hydrogen flame. The phosphorus trichloride is oxidized into phosphorus oxychloride, and hydrolyzes into gaseous phosphoric acid. The phosphoric acid gas reacts with the gaseous aluminum chloride to form aluminum phosphates:

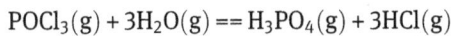

$$2H_2 + O_2 == 2H_2O(g)$$

$$2PCl_3(g) + O_2 == 2POCl_3(g)$$

$$POCl_3(g) + 3H_2O(g) == H_3PO_4(g) + 3HCl(g)$$

$$AlCl_3(g) + H_3PO_4(g) \;\rightleftharpoons\; AlPO_4 + 3HCl$$

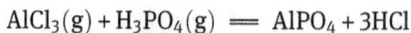

Other methods have been proposed recently for producing aluminum phosphate cement, such as microwave induction method. The microwave induction method uses sodium dihydrogen phosphate and aluminum chloride to rapidly synthesize aluminum phosphate. The crystallization time with microwave induction is about 40 min, with a microwave power of 600 W and a productivity of 95.8%.

Aluminum phosphate cement is mainly used in adhesives and as special cement. When used as adhesive, it has excellent high temperature resistance and weather resistance. It is a common binder for refractories and inorganic coating, and for siliceous, high alumina, magnesia, silicon carbide and oxide concrete.

Phosphate cement is commonly used in the lining of kilns for building materials or thermal power units. It is also used in the lining of capacitor firing furnace or open-hearth furnace in metallurgical industry. It has the advantages of high temperature resistance, low shrinkage, high strength and high wear resistance. The service life of refractory lining built with aluminum phosphate cement is normally four times longer than traditional masonry mortar and other building materials. In addition, aluminum phosphate cement can also be used for instrument manufacturing, thermal protection of instrument and heat-resistant insulating materials.

7.3 Geopolymer

Geopolymer is a kind of material with pozzolanic activity or potential hydration ability. It can react with activators and show cementitious performance. The raw material is aluminum silicate, including various types of natural silicate minerals and silicate-based industrial by-products or wastes, such as blast furnace slag, cinder, fly ash, phosphorus slag, red mud, natural volcano ash, metakaolin and so on. Compared with the production of Portland cement, the production of geopolymer does not require clinker calcination, which can greatly reduce production energy consumption and carbon dioxide emissions, and can also make full use of various industrial wastes.

The study of geopolymer materials can be traced back to 1930s. Purdon found that a small amount of sodium hydroxide has played a catalytic role in the hardening of Portland cement. The presence of sodium hydroxide helps the liberation of silica aluminum in cement, forming sodium silicate and sodium aluminate. The products will further react with calcium hydroxide, leading to the formation of silica and alumina hydrates as well as the regeneration of sodium hydroxide.

In 1957, Glukhovsky mixed gravel, pulverized boiler slag or lime and slag, then used sodium hydroxide solution or water glass solution as a paste. Geopolymers can have high compressive strength of 120 MPa and good stability, and were sometimes called "soil cement." Davidovits used metakaolin as raw material and alkali compound as an activator and obtained a mortar with compressive strength of 20 MPa after 24 h. The compressive strength of the mortar was 70–100 MP after 28 days curing. The term "geoploymers" was then for the first time proposed and patented. Pyrament geopolymer cement has been commercialized in the United States and was used in military applications. Geopolymers produced in Australia by Zerobond has been widely used in airports, roads and other projects.

Since the 1980s, China has carried out various researches on the soil polymer gel and obtained good results. Since 2005, Wuhan University of Technology carried out large number of researches and applications in bridge, structure repair and rapid construction by geopolymer cement, which has given the good effect and developed into a mature product.

7.3.1 Hydration process and hydration products of geopolymer

The hydration process of geopolymer is initiated by the breaking of covalent bond of Si–O–Si, Si–O–Al and Al–O–Al under the action of OH^- forming the ionic group or single ion group with smaller polymerization degree. The main chain of this structure is composed of –Si–O–Al–O–Si–O–, where the unbalanced charge is neutralized by K^+ and Na^+ caused by the substitution of Al^{3+} for Si^{4+}. At the same time, K^+ and Na^+ in the network structure are constrained (cage structure).

The general process of geological polymerization of metakaolin-based geopolymer is shown in Figure 7.6. According to the different materials, geopolymer cementitious materials can be classified into two categories: (1) $M_2O-M_2O_3-SiO_2-H_2O$ (alkali series) and (2) $M_2O-MO-M_2O_3-SiO_2-H_2O$ (alkaline earth series).

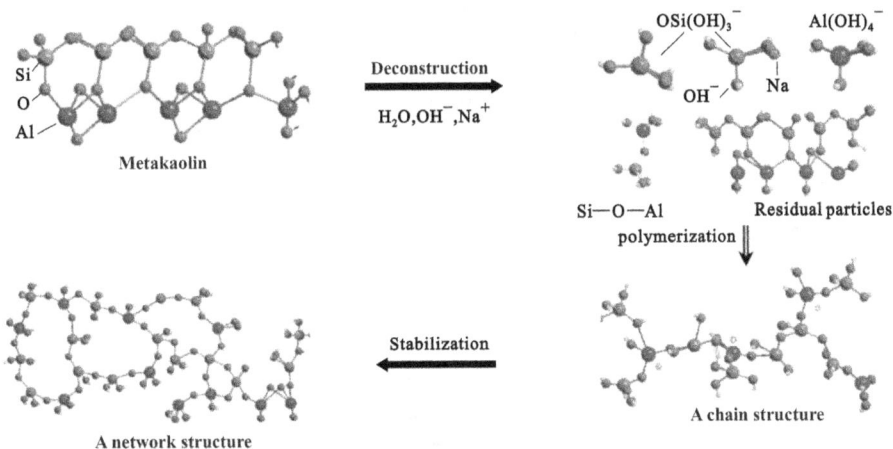

Figure 7.6: The polymerization process of metakaolin-based geopolymer.

The first class contains little calcium, mainly Si and Al, and metakaolin and fly ash are typical components for these types of geopolymers. The second kind is mainly composed of Si and Ca, which is typically based on ground granulated blast furnace slag. Because the strength of Al–O bond is higher than that of Ca–O, alkali cementitious material often needs higher alkali concentration. Compared with alkaline materials, the dissolved Ca^{2+} in alkaline earth materials react with silicate and aluminate to form precipitation, which reduces the polymerization degree. For example, the degree of polymerization of hydration products of alkali activated slag is generally between 2 and 10, while the degree of polymerization of geopolymers can reach 500–1,000 or higher. It is a three-dimensional network structure composed of silicon oxygen tetrahedron.

The main reaction product of low calcium containing geopolymer (alkali series) is a aluminosilicate inorganic polymers, while that of geopolymer rich in calcium (alkali earth series) is calcium silicate hydrates and products that highly depends on the alkali nature, alkali concentration, chemical composition and reactivity of the precursor. When NaOH was used alone as the activator, the product has hydrotalcite phase in addition to C–S–H. The XRD spectrum of the polyuresols in the kaolinite and the base of the kaolinite is the smooth diffuse peak at most diffractive angles, and the diffraction peak of the latter is moved backward. It can seen that the reaction product of kaolin based geopolymer remains amorphous.

Figure 7.7 shows the XRD pattern of geopolymer cement cured for different time. Figure 7.8 shows the XRD patterns of the metakaolin material and the geopolymer cured for 1 and 180 days at room temperature.

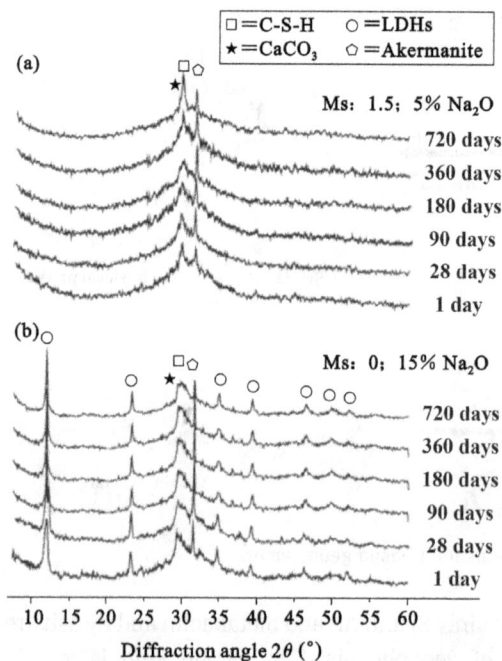

Figure 7.7: XRD patterns of geopolymer cured at different ages (a) modulus 1.5, Na_2O content 5% and (b) modulus is 0, Na_2O content 15%.

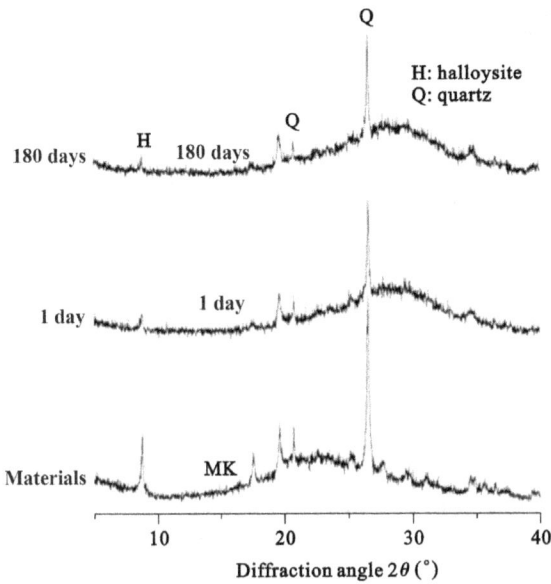

Figure 7.8: XRD patterns of raw materials and geopolymer at 1 and 180 days at room temperature.

7.3.2 Mechanical properties of geopolymers

Geopolymers have the characteristics of short setting time and high early strength. The strength development depends on the temperature and age of curing, as well as the type and dosage of the activator. Usually, metakaolin or blast furnace slag as a precursor could react at the room temperature and obtains high strength (>20 MPa after 1 day), while fly ash needs curing at elevated temperatures to accelerate the hardening. As an alkali activator, water glass could produce the silica gel in solution, which fills the pores of the paste, thus giving the highest strength (as shown in Figure 7.9).

Generally, the strength of the material has obviously positive correlation with the content of alkali, but an overdose of alkalis will aggravate the early age shrinkage and cracking.

If the modulus of water glass solution is less than 1.7, increasing the modulus increases the strength. The correlation between the modulus and the strength is not straightforward if the modulus is larger than 1.7. In the interface of traditional silicate cement, the concentration of calcium hydroxide and the transition zone of preferential orientation appear easily, resulting in the decrease of interface binding force. There is a chemical effect between limestone sand and ground polymer cementing materials, and the silica sand and the aluminum silicate composition of granite sand are also involved in polymerization. Therefore, there is no obvious interfacial transition zone between agglomerates and aggregates, which can achieve higher compressive strength.

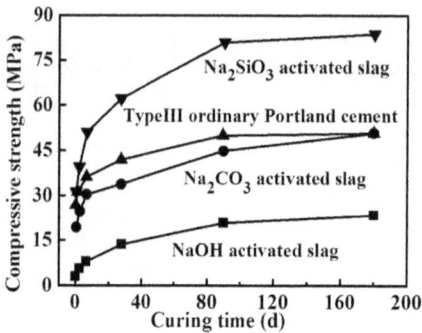

Figure 7.9: Compressive strength of geopolymers and ordinary Portland cements.

(1) Volume stability

Autogenous shrinkage and dry shrinkage of geopolymers are often larger than that of ordinary Portland cement (OPC). This phenomenon could be explained with the Laplace formula. If water is loosely absorbed, smaller pore diameters cause larger capillary force, which in turn causes greater shrinkage. Figure 7.10 shows the comparison between autogenous shrinkage and dry shrinkage of geopolymers and OPCs.

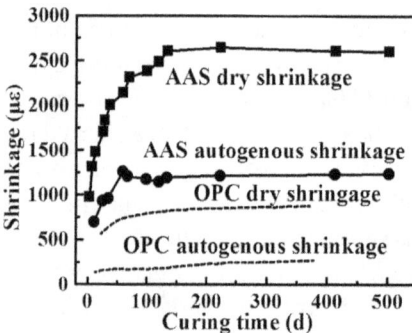

Figure 7.10: Autogenous shrinkage and dry shrinkage of geopolymers and ordinary Portland cements.

As shown in Table 7.2, the proportion of the gel pores in the geopolymers mortar is much less than that in the Portland cement mortar, and it increases with curing age. The greater shrinkage of geopolymers is also explained by the less crystalline hydration products that can act as microaggregate, pillaring the structure. Geopolymers using water glass as an alkali activator have greater drying shrinkage and autogenous shrinkage than those using sodium hydroxide. Geopolymers prepared with water glass

Table 7.2: Proportion of different pores in geopolymer mortar and OPC mortar.

Curing age (days)		3	7	28	56
Gel pores	Geopolymers	74.0%	76.0%	82.0%	81.3%
	OPC	36.4%	35.2%	32.7%	24.7%
Capillary pores	Geopolymers	16.6%	14.9%	10.4%	12.5%
	OPC	56.7%	59.6%	62.2%	69.0%
Air voids	Geopolymers	9.4%	9.1%	7.6%	6.2%
	OPC	6.9%	5.2%	5.1%	6.3%

hydrate to produce large amount of silica gel, and greater shrinkage occur when the gel dehydrates later. In addition, the geopolymers using water glass as an activator have hydration products with denser structure and refined pores.

(2) Durability

Geopolymers usually have denser hydration products and refined pore size distribution. There is no obvious interfacial transition zone in geopolymer-based concrete. These characteristics give the geopolymer-based concrete great impermeability.

The SEM morphology of the interfacial transition zone of silicate cement and metakaolin-based geopolymer concrete is shown in Figure 7.11.

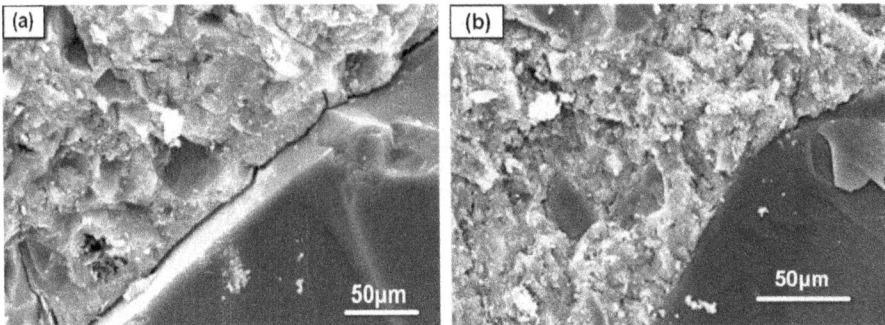

Figure 7.11: SEM images of the ITZ: (a) OPC concrete and (b) metakaolin-based geopolymer concrete.

Moreover, geopolymers show excellent resistance to chloride ingression due to the dense structure mentioned earlier, inhibiting the transfer of ions. Therefore, the geopolymer cement has good protective effect on steel reinforcement. Because of its large number of mesoporous pores, the freezing point of pore solution is significantly reduced, which greatly reduces the possibility of freezing damage caused by water icing.

The hardened pastes of geopolymer and OPC are immersed in 5% sulfuric acid and 5% hydrochloric acid solution for 60 days, and the rate of decomposition is determined (Table 7.3).

Table 7.3: Rate of decomposition in acid solution.

Type of the acid solution	5% H_2SO_4	5% HCl
OPC paste	95%	78%
Geopolymer paste	7%	6%

The results show that geopolymers have much better resistance to acid attack than Portland cements. Furthermore, geopolymer paste immersed in sodium sulfate solution has no obvious damage, but that immersed in magnesium sulfate solution shows greater damage due to the exchange of the ions and formation of the hydrated magnesium silicate (M–S–H) gel and gypsum.

Resistance of geopolymers to carbonation is normally inferior to that of Portland cement due to the absence of calcium hydroxide in the hydration products. Its carbonation rate is similar to Portland cement in the atmospheric environment. If tested in the accelerated carbonation environment, the carbonation rate of geopolymers is remarkably higher.

The alkali–silica reaction is less likely to occur in metakaolin-based geopolymer concretes due to the absence of interface transition zone. Furthermore, if there is reactive silica in aggregates, it will react with alkali oxides in binders to form new hydration products. The interfacial zone will be densified and homogeneous, which prevents the infiltration of water and further reaction of alkali–silica reactivity.

Geopolymer binders have stable volume under high temperature. Geopolymer binders have the line shrinkage in the range of 0.2–2% and retain above 60% of their original compressive strength after heating at 800 °C, which shows great mechanical behavior under high temperature and better fire resistance than OPC. The thermal conductivity is 0.24–0.38 W/(mK), which is comparable to that of lightweight refractory clay bricks (0.3–0.4 W/(mK)).

7.3.3 Applications of geopolymers

Geopolymers have broad application prospects based on their excellent performance characteristics, mainly applicable to the following aspects.

(1) Civil engineering materials and rapid repair materials
Geopolymers can greatly shorten the demold time, improving the construction speed and be used as quick repair materials for concrete structure due to high early

compressive strength and high interfacial bonding strength. Airport pavement constructed with geopolymers is ready for walking after 1 h, vehicle traffic after 4 h and aircraft taking off and landing after 6 h.

(2) Geopolymer-based coatings
Geopolymers have great water and fire resistance due to dense hydration products. Geopolymer-based coatings have a series of advantages such as acid resistant, fireproof and environmental friendly. Geopolymer-based coatings have broad application prospects as special coatings.

(3) Toxic industrial waste and nuclear waste immobilization
The typical hydration products of geopolymers are zeolite phases, which can absorb toxic chemical waste due to the three-dimensional framework (cage structure) of hydrated aluminum silicate. Thus, geopolymers are effective binders for immobilizing various chemical waste, toxic heavy metal ions and nuclear radioactive elements.

(4) Geopolymer-based composite materials
Geopolymers can be used for building panels and blocks owing to advantages including rapid hardening and early strength, high flexural strength, corrosion resistance, low thermal conductivity and high plasticity. Geopolymer products can be cured in dry condition with short curing cycle. Geopolymer-based products have great processability and have appearance of a natural stone, which are suitable for decoration applications.

(5) Chemically bonded ceramics
Geopolymers are as good as sintered ceramics but are formed at room temperature by using the cast technique. Compared to ceramics, geopolymer decorative materials or products do not need sintering process and have good integrity.

(6) Fire and high-temperature resistance materials
Geopolymers can withstand 1,200 °C and be widely used for the production of furnaces, metallurgical pipes, insulation materials and nonferrous casting in metallurgical industry.

7.4 Polymer-modified inorganic binders

Inorganic binders (Portland cement, gypsum, lime, etc.) have characteristics of great brittleness and cracking tendency. The modern engineering requires the hardened concrete possessing high ductility and toughness. Reducing the brittleness in some aspects of engineering such as tunnels, roads, hydraulic projects, buildings, fire resistance and building repaired engineering is frequently required. Adding polymers to conventional inorganic binder can enhance the toughness and tensile resistance.

Polymers are compounds with large molecular weight, which are bonded by covalent bonds between intermolecular structural units. Polymer-modified inorganic binder is a kind of new composite binder material with excellent performance. Due to the good toughness and ductility of polymers, the tensile and flexural capacity, durability and impermeability of binders can be enhanced.

Polymers have the same morphological effect as mineral admixture fly ash because the solid particles or emulsion particles of the polymers have very small particle diameters about 0.05–5 μm in general. These particles generally increase the fluidity of the fresh mixture.

Most cementitious materials are silicate materials with the unit of silica tetrahedron. Silicon and oxygen are connected by covalent bonds, while calcium and aluminum ions are combined with silicon by ionic bonds. There is almost no deformation when the covalent bonds or ionic bonds are broken and so it shows great brittleness. However, the polymer has elastic and plastic properties owing to its long molecular structure and the rotation of chain elements or segments in large molecules.

Polymers can improve the brittleness of binders when they are added to cement slurry. In addition to the characteristics of organic polymer itself, it can improve the internal structure of the paste that would reduce the formation of microcracks in the hydration process. In addition, the polymer film with small elastic modulus and large deformation can buffer the stress concentration of cracks, which would enhance the tensile and flexural strength of the hardened paste. The polymer can increase the deformability and reduce the elastic modulus of the hardened binder.

The hydrated calcium hydroxide grows along the polymer particles, which would be good for disorganizing the orientation growth of calcium hydroxide. In addition, the film formation wraps the calcium hydroxide crystals and prevents them from chemical attack such as leaching and carbonation. Ettringite in the polymer-modified Portland cement paste is shorter and thicker than that in the ordinary Portland cement paste. Polymer can improve the pore structure of the Portland cement paste, which acts as a binder and filler due to the morphological effect. The particles of polymers refine the pore structure.

Since the process of film formation of the polymer occurs during the hydration of Portland cement, the water is consumed during hydration or evaporates. The polymer can form a tough, dense network of film network structures throughout the matrix that are distributed in the paste and fill the pores, cutting off the percolated pore structure and thus enhancing the impermeability and durability of the hardened Portland cement paste.

Some polymers can react with the hydration products of the Portland cement such as the reaction between methacrylate and calcium hydroxide of Portland cement hydration. The reason is that the ester group in acrylic acid can be hydrolyzed in alkaline calcium hydroxide solution that would generate carboxylate anions that can be combined with calcium ions by ionic bonds. As a consequence, interwoven network structure of ionic bond chain macromolecule systems is formed by bridged calcium ions that would enhance the compactness of the structure.

Adding some polymers can enhance the flexural strength, fatigue resistance and impermeability of the binders. In general, the effect of polymer on the compressive strength of Portland cement concrete is marginal.

Polymer-modified inorganic binders are widely used in hydraulic engineering, tunnel engineering and other special projects owing to high permeability and durability. Polymer-modified inorganic binders are also used in pavement engineering and maintenance owing to their high tensile and flexural strength. The polymer also can be used in mass concrete because it can reduce the rate of hydration heat of Portland cement. Polymer-reinforced binders can reduce the amount of water and can increase the density without changing the fluidity. Therefore, it can also be used in high-performance concrete.

Problems

7.1 What is the modulus of water glass? What is the effect of the modulus of water glass on its technical properties?

7.2 What is the effect of water glass on the hardening process of Portland cement paste?

7.3 Why the magnesium phosphate binder is poor for water resistance?

7.4 Why the rate of shrinkage of geopolymer binder is high? How to reduce it?

7.5 What are the advantages and limitations of polymer-modified inorganic binders?

Appendix

I Relative atomic mass (A_r) in common use

Hydrogen (H)	1.0079	Phosphorus (P)	30.97376
Boron (B)	10.81	Sulfur (S)	32.06
Carbon (C)	12.011	Chlorine (Cl)	35.453
Nitrogen (N)	14.0067	Potassium (K)	39.098
Oxygen (O)	15.9994	Calcium (Ca)	40.08
Fluorine (F)	18.99840	Titanium (Ti)	47.90
Sodium (Na)	22.98977	Manganese (Mn)	54.9380
Magnesium (Mg)	24.305	Iron (Fe)	55.847
Aluminum (Al)	26.98154	Strontium (Sr)	87.62
Silicon (Si)	28.086	Barium (Ba)	137.34

II Relative molecular mass (M_r) in common use

Al_2O_3	101.96128	$MgCO_3$	84.3142
CaF_2	78.0768	$Al_2O_3 \cdot 2SiO_2 \cdot 2H_2O$	258.16128
CaO	56.0794	$3CaO \cdot Al_2O_3$	270.19948
$Ca(OH)_2$	74.0946	$12CaO \cdot 7Al_2O_3$	1386.68176
CO_2	44.0098	$CaO \cdot Al_2O_3$	158.04068
$CaCO_3$	100.0892	$CaO \cdot 2Al_2O_3$	260.00196
$CaSO_4$	136.1376	$2CaO \cdot Al_2O_3 \cdot SiO_2$	274.20488
$CaSO_4 \cdot 2H_2O$	172.168	$CaO \cdot Fe_2O_3$	215.7716
FeO	71.8464	$2CaO \cdot Fe_2O_3$	271.851
Fe_2O_3	159.6922	$4CaO \cdot Al_2O_3 \cdot Fe_2O_3$	485.97108
H_2O	18.0152	$3CaO \cdot SiO_2$	228.323
K_2O	94.1954	$2CaO \cdot SiO_2$	172.2436
MgO	40.3044	$11CaO \cdot 7Al_2O_3 \cdot CaF_2$	1408.67916
Na_2O	61.97894	$3CaO \cdot Al_2O_3 \cdot CaSO_4 \cdot 12H_2O$	622.51948
K_2SO_4	174.2536	$3(CaO \cdot Al_2O_3) \cdot CaSO_4$	610.25964
Na_2SO_4	142.03714	$K_2O \cdot 23CaO \cdot 12SiO_2$	2105.0392
P_2O_5	141.94452	$Na_2O \cdot 8CaO \cdot 3Al_2O_3$	816.49798
SiO_2	60.0848	$6CaO \cdot 2Al_2O_3 \cdot Fe_2O_3$	700.09116
SO_3	80.0582	$3CaO \cdot Al_2O_3 \cdot 6H_2O$	378.29068
TiO_2	79.8988	$3CaO \cdot Al_2O_3 \cdot 3CaSO_4 \cdot 31H_2O$	1237.08348

https://doi.org/10.1515/9783110572100-008

Bibliography

[1]   YUAN Runzhang. Cementitious Materials Science [M]. Wuhan: Wuhan University of Technology Press, 2009.
[2]   CHEN Yan, YUE Wenhai. Gypsum Building Materials [M]. 2nd edition. Beijing: China Building Materials Industry Press, 2012.
[3]   WANG Xin, CHEN Meimei. Building Materials [M]. Beijing: Beijing University of Technology Press, 2012.
[4]   XU Youhui, HE Zhanrong. Building Materials [M]. Beijing: Beijing University of Technology Press, 2012.
[5]   PANG Lufeng. Civil Engineering Materials [M]. Beijjing: China Electric Power Press, 2012.
[6]   HE Xiaoyan, HOU Yongli. Civil Engineering Materials [M]. Nanjing: Jiangsu Science and Technology Press, 2013.
[7]   CHU Jianmin, GAO Shilin. Handbook for Production Technology of Metallurgical Lime [M]. Beijing: Metallurgical Industry Press, 2009.
[8]   GONG Jianghong, GUO Hui. Building Materials [M]. Beijing: China Environmental Science Press, 2012.
[9]   GAO Hengju, WEN Xuechun. Building Materials [M]. Xi'an: Xi'an Electronic Science University Press, 2012.
[10]   LI Dongxia, WANG Tie, ZHENG Baohua. Building Materials [M]. Beijing: Beijing University of Technology Press, 2012.
[11]   ZHANG Yamei, SUN Daosheng, QIN Honggen. Civil Engineering Materials [M]. Nanjing: Southeast University press, 2013.
[12]   LIN Zongshou. Cement Technology [M]. 2nd edition. Wuhan: Wuhan University of Technology Press, 2017.
[13]   LIN Zongshou. Inorganic Nonmetallic Materials Technology [M]. 4th edition. Wuhan: Wuhan University of Technology Press, 2014.
[14]   YU Xingmin. Pandect of Operative Technology for New-dry Cement Production [M]. Beijing: China Building Materials Industry Press, 2006.
[15]   ZHU Peinan, WENG Zhenpei, WANG Tiandi. Microstructural Atlas of Inorganic Nonmetallic Materials [M]. Wuhan: Wuhan University of Technology Press, 1994.
[16]   LIN Zongshou. HundredThousand Questions of Cement: Volumes 1 to 10 [M]. Wuhan: Wuhan University of Technology Press, 2010.

https://doi.org/10.1515/9783110572100-009

Index

https://doi.org/10.1515/9783110572100-010